Silver in Healthcare
Its Antimicrobial Efficacy and Safety in Use

Issues in Toxicology

Series Editor:
Professor Diana Anderson, *University of Bradford, UK*
Dr Michael D Waters, *Integrated Laboratory Systems, Inc, N Carolina, USA*
Dr Timothy C Marrs, *Edentox Associates, Kent, UK*

Titles in the Series:
  1: Hair in Toxicology: An Important Bio-Monitor
  2: Male-mediated Developmental Toxicity
  3: Cytochrome P450: Role in the Metabolism and Toxicity of Drugs and other
     Xenobiotics
  4: Bile Acids: Toxicology and Bioactivity
  5: The Comet Assay in Toxicology
  6: Silver in Healthcare: Its Antimicrobial Efficacy and Safety in Use

How to obtain future titles on publication:
A standing order plan is available for this series. A standing order will bring delivery of
each new volume immediately on publication.

For further information please contact:
Book Sales Department, Royal Society of Chemistry,
Thomas Graham House, Science Park, Milton Road, Cambridge,
CB4 0WF, UK
Telephone: + 44 (0)1223 420066, Fax: + 44 (0)1223 420247, Email: books@rsc.org
Visit our website at http://www.rsc.org/Shop/Books/

Silver in Healthcare
Its Antimicrobial Efficacy and Safety in Use

Alan B. G. Lansdown
Charing Cross Hospital and Imperial College, Faculty of Medicine, London, UK

RSCPublishing

Issues in Toxicology No. 6

ISBN: 978-1-84973-006-8
ISSN: 1757-7179

A catalogue record for this book is available from the British Library

Published by The Royal Society of Chemistry,
Thomas Graham House, Science Park, Milton Road,
Cambridge CB4 0WF, UK

Registered Charity Number 207890

For further information see our web site at www.rsc.org

Foreword

Silver—is it a wolf in sheep's clothing? Or is it another case of the Emperor's new clothes?

Many people have their own views on this old, but recently rediscovered agent, though their views may be coloured by a number of factors. For the first time, as far as I am aware, Alan Lansdown has provided us with a focused, academically sound and balanced view of this important and increasingly valuable metal.

As a clinician I want to know before I treat my patients: is a product both safe and effective? However, my ability to make an informed judgement can, if I am not careful, be influenced by laboratory research results which may have limited translation to the clinical setting and the "*smoke and mirrors*" often seen when commercial concerns become involved in developing new products with an ability to cure all ills. I am sure the number of papers and citations about silver have increased enormously over the past decade, but how balanced are these papers? The analogies I drew at the beginning are very true in this situation and we all risk being "duped" into sitting one side of the fence or other. This book provides a comprehensive review of silver in a way that I believe will allow clinicians, academics, product developers and decision-makers who control the availability of silver-containing products to make an informed decision.

The chapters provide a logical analysis of the current situation and understanding of this element's use in healthcare. Understanding the chemistry of silver, its uptake, metabolism and toxicology and its use in medical devices are all scene setters for the main body of this work, which focuses on silver as an antimicrobial agent in wounds.

Infection causes great concerns in clinical practice. The simple question of—is this wound infected?—can not be answered simply in all cases. As a consequence, the appropriate selection of antimicrobial agents and measures of their efficacy are fraught with difficulty. For a variety of reasons there has been an explosion of interest in silver as an antimicrobial agent in wound healing in recent years. Discussions and debates on its value are still ongoing and the limited amount of high quality evidence published in the peer reviewed literature all make for a confusing picture at the present time. The interest and apparent success of many new wound treatments containing silver should mean that new data and consensus will eventually emerge.

Silver is an old element. The resurgence of interest in its use clinically has sparked considerable discussion. Alan's book should be seen as something that helps provide an informed balanced view of this element's use in clinical practice. The potential for further innovations and uses are considerable but, in keeping with the comments made at the beginning of this piece, we should not *"want to throw the baby out with the bath water"*. This work helps us all resolve questions and concerns we may have and, as such, I would recommend all who are involved in this area to read it.

Professor Keith Harding
Head of Department of Dermatology and
Wound Healing Academic
Director of Innovation & Engagement
School of Medicine, Cardiff University,
Upper Ground Floor, Room 18, Heath Park, Cardiff, UK

Preface

This book owes much to my initial training and guidance in experimental pathology and toxicology by Professor Paul Grasso and his team in the British Industrial Biological Research Association. At different times, I can look back on the fruitful discussions that I have held with Paul and other colleagues in industry and academia. I acknowledge the support that I have received from companies within the pharmaceutical and healthcare industries; this has enabled me enrich my knowledge, to pursue research, and to travel and lecture in Britain and overseas. It is noteworthy that my first research with silver was as a collaborative project with Dr Pisamai Laupattarakasem and Dr Auranut Vuttivirojana of the Faculty of Medicine, Khon Kaen University (Thailand), who visited my laboratory in Britain with support from the British Council. Their interest and enthusiasm is greatly appreciated.

This publication is the fruit of many years of experimental study on the action and interaction of metals in biological systems fostered by discussion and collaboration with clinical and academic colleagues at Charing Cross Hospital and the former Charing Cross and Westminster Medical School. I appreciate the close collaboration and friendship of Barry Sampson who has taught me much about the analysis of metals in body tissues and fluids, and fostered my interest in zinc and metal-carrier proteins as essential modulators in all living systems. Angela Williams in the Department of Vascular Surgery has been a good friend and teacher in the clinical evaluation of silver-containing wound therapies. Her nursing experience and devotion to patient care has been highly beneficial in aiding my understanding patterns of clinical wound healing. In wound care, as much depends upon the bedside manner, as upon treatment of the wound with appropriate silver therapy.

I should like to acknowledge with sincere thanks the constant support and tolerance that my wife Veronica has given me over many years. Her invaluable advice and helpful criticism has encouraged me to present my ideas on a subject that is presently topical and of wide-reaching implication in human health and welfare.

Issues in Toxicology No. 6
Silver in Healthcare: Its Antimicrobial Efficacy and Safety in Use
By Alan B. G. Lansdown
© Alan B. G. Lansdown 2010
Published by the Royal Society of Chemistry, www.rsc.org

We have sought to locate owners of all reproduced material not in our own possession. In a few cases we have been completely unsuccessful, but trust we have not inadvertently infringed any copyrights. Should we have done so inadvertently we shall of course take appropriate action for any subsequent editions.

Contents

Issues in Toxicology No. 6
Silver in Healthcare: Its Antimicrobial Efficacy and Safety in Use
By Alan B. G. Lansdown
© Alan B. G. Lansdown 2010
Published by the Royal Society of Chemistry, www.rsc.org

Acknowledgements

I should like to acknowledge the help and advice that I have received from the following colleagues, who have kindly provided me with photographs from their files to illustrate the text.

Professor Robert E. Burrell, University of Alberta, Canada
Ms Rosemary Jacobs
Dr Bruce Bouts, Blanchard Valley Medical Associates, Findley, Ohio, and Professor of Clinical Pharmacy, Ohio Northern University, USA
Professor Aldo R. Boccaccini, Imperial College, London, UK, and University of Erlangen-Nuremberg, Germany
Dr Superb Misra, Department of Mineralogy, Natural History Museum, London, UK
Dr Eva Valsami-Jones, Department of Mineralogy, Natural History Museum, London, UK
The Librarian of The Worshipful Company of Goldsmiths, London, UK
Ms Sarah Pearson, Curator of the Hunterian Museum, Royal College of Surgeons, London, UK
Dr Michael Kessler, Pharmazie-Historisches Museum, Basel, Switzerland
Mr Barry Sampson, Trace Metals Laboratory, Charing Cross Hospital, London, UK
Dr. Steve Gentleman, Reader in Neuroanatomy, Imperial College, London, UK

CHAPTER 1

Silver in Health and Disease

Silver is a lustrous white metallic element found in many parts of the world in soil and rocks, fresh and salt water, and in the atmosphere. Silver is the sixty-third most abundant element in the Earth's crust and is found naturally as the native metal, as argentite (Ag_2S) and horn silver (AgCl), and in numerous complexes with lead, zinc, copper, arsenic, mercury, tellurium and antimony. The date of its discovery is not known but silver coins, jewellery, religious icons and ornaments, and instruments have been recovered from sites of the ancient civilisations of South America, Egypt and the Middle East and China dating from 2000BC.[1] Silver and the trade of silversmithing are recorded in Biblical texts and in writings by Homer.

Silver and its alloys have a wide variety of commercial uses of which coinage, silverware, jewellery, photography, electrical contacts and electroplating, dental amalgams, batteries and medical devices are the most important nowadays. Silver in hygiene clothing is a very recent innovation.[2,3] The mention of the word "silver" conjures up visions of cups, trinkets, trophies, religious icons, *etc.*, and possession of items of value which have come to signify wealth and posterity.

Silver has a long and fascinating history in the treatment of human diseases including epilepsy, neonatal eye disease, venereal diseases, cholera, dysentery and wound infections.[4–6] Silver prostheses and silver surgical instruments were used by surgeons from the Middle Ages; the eminent paediatric neurosurgeon, Ambrose Paré (1517–1590), employed silver clips in facial reconstruction surgery.[7] Paré had extensive clinical experience in head surgery management, and his patients included Henri II King of France. He had considerable historical importance as a renaissance surgeon and his teaching extended throughout the literate world. At about the same time, the English surgeon John Woodall is also believed to have used silver nitrate in treating infections in craniofacial surgery. Later, William Halsted (1895) preferred silver wire sutures in hernia operations and used silver foil as a safeguard against post-operative infections.[8] It is unlikely that either was fully acquainted with the true nature of infectious disease or the true prophylactic action of silver which became apparent through

Issues in Toxicology No. 6
Silver in Healthcare: Its Antimicrobial Efficacy and Safety in Use
By Alan B. G. Lansdown
© Alan B. G. Lansdown 2010
Published by the Royal Society of Chemistry, www.rsc.org

the classical studies of Louis Pasteur (1822–1895) and Robert Koch (1843–1910), who provided a basic understanding of infectious diseases and the metabolism and classification of bacteria.[9,10] The powerful methodology developed by Koch in Germany heralded the "Golden Age" of medical bacteriology when such pathogens as cholera vibrio, typhoid bacillus, diphtheria bacillus, pneumococcus, staphylococcus, streptococcus, meningococcus, gonococcus and tetanus bacillus were identified, and the famous Koch's postulates formulated.[11] At the time, early evidence was emerging that metallic ions (notably Hg^{++}, Ag^+) were effective antibacterial agents at concentrations as low as one part per million (ppm),[10] but it is unclear whether Pasteur or Koch ever used silver or mercury as antibacterials. We can speculate that much of the early enthusiasm for using silver in bone prostheses, sutures, operating needles and surgical instruments, dental devices and wound therapy may have derived from its aesthetic value as a precious metal, but clinical evidence accumulating over the past 150 years has established that metallic silver and ionisable silver compounds can provide a safe and efficacious means of protecting the human body from infectious diseases.

Silver nitrate, supposedly introduced by the French ophthalmologist Credé, should possibly be regarded as the first efficacious antibiotic known to medical science. Dr. Credé (1895) claimed that silver nitrate used in vaginal douching dramatically reduced the incidence of ophthalamia neonatorum in his clinic from 10.8% to 0.2%.[12] Although the full details of his procedure are not available, it is conceivable that his successes were largely attributable to 2% phenol and not silver nitrate *per se*. Later, Lehrfeld recommended superficial sterilisation of the birth canal and thorough flushing of the eyes with boric acid followed by 0.5% silver nitrate as a prophylactic therapy.[13] Solleman cautioned against the use of silver nitrate in treating eye and mucus membrane infections on account of its irritancy and risks of corneal damage and blindness; he suggested that a "single drop of silver nitrate in a wax capsule" placed in the conjunctival sac for a few minutes and the eye rinsed in saline might be a more suitable therapy in controlling infections and irritation.[4] In the 1940s, it was common practice to use silver nitrate to treat skin infections. Clinicians considered that its local antibacterial action could be readily controlled and that its action extended "quite deeply", with silver ion complexing with albumins or precipitating as silver chloride.[14] Whereas it seems that silver nitrate formed the mainstay of antiseptics available to physicians and surgeons for many years, the safer and efficacious modern antibiotics including penicillin, sulfonamides, tetracyclines, *etc.* available nowadays are more widely accepted in clinical practice. Silver nitrate still has a place in burns unit wound clinics today and is frequently a life-saving therapy in cases of *Pseudomonas aeruginosa* infections.[15]

Older manuals of pharmacology recommend that painting the posterior pharynx with 10% solution or rinsing the mouth with 0.5% after meals is an effective antismoking measure (the foul taste experienced in the presence of cigarette smoke discouraged smoking).[4] Even these days, antismoking remedies using silver are available although the extent to which they are used is not

known.[16] In the 1940s, silver nitrate "pills" were available to treat gastrointestinal complaints, particularly peptic ulcers. Prescriptions of 10–20 mg of
silver nitrate with kaolin and petrolatum may have been effective in alleviating
the complaint, but the risks of irritancy and argyria precluded their clinical
acceptability. As discussed later, silver colloidal preparations for oral or
respiratory conditions are not regulated in UK, USA and many other countries.[17,18] Colloidal silver preparations, which may have been commonplace
throughout the earlier part of the 20th century, were superseded by safer and
more efficacious antibiotics. Nevertheless, they are still available and have been
associated with risks of argyria, one such case being the "silver man".[19]

   Advances in materials science and instrumentation have led to the inclusion
or coating of a wide range of industrial, medical and domestic devices and
appliances with silver as a means of controlling or preventing infections.[7] As an
overview, we might view the silver story in four main phases:

1. The primary phase being the long period from the initial identification of
   silver as a bright ductile and malleable material, which could be fashioned into attractive shapes and figures, but when little or nothing was
   known about its chemical or biological properties.
2. The second phase would cover the period from the mid-1800s to about
   1920 when scientists acquired an understanding of the chemical and
   biological properties of silver and its ability to control disease. Silver
   instruments were fashionable in surgery and dentistry (Figure 1.1). Silver

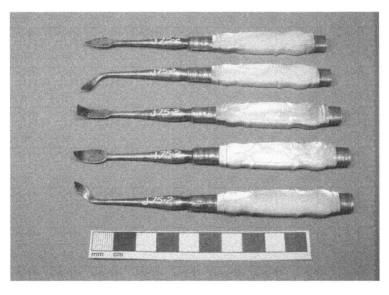

Figure 1.1   18th century silver dental instruments. *By courtesy of the Royal College of
Surgeons of England.*

nitrate became recognised as an efficacious antiseptic against many of the bacterial infections of the time.

3. Through the greater part of the 20th century, the silversmith's art became more refined; health scientists became aware of toxic risks of silver dust and fumes in the work place, and in the products released into clinical and domestic use. During this period, silver nitrate, colloidal silver preparations and silver sulfadiazine formed the major means of controlling infections, but manufacturers became aware of their legal requirements in marketing safer and more efficacious products.

4. The modern era extending from about 1990: the value of silver in healthcare has seen the introduction of sustained silver-ion release products for wound care and new technology for producing medical catheters, bone implants and cements, cardiovascular devices (stents, heart valves, *etc.*) and dental products with antibiotic protection against bacterial colonisation and biofilm formation.

Silver has a long history in water purification. The earliest evidence of this comes from evidence that silver coins were placed in the water of monarchs and nobles of the ancient dynasties of the Egypt and the Middle East and when it was fashionable to retain drinking water in silver urns and cups. Even in more recent times, the Maharajah of Jaipur (India) used massive silver urns to transport the sacred waters of the river Ganges on his trips to Europe (Figure 1.2). These days, silver–copper filters are in use in many hospitals to control risks of *Legionella* and methicillin-resistant *Staphylococcus aureus* (MRSA) in hot water systems. In each case, manufacturers strike a balance

Figure 1.2   Giant silver urn from Diwan-i-khas: preserved sacred water from the river Ganges for the Maharajah of Jaipur.

between including silver in or on their products to control health-threatening infections and the inherent safety risks of silver accumulating in the human body. Safety measures are in place to control levels of silver in drinking water in many countries.

A discussion on silver would be incomplete without recognition of the numerous advances made over the past 150 years using silver impregnation techniques in investigative anatomy and pathophysiology. Such names as Ranvier, Camillo Golgi, Santiago Ramon y Cajal, Bielschowsky and del Rio Hortega are routinely quoted in dissertations and textbooks of neuroanatomy and pathology. Ranvier developed a classical method using a silver impregnation to demonstrate nerve fibres and their nodes (Ranvier myelinated nerve fibres) in chromic acid preserved tissues.[20] The significance of the nodes of Ranvier in neuroconduction has been subject to later investigations.[21,22] Silver impregnation developed and further refined by Golgi and Ramon y Cajal has been instrumental in aiding an understanding of cells and fibres of the central nervous system;[23,24] they published more than 300 scientific papers leading to the award of the Nobel Prize for Physiology and Medicine to Ramon y Cajal in 1906.[25–27] Bielschowsky introduced the first and most reliable technique of silver impregnation using ammoniacal silver for demonstration of axons and neurofibrils (Figure 1.3); a method that is still preferred today on account of its specificity and reliability.[28,29] The true significance of Ramon y Cajal's teaching has become apparent with advent of the electron microscope.[30] Advances and

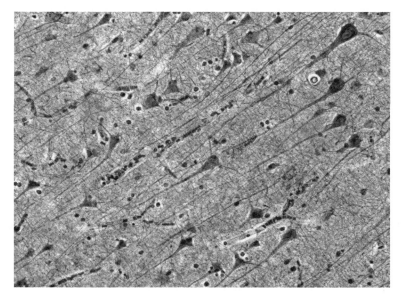

Figure 1.3 Bielschowski's silver impregnation histological method for demonstration of neurones, axons and neurofibrils. *By courtesy of Dr. S. Gentleman, Imperial College.*

modifications of silver impregnation and silver proteinate methods have provided great benefits to neuropathologists investigating degenerative changes in axons and nerve endings.[31–33] Interestingly, silver *per se* does not seem to be injurious to the brain or the central or peripheral nervous system.[34]

Silver has financial, pleasurable, clinical and industrial implications for a wide spectrum of the population in all countries. The human body is exposed to silver through drinking water, the food we eat, and through a wide range of medical and work place exposures. But how safe is silver? The lay and scientific literature is coloured by an alarming range of inaccuracies, misleading colloquialisms and speculations which tend to obscure the true value of the metal and its low toxicity relative to most other metallic elements. Internet captions have been seen quoting that "... although silver metal is not in itself toxic, most of its salts are poisonous". As far as I am aware, no comprehensive account of the toxicity of silver has yet been produced in any language and present safety thresholds published are at the best, only approximate. A toxicological profile for silver was compiled by William Roper for the Agency for Toxic Substances and Disease Registry of the US Public Health Service in 1990, but this preceded the tremendous advances in the use of silver in wide-ranging medical devices and consumer products, and fails to predict the massive advances in silver technology seen in recent years.[35]

This book is based on my experience in the biological and clinical evaluation of silver and silver products in skin wound management and subsequent researches into the use of silver in biomaterials for medical devices.[36,37] Argyria is the most commonly cited toxic manifestation of silver exposure, but although it is a cosmetically undesirable discolouration associated with exposure to silver, there is no evidence that it is life-threatening.[38] The famous "Blue Man of Barnum and Bailey's Circus" is recorded as having 90–100 g of silver in his body and was quite an attraction at the time.[39]

References

1. A. B. G. Lansdown, Silver, in *Chemistry: Foundation and Application*, ed. J. J. Lagowski, Thomas Gale, MacMillan Reference, Farmington Mills, MI, 2004, Ch. 4, p. 126.
2. D. Höfer, Antimicrobial textiles, skin borne flora and odour, in *Biofunctional Textiles and the Skin*, eds. U.-C. Hipler and P. Elsner, Karger, Freiberg, 2006, Current Problems in Dermatology, Vol. **33**, pp. 67–77.
3. U.-C. Hipler, P. Elsner and J. W. Fluhr, A new silver-loaded cellulosic fibre with antifungal and antibacterial properties, in *Biofunctional Textiles and the Skin*, eds. U.-C. Hipler and P. Elsner, Karger, Freiberg, 2006, Current Problems in Dermatology, Vol. **33**, pp. 165–178.
4. T. Sollemann, Silver, in *A Manual of Pharmacology and its Applications to Therapeutics and Toxicology*, Saunders, Philadelphia, 1942, pp. 1102–1109.
5. R. J. White, An historical overview of the use of silver in wound management, *Br. J. Nurs.* 2002, Silver Suppl., S3–S8.

6. H. J. Klasen, A historical review of the use of silver in burns, *Burns*, 2000, **26**, 117.
7. D. Simpson, Paré as a neurosurgeon, *Aust. N.Z. Med. J.*, 1997, **67**, 54.
8. W. G. MacCallum, Biographical memoir of William Stewart Halstead, *Biogr. Mem. Natl. Acad. Sci.*, 1935, **17**, Seventh Memoir, 151–169; W. S. Halstead, Practical: the radical cure of hernia, *Bull. Johns Hopkins Hosp.*, 1889, **1**, 12–13, 112.
9. W. Bulloch, *The History of Bacteriology*, Oxford University Press, London, 1960.
10. D. Davis, R. Dubeccio, H. N. Eisen, H. S. Ginsberg and W. B. Wood, Evolution of microbes and of microbiology, in *Principles of Microbiology and Immunology*, Harper and Row, New York, 1968, pp. 1–18.
11. A. S. Evans, Causation and diversity: the Henle Koch's postulates revisited, *Yale J. Biol. Med.*, 1976, **49**, 175–195.
12. K. S. F. Credé, *Die Verhütung der Augenentzündung der Neugeborenen, der häufigsten und wichtigsten Ursache de Blindheit*, Hirschwald, Berlin, 1895.
13. L. Lehrfeld, Ophthalamia neonatorum, *J. Am. Med. Assoc.*, 1935, **104**, 1468.
14. W. Lubinski, Silbernitrat oder silber eiwess, *Berlin Klin. Wochenschr.*, 1914, **51**, 1643.
15. E. J. Lowbury, Problems of resistance in open wounds and burns, in *The Rational Choice of Antibacterial Agents*, Kluwer Harrap Handbooks, London, 1977, pp. 18–31.
16. W. East, K. Boddy, E. D.Williams, D. MacIntyre and D. A. C. McLay, Silver retention and tissue silver concentrations in argyria associated with exposure to an antismoking remedy containing silver acetate, *Clin. Exp. Dermatol.*, **5**, 305.
17. National Center for Complementary and Alternative Medicine, *Colloidal Silver Products*, National Institutes of Health, Bethesda, MD, 2004, Consumer Advisory Bulletin.
18. A. B. G. Lansdown, Controversies over colloidal silver, *J. Wound Care*, 2003, **12**, 120.
19. N. S. Tomi, T. B. Kränke and W. Aberer, A silver man, *Lancet*, 2004, **363**, 532.
20. L. A. Ranvier, Contributions à l'histologie et à la physiologie des nerfs périferiques, *C.r. Hebd. Séances Acad. Sci. Paris*, 1871, **73**, 1168.
21. A. Hess and J. Z. Young, The nodes of Ranvier, *Proc. R. Soc. London, Ser. B*, 1952, **140**, 301.
22. N. Landon and P. L. Williams, Ultrastructure of the node of Ranvier, *Nature (London)*, 1963, **199**, 575.
23. A. Golgi, I recenti studi sull'istologia del sistema nervosa central, *Riv. Sper. Freniatr. Med. Leg. Alien. Ment.*, 1875, **1**, 121–260.
24. W. H. Cox, Imprägnation des centralen Nerven-systems mit Quecksilbersalzen, *Arch. Mikr. Anat.*, 1891, **7**, 123–176.34.
25. S. R. y Cajal, Structure et connexions des neurons, in *Les Prix Nobel en 1906*, Norsted and Söner, Stockholm, 1908, pp. 1–25.

26. S. R. y Cajal, *Degeneration and Regeneration of the Nervous System*, Oxford University Press, London, 1928.
27. S. R. y Cajal, *Studies on the Cerebral Cortex* (translated L. M. Kraft), Lloyd-Luke, London, 1955.
28. M. Bielschowsky, *J. Psychol. Neurol. (Lpz)*, 1904, **3**, 169.
29. M. Bielschowsky, *J. Psychol. Neurol. (Lpz)*, 1909, **12**, 135.
30. P. L. Williams and R. Warwick, in *Gray's Anatomy*, Churchill Livingstone, Edinburgh, 1980, pp. 801–1126.
31. D. Bodian, A new method for staining nerve fibers and nerve endings in paraffin sections, *Anat. Rec.*, 1936, **65**, 89.
32. G. Holmes, The cerebellum of man, *Brain*, 1939, **62**, 1.
33. E. G. Grey and R. W. Guillery, Synaptic morphology in the normal and degenerating nervous system, *Int. Rev. Cytol.*, 1966, **19**, 111.
34. A. B. G. Lansdown, Critical observations on the neurotoxicity of silver, *Crit. Rev. Toxicol.*, 2007, **37**, 237.
35. W. L. Roper, *Toxicological Profile for Silver*, Agency for Toxic Substances and Disease Registry, US Public Health Service, Atlanta, GA, 1990.
36. S. S. Bleehan, D. J. Gould and C. I. Harrington, Occupational argyria: light and electron microscopic studies and X-ray microanalysis, *Br. J. Dermatol*, 1981, **104**, 19.
37. W. R. Buckley and C. J. Terhaar, The skin as an excretory organ in argyria, *Trans. St John's Hosp. Dermatol. Soc.*, 1973, **59**, 39.
38. A. B. G. Lansdown, Silver in healthcare: antimicrobial effects and safety in use, in *Biofunctional Textiles and the Skin*, eds. U.-C. Hipler and P. Elsner, Karger, Basel, 2006, Current Problems in Dermatology, Vol. **33**, pp. 17–34.
39. O. Gettler, C. P. Rhoads and A. Weiss, A contribution to the pathology of generalised argyria with a discussion on the fate of silver in the human body, *Am. J. Pathol.*, 1927, **3**, 631.

CHAPTER 2

Silver and its Compounds, Chemistry and Biological Interactions

2.1 Silver: the Essentials

Silver is a stable, ductile and malleable transitional element found widely throughout the world. The name of the element derives from the Anglo-Saxon *seolfor* or *siolfur*, and the Latin *argentum*. It is fractionally harder than gold, but has the highest thermal and electrical conductivity of all metals in the Periodic Table and presents the lowest contact resistance. Pure silver has a brilliant white metallic lustre, but although it is stable in air it readily tarnishes on exposure to ozone, air containing sulfur or hydrogen sulfide (Figure 2.1). Commercially extracted silver is 99.9% pure.

Silver possibly accounts for 0.1 ppm in the Earth's crust and about 0.3 ppm in soils. It occurs as cubic crystals and in deposits of pure metal in the form of two isotopes, ^{107}Ag and ^{109}Ag, which occur in approximately similar proportions; ^{109}Ag exhibits greater sensitivity and is of greater value as a probe in analytical chemistry. A total of thirty five isotopes of silver are known. Of these, the radioactive silver isotope, ^{110m}Ag, is produced for neutron activation analysis (NAA) of silver in tissue samples or in air; it has a half life of 249.79 days and emits γ-radiation.[1] ^{108}Ag has a half-life of 2.4 minutes and emits γ-rays with a photopeak of 0.63 MeV.[2]

The date of discovery of silver is not known but slag dumps found in Asia Minor and on islands in the Aegean Sea suggest that man learned to separate silver from lead as early as 3000BC. The ancient Egyptians may have worked in silver at about that time, although the metal they used was probably mined in Turkey. A silver salver recovered from the Chaldean Empire (*ca.* 2850BC), and now exhibited in the Louvre in Paris,[3] may be the first tangible evidence of the use of silver in water purification, or it may have been used as a funerary urn, preserving the ashes of the dead for a future life.

Issues in Toxicology No. 6
Silver in Healthcare: Its Antimicrobial Efficacy and Safety in Use
By Alan B. G. Lansdown
© Alan B. G. Lansdown 2010
Published by the Royal Society of Chemistry, www.rsc.org

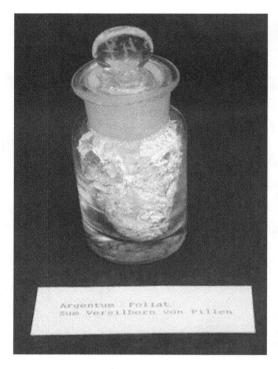

Figure 2.1 Metallic silver (enclosed within an air-tight container to prevent oxidation). *By kind permission of Dr M. Kesseler, Pharmazie-Historisches Museum, Basel.*

Early civilisations discovered that water preserved in silver vessels remained much purer and more acceptable for drinking than that held in earthenware pitchers. The disinfecting powers of silver may extend back to Babylonian times when silver containers were used to transport water for the personal use of the great kings of Persia.[4] There is evidence that silver coins and silver spoons were once placed in drinking water as a means of purification. Silver is still extensively used in water purification today and, in many hospitals, hot water systems are filtered through silver–copper devices as a protection against *Legionella* sp. and MRSA infections.[5–7]

Apart from the aesthetic and precious metal value of silver and its large-scale use in jewellery, religious icons and vessels, and silverware (mainly as silver alloys with copper and other metals), significant advances have been made in the use of silver of silver and its compounds in medicinal science and surgery.[8–10] The electro-thermal properties of silver are used in solders, electrodes, electrical contacts, batteries, *etc.* Although silver was commonly used in coinage at one time on account of its precious metal value, only few countries still mint silver in coinage today; the silver content is generally low and consequently human exposure through this route is minimal . Silver metal threads are

Table 2.1 Physico-chemical properties of silver.

Atomic number	47
Chemical symbol	Ag
Atomic weight	107.868
Isotopes	Ag^{107}, Ag^{109}, Ag^{111} *etc.*
Electronic configuration	$[Kr]5s^{1}4d^{10}$
Melting point	961.93 °C
Boiling point	2212 °C
Specific gravity (at 20 °C)	10.50
Valence states	1,2,3
Solubility/ionisation in water	$<0.1\,\mu g.\,ml^{-1}$

incorporated into paper currency as a security device in several countries, but more commonly non-precious metals such as tin are preferred.

Silver is commonly identified as a monovalent metal with chemical properties consistent with its classification as a Group II element in the periodic table (period 5d) (Table 2.1).[11] It exhibits three oxidation states [Ag(I), Ag(II) and Ag(III), *i.e.* Ag^{+}, Ag^{++} and Ag^{+++}] and potentially can form a large number of chemical compounds and complexes. A fluoride, oxide and sulfide of divalent silver are available, but most compounds of Ag^{++} and Ag^{+++} are unstable or insoluble and of minimal relevance as antimicrobial agents in medical devices and textiles.[12] The oxidation states of silver are discussed specifically with reference to microparticulate silver (nanocrystalline) silver, which has shown many beneficial properties as an antibiotic in wound dressings and medical devices.[13] Whereas the toxicology of silver and monovalent silver compounds has been the subject of intense investigation over the past 30 years, the human risks posed by Ag^{++} and Ag^{+++} in the environment are virtually unknown and many erroneous statements appear in published papers and subject reviews.

Silver has a long-standing and increasing value in medicinal applications dating from at least the Middle Ages.[14] For centuries it was observed that clinical use of silver wire in surgical sutures and silver foil in wound care seemed to promote tissue repair. A number of metals including silver exhibited anti-biotic effects,[15] but discovery of the specific properties of silver to inhibit the growth of bacteria at exceedingly low concentrations is credited to the classical studies of the Swiss botanist, Carl von Nägeli, who introduced the concept of "oligodynamic action" in 1893 to describe this action.[16] Since those days, a vast amount of research has been conducted and the capacity of silver to absorb or adsorb to metallic and non-metallic substances or surfaces for use in medical devices without losing its antibiotic properties is now well established.[17–21]

Although the true scientific value of silver as an antibiotic was not appreciated until the end of the 19th century, early surgeons and physicians recognised that silver could be inserted safely into the human body in the form of prostheses, splints and wires to aid bone repair, and that silver wire, silver sutures and silver foil were beneficial in protecting wounds from disease. Earliest records show that ancient Egyptians used silver and other metals to

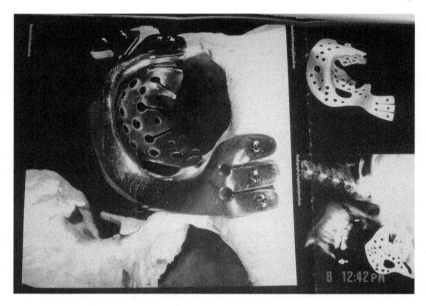

Figure 2.2   Silver in a 16th century medical prosthesis for facial reconstruction. *By courtesy of The Worshipful Company of Goldsmiths.*

reconstruct facial parts.[22] Dr Watt-Smith of the John Radcliffe Hospital in Oxford reviewed the early uses of silver in oral and maxillofacial surgery and emphasised that silver can be electroformed to create complicated shapes as in the surgical replacement of 75% of eye sockets.[22] This surgery enables a patient to retain both the eye and useful binocular vision (Figure 2.2). He noted also that, since silver eluting from these prostheses may discolour the skin, the surfaces of implants were coated with gold. In much earlier surgical studies the eminent French surgeon, Ambrose Paré, recorded the benefits derived from facial prostheses specially designed for patients and composed of paper and leather which were secured to the head with a clip and nose pieces of silver held in place with ligatures.[23] To these devices, we can now add a vast and chemically diverse range of wound dressings containing silver or ionisable silver compounds, purpose-designed medical catheters (intravascular, intra-urethral, intra-ventricular, intra-peritoneal and intra-vascular forms), silver-coated or silver-impregnated heart valves, cardiovascular prostheses, orthopaedic devices and bone cements, and silver wire and silver surgical sutures.[24]

Silver has made important contributions in X-radiography, photography and diagnostic pathology over more than 100 years and today represents a major route of human exposure to the metal.[25] X-ray films carry relatively large amounts of silver (*ca.* 40%) and radiographers are potentially at greater risk of silver toxicity than other professions. (Recovery of silver from X-ray films presents additional risk to operatives; in 1984, X-ray film stored in the USA was estimated to contain in excess of 3000 tonnes of silver worth about $3000.)

Additionally, high resolution phase lenses using silver technology are presently available for diagnostic purposes in medicine.[26,27] Silver absorbed into target tissues and visualised by photographic means provides a powerful diagnostic tool for surgeons and physicians. Recent information shows that the New York State Institute for Basic Research in Developmental Abnormalities successfully uses silver technology to identify multiple sclerosis-related antibodies in cerebrospinal fluid.[28] Numerous other applications can be illustrated where silver complexes with electrophoresis are used to detect protein disorders attributable to dietary and inherited changes.

The use of silver in textiles as a protection against bacteria and offensive fungal infections like athlete's foot (*Tinea pedis*) is a comparative new and rapidly expanding growth area.[29–33] Individuals are exposed to low concentrations of ionised silver over prolonged periods and laboratory trials have demonstrated commendable antibacterial and antifungal action.[34] Today, silver-containing textiles are marketed under the heading of "hygiene clothing", but there is increasing recognition that such textiles are potentially useful in controlling infections in elderly and immuno-compromised patients at risk of methicillin-resistant *Staphylococcus aureus* (MRSA) and other potentially fatal infections.[35,36] Advances in silver technology permit silver metal or silver alloys to be impregnated into polymers (*e.g.* polyurethane) or to applied as a surface coating to catheters to inhibit potentially fatal catheter-related bacterial or fungal colonisation and biofilm formation.[37–39] Scientific validation is still required of other recent innovations employing metallic silver and claimed to safeguard against bacteria in domestic equipment (including refrigerators).

Silver is not inert in the presence of biological materials as was thought at one time. Burrell and many other workers have established that, whereas metallic silver is largely inert, it slowly ionises in the presence of moisture, body fluids and secretions to release biologically active Ag^+. This ion readily binds to proteins, inorganic anions (*e.g.* chloride in sweat) and cell surface receptors on prokaryotic (bacteria, yeasts, fungi) and eukaryotic cell membranes.[13,40–42] Whereas silver has no known role as a nutrient in the human body, low concentrations are found in the serum and tissues of most people ($<2.3\,\mu g\,L^{-1}$) through ingestion of silver or soluble silver compounds in food and drink, inhalation or occupational exposures.[42,43] Silver absorbed into the human body is readily metabolised to bone and most soft tissues.[2,44] Whereas it is possible to control human silver intake in food, water and medicinal exposures, there is increasing concern in certain parts of the world (*e.g.* the San Francisco Bay area of the USA) of the risks of silver contamination through mining, metal refineries and metal industries and its elution into surface water, rivers and marine deposits, ultimately entering human food chains.[45–47]

Blood silver (argyraemia) and urine silver excretion are useful indices of human silver exposure to silver from all sources.[42,48–51] Uptake of silver by mammalian and prokaryotic cells is mostly by carrier mediated or phagocytic events.[26,52] In the body, Ag^+ binds strongly with and "inactivates" electron donor moieties containing oxygen, sulfur and nitrogen as in sulfydryl-, amino-, imidazole-, carboxylate or phosphate groups,[53] but like many heavy metals, is

excreted in the keratins in skin and hair.[53–55] Silver is not known to be a
cumulative toxin but does interact with and displace essential metal ions like
Ca^{++} and Zn^{++} in hydroxyapatite in bone.[56–58] Excessive silver intake can lead
to a long-term accumulation of insoluble precipitates in the skin and eye.
Albumins and macroglobulins in wound exudates and body fluids exert a
protective effect by binding excess silver ion liberated devices.[59] Micro-auto-
radiographical methods demonstrate that three molecules of silver bind a single
molecule of albumin to form a colourless and non-toxic precipitate.[60] Additionally, Ag^+ binds receptor groups on the surfaces of cells, tissue debris (in
wounds) and bacterial and fungal/yeast cell envelopes.[41,60]

2.2 Silver in the Environment and Ecosystem

Silver ores found naturally include sulfide, sulfate, bicarbonate, telluride and
complexes with a number of other metals such as lead, zinc, copper, gold, mercury, arsenic, antimony, arsenic, molybdenum and tellurium.[61] In addition to the
numerous deposits of pure silver metal, the principal mineral ores of silver include
argentite (silver sulfide), proustite (complexed with arsenic and sulfur), stephanite
and pyrargyrite (complexed with antimony and sulfur), hessite (complexed with
tellurium) and cyrargyrite (silver chloride). The principal silver mining areas of
the world include the USA (mainly in Utah, Montana, Idaho, Colorado, Arizona
and Nevada), South and Central America, Canada, Japan and Australia. The
principal mining sites in the USA are located in the Coeur d'Alene region of
northern Idaho, which at one time accounted for 71% of national production.
Smaller silver deposits occur in Europe (including Britain) and Middle Eastern
countries. Silver mining in Britain is recorded from Roman times when extraction
of the metal at Charterhouse-on-Mendip (Somerset), Machen (mid-Glamorgan),
Pentre (Flintshire), Lutudarum (Derbyshire) and the southern regions of the
Pennines was strictly controlled by the military administration.

Thornton reviewed the facts and misconceptions of metals in the global
environment and discussed the distribution of silver in organic rich black shales
and hydrothermal sulfides.[61] Natural deposits are exposed, eroded, transported
and redistributed by geophysical phenomena, glacial changes, temperature,
volcanic action and human activity. Mining, metallurgy, industrial processing
and metal recycling can influence local distribution, chemical form and bioavailability of metals and metal–metal ratios in the biosphere. In addition,
organic and inorganic deposits of metals and metal complexes can be "trans-
formed" by micro-organisms, thereby influencing their bioavailability and
dispersion in aquatic and soil ecosystems and their movement into food chains
and concentration in plants and animals.[62,63] Distribution and redistribution
patterns are influenced by the acidity of the local environment, oxidation–
reduction reactions, the clay content of the substratum, and exposure to water.
Minute silver particles and many other metals are released into the atmosphere
through progressive weathering and erosion of the rocks and soils.[63] Control of
such emissions is rarely possible, but inhalation of silver from particles in the
air and industrial dusts represents a major route of human exposure to silver

and a potential cause of argyria and related conditions.[26,61,64,65] Environmental risk assessments are made using "data that are relevant to the situation or region", rather than on the basis of data generated by basic laboratory studies. The experimental practice of "seeding clouds with silver iodide to promote rainfall" poses a potential hazard but the true risk to human and animal populations is not fully appreciated.[66,67] Silver iodide has a very low solubility in water.

Roper reviewed the human and environmental implications of airborne silver and silver contamination in industrial waste in studies conducted up to 1990 for the US Public Health Service.[68] In his opinion, data available then indicated that silver will bioaccumulate to a "limited extent" in algae, mussels, clams and other aquatic species, but that published studies did not conform to current state-of-the-art requirements and did not reflect levels of human exposure accurately. The ecotoxicological data reviewed showed that marine algae accumulate silver more than shellfish and fish, and exhibit bioconcentration factors ranging from 13 000 and 66 000.[69] In controlled exposure studies, bluegill (*Lepomis macrochirus*) and large-mouth bass (*Micropterus salmoides*) were shown to bioaccumulate exceedingly low amounts of silver from dilute silver nitrate in four months.[70] Bioaccumulation studies in algae (*Scenedesmus* sp.), water flea (*Daphnia magna*), mussels (*Legumia* sp. and *Margaritifera* sp.) and fat-head minnow (*Pimephales promelas*) exposed for ten weeks to silver thiosulfate indicated very limited increase in silver residues in the aquatic food chain.[71,72] Silver absorbed from sewage sludge, and factory waste disposal sites by algae and sediment-feeding animals is proportional to local concentrations, but the biological half-life for elimination of the metal varies greatly according to species.[73,74]

More reliable estimates of the ecotoxicology of silver are possible through analysing recent surveys such as that conducted in the so called "Silver Estuary" region of San Francisco Bay.[74,75] At one time, the Bay was reported to have the highest known levels of silver in its sediments and biota in the world, and was the only area with accurately measured levels of silver in solution. Using newly developed silver analysis of inductively coupled plasma atomic absorption spectrometry, scientists have measured silver residues in sediments, sea water, algae and shellfish and monitored changes in the temporal and regional distribution of the metal. In keeping with previous studies, they noted that airborne particles of metallic silver or silver compounds released by silver smelters and extraction plants slowly ionised in surface water and were eluted into streams, rivers lakes and estuaries and into seas and oceans where they concentrated in algae, shellfish, fish and crustateans.[76–78] The mobility of silver/silver ion in the ecosystem is strongly inhibited in the presence of organic matter (peat, boggy soils and in marshes), which binds silver ion prior to its uptake by plant life as a preliminary to entering human food chains through food animals grazing these pastures.[79–81]

2.3 Analysis of Silver in Biological Materials

Legislative and regulatory authorities, manufacturers, medical scientists and toxicologists are increasingly required to monitor the silver content in biomaterials, medical devices and antibiotic therapies, occupational exposures, industrial waste and ecosystems as quality control measures and safety requirements. Occupational limits and guidelines for silver reflect the sensitivity of analytical procedures available at the time and current knowledge on the inherent risks of silver exposure. In 1986, the American Conference of Governmental Industrial Hygienists (ACGHI) recommended that threshold limit values for human exposure to metallic silver should be $0.1\,mg\,m^{-3}$ and $0.01\,mg\,m^{-3}$ (soluble silver compounds).[82] Other US authorities including the Occupational Safety and Health (OSH) and the National Institute for Occupational Safety and Health (NIOSH) have set $0.01\ mg\,m^{-3}$ as permissible exposure limits.[81]

Improved technology and equipment available these days have enabled more accurate and metal-specific analyses in natural and synthetic materials. Whereas in materials science, quantification of the silver content of alloys, coatings and impregnates is necessary as a quality control measure, in biological materials the accent is on how much silver, where is it concentrated, what are the bioaccumulation patterns, and how long does it stay there. Analyses are generally tailored to the circumstances of the investigation (*i.e.* levels of human risk), the equipment and technical expertise available. Their aim can include determination of:

- total silver content—in biomaterials, plant and animal tissues, industrial wastes and ecosystems;
- patterns of release of silver release from specified products and emissions;
- release of silver from antibiotic dressings, coatings and impregnates in relation to its antibiotic efficacy (a measure of the bactericidal or fungicidal properties of a product;
- metabolic pathways and elimination patterns in individuals exposed to silver occupationally, in therapeutic or prophylactic medicinal, or through food chains and environmental exposures. Evaluations cover tissue-specific capacity to concentrate metallic silver or accumulate silver in the form of insoluble precipitates such as silver sulfide or silver selanide (as in cases of argyria).[26,27]

Methods available for quantitation of silver in biological tissues, fluids and environmental media have improved greatly in sensitivity/detection limits and reproducibility over the past 40 years. Whereas standard textbooks of the 1960s documented gravimetric, electrolytic, titrimetric and compleximetric means for quantifying silver in biological samples,[83] these methods are insufficiently sensitive or accurate nowadays for legislative control and have been largely replaced by atomic absorption spectrometry supplemented by qualitative analyses using high resolution electron microscopy and accurate methods with modifications to eliminate inaccuracies attributable to other metals or non-metal contaminants of the media. Although emission spectrographic analysis, flame photometric and atomic absorption spectroscopy were available in the early 1960s for the determination of precious metals like silver, gold, platinum, palladium and rhodium, equipment was expensive and not widely available for routine analyses.

Roper carried out a review for the US Public Health Service of analytical methods used for detection of silver presented in 21 major publications from 1973 to 1987.[68] These methods included:

- flame atomic absorption spectroscopy (FAAS);
- graphite furnace (flameless) atomic absorption spectroscopy (GFAAS);
- neutron activation analysis (NAA);
- high frequency torch-atomic emission spectroscopy (HFT-AES);
- micro-cup atomic absorption spectroscopy (MCAAS).

These methods are suitable for estimating silver absorption by human tissues and body fluids from wound care products, medical devices, textiles and other products containing silver for antibiotic purposes. Recent literature shows that FAAS equipped with atomisers to improve specificity is the most widely used procedure to analyse trace levels of silver in biological tissues and fluids. We have used FAAS in evaluating silver uptake from silver nitrate or silver sulfadiazine in experimental skin wounds and in patients with chronic venous leg ulcers treated with sustained silver-release wound dressings.[60,84] In each case, samples of 0.5 g wet tissue (or 0.5 mL wound fluid or plasma) were analysed with silver detection limits of $0.002\,\mu g\,g^{-1}$. In the clinical study, the silver content of wound exudates and wound debris and levels of silver deposited in tissue debris correlated closely with the amount of silver appreciated histologically as black granules (Figure 2.3) and patterns of silver ion release from

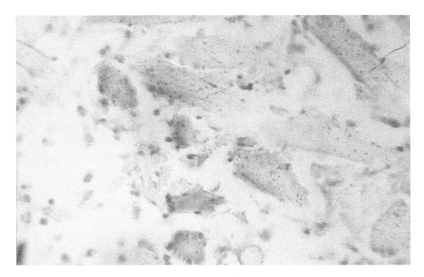

Figure 2.3 Deposition of silver sulfide granules in wound exudate from a patient treated with Aquacel® Ag. Note remains of carboxymethylcellulose fibres.

wound dressings. Wound exudates ranging from a thick mucoid consistency to moderate watery state exhibited silver levels from 0.74 to 417 μg g^{-1}. Atomic absorption spectroscopy offers high sensitivity (sub-nanogram to gram levels) and can be used with relatively small samples.[68]

GFAAS with a deuterium continuous light source to eliminate background emissions due to contaminants is capable of measuring silver at concentrations of 2 × 10^{-5} μg g^{-1} sample weight. Compared to the older gravimetric methods, which at best would detect silver with an accuracy of 10^{-2} mg g^{-1}, Segar and Gilio[85] and DiVincenzo et al.[86] used GFAAS to measure silver in biological tissues digesting samples in concentrated nitric acid (with or without addition of EDTA as an anticoagulant or ammonium hydrogen sulfate buffer as a modifier for matrix components) to measure silver content in whole blood, hair, faeces and miscellaneous biological extracts and recorded near 100% silver recovery; the accuracy of the technique is influenced by the tissue to be analysed and its preparation.

Wan et al.[42] and Coombs et al.[87] critically evaluated methods available for silver analysis in human tissues and body fluids, and refined the GFAAS method. They claimed that their method using a Perkin Elmer spectro-photometer equipped with a deuterium background correction, attached graphite furnace unit, and a silver hollow-cathode lamp with a wavelength of 328.1 nm and a high density graphite carbon tube without platform, was sufficiently sensitive to measure silver in blood, urine, liver and kidney of subjects without known industrial or medical exposure at < 2.3 μg L^{-1}, 2.0 μg d^{-1} and 0.05 μg g^{-1} (wet tissue) and 0.05 μg g^{-1} (wet tissue), respectively, with a detection limit of 0.4 μg L^{-1} and precision recovery of 94–99%. In a group of ten patients treated with silver sulfadiazine (SSD) for wound sepsis (> 5% total body surface area), blood silver increased rapidly to 69 μg L^{-1} after 6 hours to a maximum or 310 μg L^{-1} after about 10 days. Urinary silver in nine patients increased to 11 μg L^{-1} after 24 hours to a maximum of 558 μg L^{-1} after 7 days. Tissue silver analysis in an 81-year-old patient dying of renal failure after eight days SSD therapy showed levels of 970 μg L^{-1} (cornea), 14 μg L^{-1} (liver) and 0.2 μg L^{-1} (kidney) on a wet tissue weight basis. The authors pointed out that their determinations were significantly lower and appreciably more accurate that those published in an earlier study of thermal injury patients treated with SSD where flame atomic absorption spectrometry had been used.[50,86] Wan et al.[42] paid particular attention to minimising in-run and between-run precision, and controlling critically the influence of temperatures on analyses (lacking in most other reported studies).

A second illustration of the value of GFAAS is provided by a clinical case of a lady who suffered a massive accumulation of silver in her hip joint following total hip arthroplasty and use of silver-impregnated bone cement as prophylaxis against deep infections.[88] GFAAS using a Zeeman effect background correction and ammonium phosphate as a matrix modifier provided a detection level of < 0.03 μg L^{-1}.[89] The level of silver seen in her hip joint fluid (103.2 μg L^{-1}) was approximately 1000 times higher than present in the plasma of 12 non-silver exposed individuals (0.08 μg L^{-1}). Although fluid silver declined dramatically within 2–24 months after removal of the silver bone cement, it remained 15–20 times higher than normal, indicating a slow elimination of silver from the body in this patient at least. Histopathological examination of the soft tissue of the joint confirmed a silver sulfide discolouration characteristic of argyria along elastic fibres and in connective tissue.

GFAAS (using a Zeeman background correction) and/or high resolution magnetic sector inductively coupled plasma mass spectrometry have been used extensively to determine elemental silver in the large scale ecotoxicological studies of factory waste in the San Francisco Bay area.[75,79,90] On the pretext that silver is "one of the most toxic metals known to aquatic organisms in laboratory testing",[76,77,91] Flegal *et al.* quantified temporal changes in silver contamination in sea water and sediment in Bay "cores" with estimated dates of pre-1850, the 1950s period and 1970 by GFAAS (using a Perkin Elmer 4100 spectrophotometer). Pronounced vertical gradients in silver concentration provided unequivocal evidence of a persistent flux of silver into Bay water from sludges and factory effluents in recent years and its propensity to concentrate in shellfish (bivalves and clams) at concentrations ranging from 0.001 μg g^{-1} to 51.7 μg g^{-1}.[74]

In recent years, considerable emphasis has been placed upon the evaluation of silver released from sustained silver-release wound care dressings as a means of assessing their relative antibiotic efficacy and safety in use. A so-called "Static Silver Dissolution Test" developed by the Westaim Biomedical Corporation in Canada as part of its Silcryst™ technology programme and production of Acticoat® dressings (Smith & Nephew)[92] employed atomic absorption spectrometry to measure patterns of silver release from wound dressings immersed in water or simulated wound fluid in a continuous flow model. Following digestion in concentrated nitric acid (acid-soluble silver) or ammonia (soluble silver), the silver content of dressings was assayed analysed by AAS using a wavelength of 328.1 nm.

X-ray fluorescence is claimed to be an extremely useful *in vivo* technique for measuring heavy metals in soft tissues life the skin.[93] The system relies upon the ability of an 1251 irradiation source to excite K X-rays from silver atoms with quantitation of the emissions using a suitable detector. Experiments with "silver-doped" skin phantoms indicated that a minimum detectable silver concentration of 3–4 μg g^{-1} is possible in a period of 10–20 minutes. The technique provides a safe and non-invasive means of monitoring populations exposed to silver and allows accurate quantitation of silver concentrations in the skin of patients with occupational argyria.

The NAA method is based upon activation of silver atoms with a source of neutron irradiation and the creation of the radionuclide ^{110m}Ag which emits γ-radiation. Quantification of γ-emissions using a high resolution drifted-germanium detector or other suitable instrument provides a detection limit for silver in human hair of 0.69 μg g^{-1}. ^{110m}Ag has a half-life of 250.4 days providing sufficient time for counting after irradiation and cooling.[94] The NAA method has the advantage of measuring silver concentrations in environmental samples containing other metallic elements.[1,95,96] Good correlations have been reported between the silver content of tissue samples and the level of argyrotic changes.

East and Boddy adapted the NNA method of silver analysis to determine whole body concentrations of silver in patients using silver acetate therapies as an antismoking remedy.[2,97] Their patients were irradiated with low emission neutron radiation (14 MeV) equivalent to 1 rem at the body surface to excite silver atoms in the body. The resulting radioactive nuclide, $^{107}Ag(n,\gamma)^{108}$, with a half life of 2.4 minutes emitted γ-rays which were measured using a suitable whole body counter, silver being distinguished from other trace metals by its distinctive γ-ray emission profile. Alternatively, these authors monitored body silver retention using a low level radioactive silver tracer such as ^{110m}Ag, which was administered either in the form of a radiolabelled lozenge or as an instillate onto the tongue.[2] They claim that this isotope with a half-life of 253 days and emitting γ-rays with peaks at 0.658, 0.764 and 0.885 MeV provides a safe and convenient technique for measuring blood and urine as a means of investigating silver metabolism and elimination from the body.[2,98] In an experimental situation, the radioactive tracer ^{110m}Ag has been used to monitor the uptake of silver in mice given silver nitrate solutions (0.03 mg L^{-1}) in their drinking water.[99] Animals were dosed for one or two weeks with ^{110m}Ag (specific radioactivity 30 MBq g^{-1}), and silver deposition in soft tissues (notably the brain) and urinary excretion quantified using a 658 kV Ag- peak and Ge-detector connected to a multichannel analyser. The detection limits of silver in blood were 0.96 ng g^{-1}, kidney 1.04 ng g^{-1} and musculus soleus 26.10 ng g^{-1}, but levels of silver in faeces and urine were below the detection limits of the methodology.

Electrochemical techniques are deemed to be relatively inexpensive and accurate means of measuring silver ion (and not total silver content) in biomaterials. Whereas voltammetric methods (VAM) are less widely reported in the estimation of silver ion in tissues and fluids, they have been developed as a means of determining silver released from biomaterials used in urinary catheters.[37,100,101] Silver released from impregnated polyurethane catheters can be measured in both static and dynamic models to establish the longevity of the antibiotic action of the catheter material and its ability to release silver into a patient's body. The method involves electric cells fitted with either a single rotating glassy carbon electrode or a three electrode system comprising a platinum counter electrode, a carbon electrode and a 0.1N potassium chloride calomel reference electrode. In each case, the equipment was designed to measure changes in current due to silver ions eluted into the electrolyte. Changes in current attributable to silver are detected using a suitable potentiometer and compared to reference standards. The prototype equipment constructed in the authors' own workshop was housed in a protective Faraday-type cage to eliminate any external interference. Samuel and Guggenbichler employed a voltammetric method to evaluate the silver eluting from intraurethral catheters containing nanoparticulate silver using a continuous flow apparatus in which catheters were perfused with a simulated human urine.[37] They claimed that their method was sensitive enough to monitor silver elution of 250 ng Ag$^+$ per day. The anode stripping voltammetric (ASV) method developed by Kumar and the Münstedt is available for measuring silver release from biomaterials and in monitoring the identity and concentration of analytes in aqueous and non-aqueous media.[101]

Visual appreciation and qualitative estimation of silver in human and animal tissues is traditionally achieved using light microscopy. In cases of argyria, silver deposits that have been characterised as silver sulfide or silver selenide precipitates are localised intracellularly and in extracellular sites as spherical electron-dense black-brown granules.[26,102,103] These spherical, ovate or irregularly shaped granules are intensely refractile with dark-field illumination, and X-ray microanalysis has shown them to be associated with other metal elements including mercury, titanium and iron. Electron microscopy of silver sulfide/silver selenide deposits in the skin following occupational silver exposures has shown the granules to be 30–100 nm in diameter and most numerous in the region of the basal lamina, dermal elastic fibres, hair follicles and small blood vessels, but not in epidermal tissues. There is good histological evidence of silver deposits in macrophages, both in occupational exposure and in the use of silver impregnated in dwelling catheters, acupuncture needles and bone cements.[88,104,105]

Argyria is the commonest of the complications seen with occupational silver exposure and many authors have attempted to further examine the structure of the granules using scanning and transmission electron microscopy, X-ray probe energy dispersive microanalysis and histochemical techniques. In each case the observations are complementary, the granules are generally too dark and heavily blackened to detect any internal structure. Buckley *et al.* demonstrated that the granules of silver sulfide/silver selenide could be removed by treating histological sections with ferricyanide bleach (2% potassium ferricyanide + 1% potassium bromide in water) for five minutes.[103] In their opinion, a small granule of about 25 Å diameter would contain 150 molecules and have a calculated mass of 6×10^{-20} g (assuming that the granules are approximately spherical in shape and that the density of finely dispersed particles is the same as that of metallic silver or silver sulfide in bulk). Silver sulfide granules of 20, 300 or 400 Å would have calculated masses of 3.7×10^{-17}, 1.035×10^{-16} and 2.45×10^{-16} grams, respectively.

Electron microscopy supported by microchemical and histochemical techniques may have advantages over electron probe analyses in identifying minute deposits of silver sulfide in tissues. Buckley *et al.* considered that electron probe X-ray analysis is capable of detecting silver sulfide deposits in the skin at least at levels of 1×10^{-14} g μ^{-2}, assuming that the sections were of 5μ in thickness.[103] With electron probe analysis, silver sulfide granules displayed prominent emission peaks characteristic of silver at K_α 22.10 keV, K_β 24.9 keV, L_α 2.98 keV and M 0.75 keV.[2,97,98]

Danscher *et al.* developed a silver "amplification" technique identified as autometallography for detection of silver in electron microscopic sections.[106] Their technique is based on the capability of metallic silver, silver sulfide and silver selenide to catalyse reduction of silver ions to metallic silver.[107] To ensure reduction of silver ions not bound to sulfur or selenium anions, histological or electron microscopy sections were exposed to ultraviolet (UV) light before development with a reagent such as hydroquinone. In their opinion, physical development amplifies trace deposits of silver in tissues until they reach a size

Figure 2.4 The Trace Metals Laboratory at Charing Cross Hospital, London, showing the mass spectrometer used for analysis of silver and other metals in tissues and body fluids. *By courtesy of Mr B. Sampson.*

that is detectable in light and electron microscopy sections. It is unclear how sensitive this method is, but to ensure specificity and reliability, it may well be supplemented by use of X-ray microanalysis or other quantitative means of measuring and visualising silver in tissue sections.[27,108]

Mass spectrometry is used increasingly these days for silver determination in body tissues and fluids (Figure 2.4). The technique is accurate to levels of at least $0.1 \, \mu g \, g^{-1}$ but the biological relevance of minute levels of silver is questionable (B. Sampson, personal communication).

2.4 Nanotechnology

Various technologies developed in recent years have sought to maximise ionisation of metallic silver or silver compounds and release of high levels of Ag^+ for antibiotic action.[13] Ionisation of metallic silver in the presence of moisture is a temperature, pH and time related process, but at room temperatures the solubility/ionisation of silver is low ($<1 \, \mu m L^{-1}$). The ionisation rate of metallic silver is proportional to the surface area of metal area/particle size exposed.[13] Spadaro and Becker pioneered electrolytic means of enhancing ionisation of silver and applied their technology in the production of antibiotic bone cements (polymethyl methacrylate), orthopaedic infections and bone repair.[109–111] However, the new concept of "sustained silver ion-release technology" has been beneficially developed over the past 20 years to maximise release of silver ions into a wound to provide maximal antibiotic efficacy without compromising patient comfort or safety. The release of biologically active silver ion (Ag^+) and antimicrobial capacity of a product is proportional to the surface area of particles exposed to moisture.[112]

Nanotechnology represents a major advance in silver science and production of devices with antibiotic action. Technology involving an atomisation process introduced by the Nucryst Pharmaceuticals division of the Westaim Corporation (Canada) facilitates production of silver ions of 5–20 nm (20 nm diameter) or less for rapid action wound dressings such as Acticoat® (Smith & Nephew Healthcare) which are designed for disinfection of ulcers and skin wounds with moderate to high bacterial colonisation.[113] Nucryst Pharmaceutics claims[92] that nanocrystalline silver particles are appreciably more soluble and ionise more rapidly than silver foil or silver compounds used in wound management. The Silcryst™ Technology affords appreciably higher microbicidal action than materials based on silver nitrate, silver sulfadiazine or polyurethane film products impregnated with ionisable silver salt.[13,113] Atomisation is considered to alter the physical and chemical properties of silver particles by decreasing their size and increasing the "proportion of exposed atoms" on their surface (Figure 2.5). Birringer suggested that the grain boundary region may represent a "new state of solid matter".[114] Burrell indicated that the nanocrystalline particles exhibit unique dissolution properties, ionising to release Ag^+ and $Ag°$ ions. The dissolution process is sensitive to temperature and reaches a steady state with silver concentrations of 70–100 µg mL^{-1}.[13,114] The ionisation of nanoparticulate silver is 70–100 times greater than with silver foil, wire or powder.

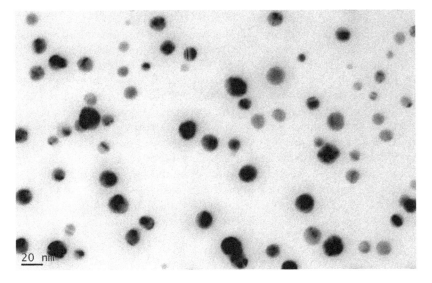

Figure 2.5 Nanocrystals of metallic silver < 20 nm diameter. The high surface to volume ratio of nanocrystals greatly increases their capacity for ionisation-reflecting special features—the "grain boundary phenomenon".[13,114] *By courtesy of Dr S. Misra, Department of Mineralogy, Natural History Museum, London.*

Antimicrobial technologies involving silver developed in recent years in relation to the sustained silver-release wound dressings include the adsorption of metallic silver to charcoal, impregnation of nanoparticulate silver into polymers for use in wound dressings and catheter materials, and silver ion beam deposition to provide a silver coating for biomaterials.[19,38,115] A novel silver hydrogel-impregnated catheter designed to reduce catheter-associated urinary tract infections developed by Maki *et al.* in 1998 was claimed to be effective against a range of infections resistant to common antibiotic therapy, including infections associated with biofilm formation.[116]

2.5 Silver Compounds in Antibiotic Therapy and Medical Devices

Metallic silver, silver sulfadiazine and a wide range of other silver compounds and complexes are now used to provide antibiotic properties to:

- wound dressings;
- catheters for urethral drainage;
- intravascular insertion;
- intracerebral cannulation and intraperitoneal use;
- cardiac devices (heart valves and prostheses);
- bone cements and orthopaedic devices;
- water purification;
- textiles and domestic appliances.

In each case, the form of silver selected is based upon its stability, ionisation in the presence of moisture and body fluids, and biocompatibility with surrounding tissues (Table 2.2).

Although 0.5% silver nitrate and 1% silver sulfadiazine have been used for many years as the principal delivery vehicles for silver in wound care and treatment of topical infections, they are less than ideal as antibiotics because of the need to provide large excesses of silver to compensate for losses due to chemical inactivation, the presence of undesirable counter-ions and components of the carrier creams and the variety of other inorganic and organic complexes introduced. Recent research has established that only a proportion of "free" ion released from a silver product is actually available to interact with and kill bacteria, fungi and protozoal infections. The antibiotic efficacy of a silver compound (A) can be represented by the equation:

$$A = S^t - (B^1 + B^2 + B^3)$$

where S is the total amount of silver ion released at a given temperature t, and B^1, B^2, and B^3 represent the amount of silver ion bound to anions (*e.g.* chloride), protein residues and tissue debris, respectively.

Table 2.2   Silver Compounds as Antibiotic Agents.

Compound	Formula
Chloride	AgCl
Iodide	AgI
Sulfate	Ag_2SO_4
Carboxymethyl cellulose	Ag-complex
Zeolite	Ag-complex
Oxide	Ag_2O
Phosphate	Ag_3PO_4
Carbonate	Ag_2CO_3
Zirconium salt[a]	Ag-complex
Lactate	Ag-lactate
Metallic silver	Ag^0
Nanocrystalline silver	Ag^0
Allantoinate	Ag-complex
Proteinate	Ag-proteinate
Citrate	$AgC_6H_7O_7$, H_2O_2-Hydroxy propane tricarboxylic acid, silver salt
Alloys used in catheters	Complex

[a]silver zirconium lactate.

Current technology aims to produce silver-containing wound dressings and medical devices that release silver ions in sufficient amounts to maximise the concentrations available for antibiotic action without compromising the health of patients or exposed personnel over the product's expected lifetime. Secondly, silver ion release is prolonged to give longer "active life" and more economical use of products. The cost efficacy balance becomes increasingly important as healthcare budgets decrease.[117,118] New technologies aim to improve and regulate silver ion release with respect to expected levels of infection and are increasingly tailored towards intended use of a product.

2.5.1   Silver Nitrate

Silver nitrate **1** has a molecular weight of 169.98 and is the salt of silver and a strong acid. As an analytical reagent, it is obtained in a very pure form (99.9–100%) and its chemical properties well known.[119]

$$Ag^+ \quad \begin{array}{c} O^- \\ | \\ N^+ = O \\ | \\ O^- \end{array}$$

1

Silver nitrate is the most soluble of all inorganic silver salts and readily dissolves in water ($> 2000 \, g \, L^{-1}$) at room temperature to give a colourless, odourless and stable solution which discolours on exposure to solar radiation.

It is the simplest and most efficient carrier for releasing silver ions into a moist environment for antibiotic action and has been in use as a chemical antibiotic, astringent and caustic agent for at least 100 years.[8,120] (Mercuric salts were also highly effective antibiotics in the late 1800s but withdrawn in view of their serious of health risks.)

Medical applications of silver nitrate documented in pharmacopoeias list local antiseptic, astringent, corrosive agent for removal of calluses, excessive granulations and stimulation of indolent wounds. The preparations differ greatly in concentration (and acidity), expected duration of action and mode of application. Records show that silver nitrate solutions [$AgNO_3$, United States Pharmacopeia (USP)] were introduced into pharmacopoeias in the 19th century as *Argenti nitras* or lunar caustic as astringents, strong antiseptics and caustic agents for ablation of warts and stimulation of ulcers and granulations. Silver nitrate in the form of pencils (lunar caustic) sometimes mitigated with potassium nitrate was moistened before application to disfiguring skin conditions.[121] Painting with a 10–20% solution achieved the same objective, but these preparations were a serious irritant to the skin. In surgery and in wound clinics, silver nitrate pencils or sticks are still available for treatment of calluses, warts and cosmetically unsightly granulations. In our clinic, Toughened Silver Nitrate (Avoca, Bray) was used to remove a troublesome skin lesion complicated by a bone fragment.[24] In this case, the silver ion provided an antibiotic action whilst the nitric acid acted on unwanted callus tissue. Silver ion precipitates as silver sulfide with skin keratins and many other proteins, but the nitric acid released is highly corrosive. Other prescribed uses for strong or Toughened Silver Nitrate include treatment for warts (not facial or anogenital warts) and verrucae, with recommendations for neutralising with physiological saline.[121]

Apart from their ability to evoke severe irritancy to the skin and mucous membranes, silver nitrate pencils, sticks or solutions used in topical therapy exhibit the major disadvantage of discolouring the tissue and everything in contact black-brown. Whereas the disfigurement is usually temporary and largely superficial, it may be a cause of social embarrassment. In time, discoloured tissues tend to wear away by continual environmental contact and wear-and-tear. In contrast, argyria-like symptoms frequently associated with occupational exposure to silver or colloidal silver preparations used for gastrointestinal or rhino-laryngeal infections are long-lasting and seriously disfiguring as in the case of the "silver man".[122] Suggested methods for reducing superficial skin stains due to silver sulfide precipitates include rubbing with potassium iodide crystals or paining with 10% potassium iodide solution.[123] Other, less safe remedies for silver nitrate discolouration are solutions of 10% potassium cyanide or 10% mercuric chloride in 10% ammonium chloride,[124] neither of which are listed in modern pharmacopoeias.

Although silver nitrate was possibly used first by John Woodall in the 16th century as "the surgeon's mate", silver nitrate solution was possibly first recognised as an antiseptic by the French neonatologist, Credé who in 1884 reported using 1% solutions to alleviate neonatal eye infections.[125] Although he claimed that the incidence of infection declined from 10.8 to 0.2% in his clinic, later statistical analyses suggest that about 2% of Credé's therapy included vaginal douching with 2% phenol. Later investigators found that 0.25% and 0.01% silver nitrate was efficacious in alleviating typhoid and anthrax bacilli respectively.[126] Lehrfeld recommended superficial sterilisation of the birth canal with boric solution followed by 0.5% silver nitrate for three days followed by boric solution as a prophylactic for neonatal eye infections.[127] In his *Manual of Pharmacology*, Sollemann cautioned in using silver nitrate in treating any eye infections or corneal lesions on account of its propensity to cause blindness; he advised a single drop of 1% silver nitrate as marketed then in wax capsules to be placed in the conjunctival sac at birth after cleansing the eyelids, followed by a saline rinse.[120]

Other uses of silver nitrate in the 1940s included antismoking remedies and "pills" of silver nitrate as remedies for gastric ulcer.[120] Although East *et al.* have discussed at length the merits of silver acetate as an antismoking remedy,[2] silver nitrate and other silver preparations are not presently licensed for oral use or therapy for gastrointestinal complaints in many countries including the European Union, USA and Australia. Pills of silver nitrate (10–20 mg) containing kaolin and petrolatum exhibit only slight efficacy against peptic ulcer but carry unacceptable risks of irritancy and argyria.[120]

Silver nitrate still retains important clinical value in burn wound clinics, where infection with *Pseudomonas aeruginosa*, *Staphylococcus aureus*, *Streptococcus pyogenes* and occasionally *Escherichia coli* can be life threatening.[128,129] Edward Lowbury conducted extensive comparative studies on silver antibiotics between 1972 and 1980 and emphasised that silver nitrate is still preferred to most other antibiotics as bacteria did not develop resistance to silver. In his clinic, the frequency of isolation of *Pseudomonas aeruginosa* in burns patients was reduced from 70% to 3% in a trial using 0.5% silver nitrate compresses; this was associated with improved clinical results and reduced mortality.[129] Later studies showed that silver sulfadiazine cream developed by Charles Fox[130] appeared to have outstanding promise as an alternative to silver nitrate solution, and that controlled trials in the Industrial Injuries and Burns Unit of Birmingham Accident Hospital showed that its prophylactic value on extensive burns compared favourably with 0.5% silver nitrate.[131]

2.5.2   Silver Sulfadiazine (Flamazine, Flammazine)

Silver sulfadiazine (SSD) is possibly the most widely used of the silver antibiotics is wound therapy and medical devices.[129,132] It was developed at a time when hypotonic (29.4 mM L^{-1}) silver nitrate therapy was contraindicated on account of its irritancy and propensity to cause disturbances in body sodium, potassium and chloride electrolyte balances. Mafenide, which was commonly used in the 1950s–1960s to treat life-threatening *Pseudomonas aeruginosa* infections, also evoked electrolyte imbalances but as a profound inhibitor of carbonic anhydrase; it led to hyperpnea and hyperchloraemic acidosis.[133,134] SSD was developed in 1968 by Charles Fox, a microbiologist who recognised the major contribution made by sulfonamides in the prophylaxis of wound infections in World War II and was aware that there was a reappraisal of the potential clinical benefits to be gained by using dilute silver nitrate.[135] By combining the weak acidic properties of sodium sulfadiazine with silver nitrate, Fox produced a white, fluffy, complex salt that was virtually insoluble in water (1 in 10 000).[136] Fox concluded from his initial experiments that, unlike silver nitrate, SSD did not precipitate chloride from body fluids as silver chloride, was a very low irritant and did not stain the tissues,[137] but retained the oligodynamic properties of silver ion proposed by von Nägeli in 1893,[16] *i.e.* the ability of silver sensitive bacteria to concentrate silver ion from dilute solutions.[138] In effect, Fox had combined the antibiotic potency of Ag^+ with the acclaimed antibacterial action of the sulfonamide moiety. With a relatively low health risk threshold, SSD has become the most popular antibiotic for topical application in burn and scald clinics with particular efficacy in treating *Pseudomonas aeruginosa*.[129] In his preliminary experiments, Fox claimed that 30 mM SSD was not toxic in mice (burn wound model) or human patients subject to thermal injuries, and was not likely to sensitise. This latter claim has been contested by later clinicians.[139,140]

The chemistry of SSD is complex, but the white microcrystalline powder comprising silver atoms co-ordinated to three sulfadiazine molecules has exhibited antibiotic efficacy against more than 95% infections found in human wounds. Using X-ray crystal structure analysis, Cook and Turner investigated the chemical configuration of the "monolithic" crystals and identified silver atoms co-ordinated to three sulfadiazine molecules by a distorted tetrahedral arrangement of three nitrogen atoms and one oxygen atom.[141] Other studies demonstrated that silver is ionically bound to the amide nitrogen atoms and that oxygen atoms are not co-ordinated to the silver ion, in contradiction to Fox who had claimed that SSD was analogous to silver amines.[142] Chemically, SSD has the formula $AgC_{10}H_9N_4O_2S$ and a molecular weight of 357.137 (Figure 2.6).[137]

Application of SSD to burn wounds is considered to be like treating a wound with a "reservoir of silver ions".[143] During its continuous dissociation in the presence of body secretions and burn wound exudates, much of the silver ion released binds avidly with albumins and macroglobulins leaving the remainder to provide antibiotic function. Estimates vary concerning the uptake and metabolism of the sulfadiazine moiety, but early estimates suggested that up to 10% is absorbed into the human body to bind to plasma proteins to give a circulating concentration of $10-50\,mg\,L^{-1}$.[137,144] Exceptionally, plasma sulfadiazine levels of $150\,mg\,L^{-1}$ have been recorded. Sulfadiazine is metabolised in the liver by N_4-acetylation and by 5-hydroxylation,[144] but as much as 50% of absorbed material is excreted as unmetabolised drug and 40% as an acetylated metabolite (Figure 2.7). Boosalis *et al.* demonstrated that silver excretion in patients with burns extending over more than 60% of their body (TBSA) was

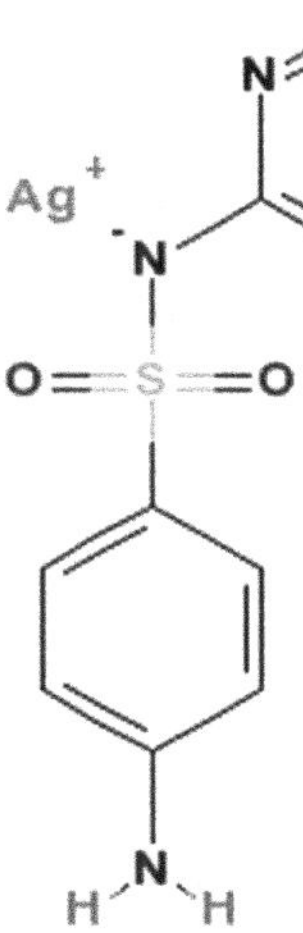

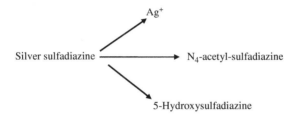

Figure 2.6   Silver sulfadiazine-4-amino-*N*-2-pyrimidinylbenzenesulfonamide mono-silver (1 +) salt, monosilver 2-sulfanilamidopyrimidine.

Figure 2.7   Metabolism of silver sulfadiazine (Ag$^+$ is released in wound sites and the sulfadiazine moiety is metabolised in the liver).

greatly increased following SSD therapy; mean peak concentrations of 1100 µg in 24 hours illustrate the high levels of silver absorption through burn wounds.[49] Urinary silver excretion in untreated patients was less than 1 µg in 24 hours. Further, these authors considered that urine silver excretion could serve as a sensitive index of silver uptake by cutaneous absorption. Wang *et al.* also evaluated percutaneous uptake of silver sulfadiazine in extensive burns extending to >20% TBSA and demonstrated silver deposits in the tissues, but this argyria-like staining diminished within one year.[145]

SSD (1%) is incorporated into sustained silver ion release wound dressings and other devices, and has been shown to be clinically efficacious as an antimicrobial agent whilst being entirely compatible with hyaluronic acid, epidermal growth factor, lipocolloids, hydrocolloids, chlorhexidine sulfate and polymers used in catheter materials without losing its antibiotic potency. In a new product designed for burn wound therapy, SSD was found compatible with a drug delivery system containing polyethylene glycol 400, poly-2-hydroxyethyl-1-methacrylate.[146] It was effective in clinical trials and found to be safe in use. Other novel SSD delivery systems include liposome encapsulation for soft tissue wounds therapy; preliminary trials have shown this formulation effective against *Pseudomonas aeruginosa*.[147] Technologies developed have incorporated SSD and other silver compounds in coatings or impregnates for medical catheters. Whilst claims have been made as to their antimicrobial efficacy in *in vitro* systems, they exhibit limited capacity to inhibit bacterial or fungal colonisation and biofilm formation *in vivo*.[148] This does not reflect molecular changes in the SSD, but is more likely to be attributable to free silver ion released combining with anions, proteins and other non-specific moieties in the environment. As discussed in more detail later, bacteria in biofilms are particularly resistant to antibiotic including silver. It is possible that some free silver ion binds to the mineralised calyx surrounding the biofilm.

The principal advantages of SSD as an antibiotic over silver nitrate can be summarised as follows:

- safety in use;
- improved stability in light, with longer shelf life;
- low capacity to discolour the tissues or wound dressings, bandages, *etc.*;
- slower release of silver ion and longer lasting therapeutic and prophylactic action;
- wider sphere of antibiotic efficacy than either Ag^+ or sulfadiazine alone;
- compatibility with other materials used in wound care and medical devices.

2.5.3   Colloidal Silver

Colloidal silver products are the most contentious of the silver products introduced as antibiotics in human and veterinary medicine. They appeared in pharmacopoeias more than 100 years ago when the antibiotic action of silver nitrate solutions ($> 5\%$) were contraindicated on account of their profound irritancy and corrosive action.[120] Colloidal silver products were based on the dispersal and suspension of ultrafine particles of silver or an insoluble silver salt [*e.g.* silver chloride, silver oxide (Ag_2O) or silver proteinate] in water or water plus denatured protein. In each case, the fine particulate materials were presumed to become sufficiently miscible with the water to form long lasting "solutions". Correctly, they were ultramicroscopic suspensions of insoluble materials in what some authors refer to as a "continuous medium". Colloidal products were phased out of national formularies in Europe (Medical Devices Agency), USA (United States Pharmacopoeia, National Formulary) and Australia (Therapeutic Goods Administration) at least 30 years ago as more efficacious, better defined and safer antibiotics such as silver sulfadiazine, sulfonamides and penicillin became available.[149–151]

At least 50 colloidal silver preparations have been produced in the past 100 years, possibly following the early philosophy of Dr Henry Crooks and others who reportedly claimed that:

"*... certain metals when in a colloidal state exhibit a highly germicidal action but are quite harmless to human beings. In fact, no microbe is known that is not killed by colloidal silver in laboratory experiments in six minutes*".[i]

Dr Torald Sollemann in his *Manual of Pharmacology* provided an insight into the colloidal silver preparations available in the USA in 1942.[120] At that time, colloidal silver products were classified under "strong" silver proteins (Protargol type) or "mild" silver proteins (Argyrol®, Collargol, *etc.*) on the basis of their "therapeutic action" (*i.e.* the level of silver ion released) rather than the total amount of silver present. Protargol is the best example of strong silver proteins; it contained low concentrations of metallic silver (7–8.5%), which were diluted to 0.05 to 10% for local use, but most of the silver content ionised in solution to provide strong antibiotic action. The solutions were supposedly less irritant (?one tenth) than equivalent concentrations of silver nitrate.[152] Argyrol®, Collargol, Silvol, Cargentos and Solargentum are examples of mild silver proteins. These contained higher total concentrations of silver (19–23% Ag) but, on dilution to 0.1 or 5% for topical application, released lower levels of silver ion and were less of an irritant. Mild silver proteins were supposedly soothing to the skin and provided a mild antibiotic action, assuming that they were freshly prepared or containing a suitable stabiliser. They were dispensed in amber coloured bottles to avoid premature oxidation.[152,153] Other colloidal silver products listed in earlier US National Formularies included silver nitrate containing 28.5–30.5% silver were recommended for use without dilution. Colloidal silver antibiotics containing the minimally soluble compounds silver iodide or silver chloride were marketed and, according to Sollemann, they did not differ in principal from other colloidal silver products (some are more ionised, some less). Colloidal silver iodide solutions contained 18–22% silver, but this was diluted to 0.05 to 10% to avoid irritancy.

More recently, manufacturers and suppliers operating through clandestine channels and retail outlets (internet communications, *etc.*) have sought to re-introduce colloidal silver products with far-reaching and largely unsubstantiated claims for their efficacy against a wide range of human infectious diseases and medical conditions. Commentaries and product promotion documents claim that these new colloidal silver products are efficacious against at least 650 human infections and diseases (including many that have no known underlying infective aetiology), but claims lack verification by competent clinical laboratories. It is unfortunate that some colloidal silver products have been promoted using misquoted information and scientifically flawed arguments citing respectable publications. One such irrelevance states that a colloidal silver product is

---

[i] H. Crooks, 1910, "The uses of colloidal silver", London. Cited by A. B. Searle, *Br. Med. J.*, 1913, p. 70.

believed to act as a "systemic disinfectant" and works like a "secondary immune system". Some colloidal products have been marketed under nutritional products, even thought silver has long been recognised as having no trace metal value or known physiological role in the human body.[42,150] The misleading expression "silver deficiency disease" is not scientifically recognised or medically valid. Of greatest concern is the high incidence of argyrias and argyroses experienced by patients who have consumed colloidal silver proteins for gastrointestinal or respiratory infections without appropriate medical supervision. Occasionally, these discolorations are profoundly disfiguring and are permanent (B. A. Bouts, personal communication).

Strong silver proteins like Protargol would have been prepared by precipitating a "peptone" (albumose) solution with silver nitrate or moist silver oxide, dissolving the resulting silver peptonate in an excess of protalbumose and drying in a vacuum.[154] Proargentum, on the other hand, was produced by reacting silver nitrate with gelatine in alkaline medium followed by precipitation by alcohol and drying. In contrast, mild silver proteins like Silvol, Cargentos and Solargentum were prepared by interaction of an alkaline protein with moist silver oxide and drying in a vacuum. Collargol, containing as much as 78% silver, was prepared by reducing silver nitrate with ferric citrate and then stabilising the solution with alkaline albumin.[155] The 1942 pharmacopoeia identified a preparation of Protargol for intravenous injection as provoking a so-called "colloidal reaction" in which the proteins play a major role.[120]

More recent methods of producing colloidal silver solutions rely on electrolytic processes in which direct or alternating currents are passed between silver electrodes in water. The resulting solutions are frequently too dilute, unstable or insufficiently ionised for antibiotic purposes.[155] Alternatively, colloidal silver has been prepared using tannic acid and sodium carbonate to reduce silver nitrate and produce clear tea-coloured colloidal solutions. A modification of an earlier method involved heating an aqueous solution of silver nitrate to boiling point and then adding sodium citrate to produce a cloudy greyish suspension. Van Hassault *et al.* evaluated all three methods of producing colloidal silver solutions and verified the ultrafine silver particles using transmission microscopy [10–30 nm (tannic acid method) and 20–50 nm (sodium citrate)].[155,156] However, in their comparative microbiological studies, no significant antibacterial action was evident for any of the colloidal silver products in laboratory cultures. Further, a commercial product supposedly containing 22 ppm silver in water and having silver particles of 10 nm was without identifiable action on common strains of pathogenic bacteria (*Pseudomonas aeruginosa, Staphylococcus aureus, Escherichia coli, etc.*) seen in skin wounds.

The official status of colloidal silver products proposed by the US Food and Drug Administration (FDA) in 1996 classified:

"all over the counter (OTC) products containing colloidal silver ingredients or silver salts for internal or external use as not generally recognised as safe (GRAS) and are misbranded".[150]

The FDA issued this proposal as a response to the marketing of products labelled as containing colloidal silver or silver salts with recommendations for treating numerous human diseases. At the time, FDA was not aware of any substantial evidence supporting manufacturers' claims. A similar view was taken by the Therapeutic Goods Administration (Australia), which sought to limit the illegal importation of colloidal silver products.[151]

2.5.4 Miscellaneous Silver Compounds

A wide range of silver compounds and complexes with a capacity to liberate biologically active silver ions, solubility and compatibility with human body fluids and secretions are employed as antibiotics in wound dressings, medical devices, textiles or domestic products (Table 2.2). Occasionally, the silver content of a dressing or other product is identified only as "ionic silver" or "Ag^+ "with the true identity of the silver complex undefined. In each case, the antibiotic properties of a silver containing or coated wound dressing, catheter, bone cement, *etc.* are directly proportional to the levels of "free Ag^+" available for interaction with bacterial or fungal cell walls as a preliminary to microbicidal action.[41]

References

1. R. Dams, J. A. Robbins and K. A. Rahn, Non-destructive neutron activation analysis of air pollution particulates, *Anal. Chem.*, 1970, **42**, 861.
2. B. W. East, K. Boddy, E. D. Williams and D. MacIntyre, Silver retention, total body silver and tissue silver concentrations in argyria associated with exposure to anti-smoking remedy containing silver acetate, *Clin. Exp. Dermatol.*, 1980, **5**, 305.
3. Ministry of Natural Resources, 1985, reference to Chaldean silver urn in Louvre Museum, Paris.
4. G. Sykes, Disinfection of viruses, in *Disinfection and sterilization*, Spon, London, 2nd edn., 1965, pp. 291–306.
5. A. Hambidge, Reviewing the efficiency of alternative water treatment techniques, *Health Estate*, 2001, **55**, 23.
6. Z. Liu, J. E. Stout, L. Tedesco, M. Boldin, C. Hwang, W. F. Diven and V. L. Yu, Controlled evaluation of copper–silver ionization in eradicating *Legionella pneumophila* from a hospital water distributing system, *J. Infect. Dis.*, 1994, **169**, 919.
7. U. Rohr, M. Senger and F. Selenka, Effect of silver and copper ions on survival of *Legionella pneumophila* in water, *Zentralbl. Hyg. Umweltmed.*, 1999, **198**, 514.
8. R. J. White, A historical overview of the use of silver in wound management, *Br. J. Nurs.*, 2001, **10**(Suppl), S3.
9. H. Klasen, Historical review of the use of silver in the treatment of burns, Part 1. Early uses, *Burns*, 2000, **26**, 117; Part 2. Renewed interest for silver, *Burns*, 2000, **26**, 131.

10. A. B. Searle, *The Use of Metal Colloids in Health and Disease*, E. P. Dutton, New York, 1919, p. 75.
11. M. J. Winter, Silver: the essentials, in *WebElements: The Periodic Table on the WWW*, www.webelements.com/silver, accessed 11 November 2009.
12. S. Patel and Z. Rappaport, *The Chemistry of the Organic Derivatives of Gold and Silver*, Wiley, Chichester, 1999.
13. R. E. Burrell, A scientific perspective on the use of topical silver preparations, *Ostomy Wound Manage.*, 2003, **49**(Suppl. 5A), 19.
14. R. O. Becker, Silver ions in the treatment of local infections, *Met. Based Drugs*, 1999, **6**, 311.
15. L. J. Ming, Structure and function of "metalloantibiotics", *Med. Res. Rev.*, 2003, **23**, 697.
16. K. W. von Nägeli, Leben die oligoynamischen Erschreinungen an lebenden Zellen, *Denkschr. Schweiz. Naturforsch. Ges.*, 1893, **33**, 174.
17. H. Shintani, Modification of medical device surface to attain anti-infection, *Trends Biomater. Artif. Organs*, 2004, **18**, 1.
18. T. Elliott, The role of antimicrobial central venous catheters for the prevention of associated infections, *J. Antimicrob. Chemother.*, 1999, **43**, 441.
19. E. J. Tobin and R. Bambauer, Silver coating of dialysis catheters to reduce bacterial colonisation and infection, *Ther. Apher. Dial.*, 2003, **7**, 504.
20. D. Langaki, M. F. Ogle, J. D. Cameron, R. A. Lirtzman, R. F. Schroeder and M. W. Mirsch, Evaluation of a novel prosthetic heart valve incorporating anticalcification and antimicrobial technology in a sheep model, *J. Heart Valve Dis.*, 1998, **7**, 633.
21. J. A. Spadaro, D. A. Webster and R. O. Becker, Silver polymethyl methacrylate bone cement, *Clin. Orthop. Relat. Res.*, 1979, **143**, 266.
22. S. R. Watt-Smith, *The Surgeon and the Goldsmith*, The Worshipful Company of Goldsmiths, London.
23. J. F. Malaigne, The works of Ambrose Paré (1510–1590), The Cowlishaw Collection, London, 1649.
24. A. B. G. Lansdown, Pin and needle tract infections: the prophylactic role of silver, *Wounds UK*, 2006, **2**, 51.
25. L. F. Siebert, Silver in radiology: problem or asset? *Med. Electron.*, 1988, **19**, 108.
26. S. S. Bleehan, D. J. Gould, C. I. Harrington, T. E. Durrent, D. N. Slater and J. C. Underwood, Occupational argyria: light and electron-microscopic studies and X-ray microanalysis, *Br. J. Dermatol.*, 1981, **104**, 19.
27. S. Sato H. Sueki and A. Nishimura, Two unusual cases of argyria: the application of an improved tissue processing method for X-ray microanalysis of selenium and sulphur in silver-laden granules, *Br. J. Dermatol.*, 1999, **14**, 158.
28. P. D. Mehta, Diagnostic usefulness of cerebrospinal fluid in multiple sclerosis, *CRC Crit. Rev. Clin. Lab. Sci.*, 1991, **28**, 233.

29. U.-C. Hipler, P. Elsner and J. W. Fluhr, A new silver-loaded cellulosic fiber with antifungal and antibacterial properties, *Curr. Probl. Dermatol.*, 2006, **33**, 165.

30. E. A. Deitch, A. A. Marino, V. Malakonok and J. A. Albright, Silver nylon cloth: *in vitro* and *in vivo* evaluation of antimicrobial activity, *J. Trauma*, 1983, **27**, 301.

31. S. Zikeli, Production process of a new cellulosic fiber with antimicrobial properties, *Curr. Probl. Dermatol.*, 2006, **33**, 110.

32. A. Gauger, Silver-coated textiles in the therapy of atopic eczema, *Curr. Probl. Dermatol.*, 2006, **33**, 152.

33. U. Wollina, M. B. Abdul-Naser and S. Verma, Skin physiology and textiles—considerations of basic interactions, *Curr. Probl. Dermatol.*, 2006, **33**, 1.

34. J. W. Fluhr, D. Kowatski, A. Bauer, P. Elsner and C. Hipler, Silver-loaded cellulose fibres with antibacterial and antifungal activity *in vitro* and *in vivo* on patients with atopic eczema, 5th World Textiles Conference AUTEX 2005, Portorož, Slovenia, 2005, p. 133.

35. G. Ricci, A. Patrizi, F. Bellini and M. Medri, Use of textiles in atopic dermatitis, *Curr. Probl. Dermatol.*, 2006, **33**, 127–143.

36. M. Juenger, A. Ladwig, S. Staecker, A. Arnold, A. Kramer, G. Daeschlein, E. Panzig, H. Haase and S. Heissing, Efficacy and safety of silver-textile in the treatment of atopic eczema, *Curr. Res. Med. Opin.*, 2006, **22**, 739.

37. U. Samuel and J. P. Guggenbichler, Prevention of catheter-related infections: the potential of a new nano-impregnated catheter, *Int. J. Antimicrob. Agents*, 2004, **23**(Suppl. 1), S75.

38. S. Saint, R. H. Savel and M. A. Matthey, Enhancing the safety of critically ill patients by reducing urinary and central venous catheter-related infections, *Am. J. Respir. Crit. Care*, 2002, **165**, 1475.

39. V. Edwards-Jones, Antimicrobial and barrier effects of silver against methicillin-resistant *Staphylococcus aureus*, *J. Wound Care*, 2006, **15**, 285.

40. A. Gupta, K. Matsui, J. F. Lo and S. Silver, Molecular basis for resistance to silver cations in *Salmonella*, *Nature Medicine*, 1999, **5**, 183.

41. M. Walker, Cochrane, P. G. Bowler, D. Parsons and P. Bradshaw, Silver deposition and tissue staining associated with wound dressings containing silver, *Ostomy Wound Manage.*, 2006, **52**, 42–46.

42. A. T. Wan, R. A. Conyers, C. J. Coombs and J. P. Masterton, Determination of silver in blood, urine and tissues of volunteers and burn patients, *Clin. Chem.*, 1991, **37**, 1683.

43. E. A. Thomson, S. N. Luoma, C. E. Johansson and D. J. Cain, Comparison of sediments and organisms in identifying sources of biologically available trace metal contamination, *Water Res.*, 1984, **18**, 755.

44. J. E. Furchner, C. R. Richmond and G. A. Drake, Comparative metabolism of radionuclides in mammals, IV. Retention of silver-110m in the mouse, monkey and dog, *Health Phys.*, 1968, **55**, 398.

45. F. Parchaso and J. K. Thompson, Influence of hydrologic processes on reproduction of the introduced bivalve *Potamocorbula amurensis* in northern San Francisco Bay, California, *Pac. Sci.*, 2002, **56**, 329.

46. A. R. Flegal, C. L. Brown, S. Squire, J. R. M. Ross, G. M. Scelfo and S. Hibden, Spatial and temporal variations in the silver contamination and toxicity in San Francisco Bay, *Environ. Res.*, 2007, **105**, 34.

47. N. Williams and I. Gardner, Absence of symptoms in silver refiners with raised blood silver levels, *Occup. Med.*, 1995, **45**, 205.

48. W. R. Hill and D. M. Pillsbury, *Argyria: The Pharmacology of Silver*, Williams and Wilkins, Baltimore, MD, 1939.

49. M. G. Boosalis, J. T. McCall, D. H. Ahrenholz, L. D. Solem and C. J. McLain, Serum and urinary silver levels in thermal injury patients, *Surgery*, 1987, **101**, 40.

50. S. Maitre, K. Jaber, J. L. Perrot, C. Guy and F. Cambazard, Increased serum and urinary levels of silver during treatment with topical silver sulfadiazine, *Ann. Dermatol. Venereol.*, 2002, **129**, 217.

51. M. Westhofen and H. Schafer, Generalised argyrosis in man: neurotological, ultrastructural and X-ray microanalytical findings, *Arch. Otorhinolaryngol.*, 1986, **243**, 260.

52. J. M. Schierholz, L. J. Lucas, A. Rump and G. Pulverer, Efficacy of silver-coated medical devices, *J. Hosp. Infect.*, 1998, **40**, 257.

53. A. B. G. Lansdown, B. Sampson, P. Laupattarakasem and A. Vuttivirojana, Silver aids healing in the sterile skin wound: experimental studies in the laboratory rat, *Br. J. Dermatol.*, 1997, **132**, 728.

54. S. N. Kales and D. C. Christian, Hair and metal toxicity, in *Hair in Toxicology: An Important Biomonitor*, ed. D. J. Tobin, Royal Society of Chemistry, Cambridge, 2005, p. 125.

55. A. B. G. Lansdown, Metallothioneins: potential therapeutic aids for wound healing in the skin, *Wound Repair Regen.*, 2002, **10**, 130.

56. R. Tupling and H. Green, Silver ions induce Ca^{2+} release from the SR *in vitro* by acting on the Ca^{2+} release channel and the Ca^{2+} pump, *J. Appl. Physiol.*, 2002, **92**, 1603.

57. A. B. G. Lansdown, Cartilage and bone as target tissues for toxic materials, in *General and Applied Toxicology*, ed. B. Ballantyne, T. C. Marrs and P. Syversen, John Wiley, Chichester, 3rd edn, 2009, ch. 62.

58. G. W. Gould, J. Colyer, J. M. East and A. G. Lee, Silver ions trigger Ca^{2+} release by interaction with $(Ca^{2+}\text{-}Mg^{2+})$ATPase, *J. Biol. Chem.*, 1987, **262**, 7676.

59. R. L. Williams and D. F. Williams, Albumin absorption on metal surfaces, *Biomaterials*, 1988, **9**, 206.

60. A. B. G. Lansdown, A. Williams, S. Chandler and S. Benfield, Silver absorption and antibacterial efficacy of silver dressings, *J. Wound Care*, 2005, **14**, 161.

61. I. Thornton, *Metals in the Global Environment: Facts and Misconceptions*, International Council on Metals and the Environment (ICME), Ottawa, 1995.

62. B. H. Olsen, Microbial mediation of the biogeochemical cycling of metals, in *Applied Environmental Geochemistry*, ed. I. Thornton, Academic Press, London, 1983, p. 201.
63. M. Grayson, Silver and silver alloys silver and silver compounds, in *Kirk-Othmer Encyclopaedia of Chemical Technology*, 3rd edn, 1978, Vol. **21**, pp. 1–32.
64. K. D. Rosenman, A. P. Moss and S. Kon, Argyria: clinical implications of exposure to silver nitrate and silver oxide, *J. Occup. Med.*, 1979, **21**, 430.
65. A. P. Moss, A. Sugar, N. A. Hargett, A. Atkin, M. Wolkstein and K. Rosenman, The ocular Manifestations and functional effects of occupational argyria, *Arch. Ophthalmol.*, 1979, **97**, 906.
66. C. F. Cooper and W. C. Jolly, Ecological effects of silver iodide and other weather modification agents, *Water Resour. Res.*, 1970, **6**, 88.
67. H. G. Petering, Pharmacology and toxicology of heavy metals: silver, *Pharmacol. Ther.*, 1976, **1**, 127.
68. W. L. Roper, *Toxicological Profile for Silver*, Agency for Toxic Substances and Disease Registry, US Public Health Service, Atlanta, GA, 1990.
69. N. S. Fisher, M. Bohe and J.-L. Teyssie, Accumulation and toxicity of Cd, Zn, Ag and Hg in four marine phytoplankters, *Mar. Ecol. Prog. Ser.*, 1984, **18**, 201.
70. R. L. Coleman and J. E. Cearly, Silver toxicity and accumulation in largemouth bass and bluegill, *Bull. Environ. Contam. Toxicol.*, 1974, **12**, 53.
71. C. J. Terhaar, W. S. Ewell and S. P. Dziuba, A laboratory model for evaluating the behaviour of heavy metals in an aquatic environment, *Water Res.*, 1977, **11**, 101.
72. A. Calabrese, J. R. MacInnes and D. A. Nelson, Effects of long-term exposure to silver or copper on growth, bioaccumulation and histopathology in the blue mussel *Mytilus edulis*, *Mar. Environ. Res.*, 1984, **11**, 253.
73. G. Pesch, B. Reynolds and P. Rogerson, Trace metals in scallops from within and around two ocean disposal sites, *Mar. Pollut. Bull.*, 1977, **8**, 224.
74. A. R. Flegal, I. Rivera-Duarte and S. A. Sañudo-Wilhemy, Silver contamination in aquatic environments, *Rev. Environ. Contam. Toxicol.*, 1996, **146**, 139.
75. G. W. Bryan, The effects of heavy metals (other than mercury) on marine and estuarine organisms, *Proc. R. Soc. London, Ser. B*, 1971, **177**, 389.
76. G. W. Bryan and W. J. Langston, Bioavailability, accumulation and effects of heavy metal in sediments with special reference to the United Kingdom estuaries: a review, *Environ. Pollut.*, 1992, **76**, 89.
77. R. Eisler, Silver hazards to fish, wildlife and invertebrates: a synoptic review, in *Contaminant Hazard Reviews*, US National Biological Service, Laurel, MD, 1996, Patuxent Wildlife Research Center Report No. 32.

78. S. N. Luoma, Processes affecting metal concentrations in estuarine and marine coastal sediments, in *Heavy Metals and the Marine Environment*, eds. P. Rainbow and R. Furness, CRC Press, Boca Raton, FL, 1990, p. 51.

79. R. W. Boyle, *Geochemistry of Silver and its Deposits, with Notes on Geochemical Prospecting for the Element*, Department of Energy, Mines and Resources, Ottawa, Ontario, 1968, Geological Survey of Canada Bulletin 160.

80. N. I. Ward, E. Roberts and R. P. Brooks, Silver uptake by seedlings of *Lolium perene* and *Tripolium repens*, *New Zeal. J. Sci.*, 1979, **22**, 129.

81. P. L. Drake and K. J. Hazelwood, Exposure-related health effects of silver and silver compounds, *Ann. Occup. Hyg.*, 2005, **49**, 575.

82. American Conference of Governmental Industrial Hygienists, *Documentation of the Threshold Limit Values and Biological Exposure Limits*, ACGIH, Cincinnati, OH, 5th edn, 1986.

83. A. I. Vogel, *A Text-Book of Quantitative Inorganic Analysis including Elementary Instrumental Analysis*, Longmans, London, 3rd edn., 1961.

84. A. B. G. Lansdown, M. Severin and J. J. Blaker, Punch biopsy wounds in the rat: development of a new model for evaluation of a sustained silver release dressing, presented at European Tissue Repair Society, Annual Conference, Amsterdam, 2003.

85. D. A. Segar and J. L. Gilio, The determination of trace transition elements in biological tissues, using flameless atom reservoir atomic absorption, *Int. J. Environ. Anal. Chem.*, 1973, **2**, 291.

86. G. D. DiVincenzo, C. J. Gordiano and L. S. Schriever, Biologic monitoring workers exposed to silver, *Int. Arch. Occup. Environ. Health*, 1985, **56**, 207.

87. C. J. Coombs, A. T. Wan, J. P. Masterton, R. A. J. Conyers, J. Pedersen and Y. T. Chia, Do burn patients have a silver lining?, *Burns*, 1992, **18**, 179.

88. E. Sudemann, H. Vik, M. Rait, K. Todnem, K. J. Andersen, K. Julsham, O. Friesland and J. Rungby, Systematic and local silver accumulation after total hip replacement using silver-impregnated bone cement, *Med. Prog. Technol.*, 1994, **20**, 179.

89. K. J. Anderson, A. Wikshaland and A. Utheim, Determination of silver in biological samples using graphite furnace atomic absorption spectrometry based on Zeeman effect background and matrix modification, *Clin. Biochem.*, 1986, **19**, 166.

90. C. L. Brown and S. N. Luoma, Use of euryhaline bivalve *Potamocorbula amurensis* as a biosentinel species to assess trace metal contamination in San Francisco Bay, *Mar. Ecol. Prog. Ser.*, 1995, **124**, 129.

91. H. T. Ratte, Bioaccumulation and toxicity of silver compounds: a review, *Environ. Toxicol. Chem.*, 1999, **18**, 89.

92. J. B. Wright, D. L. Hansen and R. E. Burrell, The comparative efficacy of two antimicrobial dressings: *in vitro* examination of two controlled release silver dressings, *Wounds*, 1998, **10**, 179.

93. S. A. Graham and J. M. O'Meara, The feasibility of measuring silver concentrations *in vivo* with X-ray fluorescence, *Phys. Med. Biol.*, 2004, **49**, N259.

94. T. Dutkiewicz, W. Paprotny and D. Sokolowska, Trace element content of human hair determined using neutron activation analysis as a monitor of exposure effects to environmental metals, *Chem. Analityczna*, 1978, **23**, 261.

95. J. Bogen, Trace elements in the atmospheric aerosol in the Heidelberg area, measured by instrumental neutron activation analysis, *Atmos. Environ.*, 1973, **7**, 1117.

96. R. Dams, J. Billet and J. Hoste, Neutron activation analysis of F, Sc, Se, Ag, Hf in aerosols using short-lived isotopes, *Int. J. Environ. Anal. Chem.*, 1975, **4**, 141.

97. K. Boddy, A high sensitivity shadow-shield whole body monitor with scanning bed and tilting geometrics incorporating mobile laboratory, *Br. J. Radiol.*, 1968, **40**, 631.

98. K. Boddy, A simple facility for total body in vivo activation analysis, *Int. J. Appl. Radiat. Isot.*, 1973, **24**, 428.

99. K. H. O. Pelkonen, H. Helvi-Tanski and O. O. P. Hänninen, Accumulation of silver from drinking water into cerebellum and musculus soleus in mice, *Toxicology*, 2003, **186**, 151.

100. R. M. Joyce-Wöhrman and H. Müenstedt, Determination of the silver release from polyurethanes enriched with silver, *Infection*, 1999, **27**(Suppl. 1), S46.

101. R. Kumar and H. Müenstedt, 2005 Silver ion release from antimicrobial polyamide/silver composites, *Biomaterials*, 1999, **26**, 2081.

102. W. R. Buckley and C. J. Terhaar, The skin as an excretory organ in argyria, *Trans. St John's Hosp. Dermatol. Soc.*, 1973, **59**, 39.

103. W. R. Buckley, C. F. Oster and D. W. Fassett, Localised argyria. II The chemical nature of the silver-containing particles, *Arch. Dermatol.*, 1965, **92**, 697.

104. H. Suzuki, S. Baba, S. Uchigasaki and M. Murase, Localised argyria with chrysiasis caused by implanted acupuncture needles, *J. Am. Acad. Dermatol.*, 1993, **29**, 833.

105. H. Steininger, E. Langer and P. Stommer, Generalised argyria, *Deutsch. Med. Wochenschr.*, 1990, **115**, 657.

106. G. Danscher, M. Stoltenberg and S. Juhl, How to detect gold, silver and mercury in human brain, and other tissues by autometallographic silver amplification, *Neuropathol. Appl. Neurobiol.*, 1994, **20**, 454.

107. . Rungby and G. Danscher, Localisation of exogenous silver in brain and spinal cord of silver exposed rats, *Acta Neuropathol. (Berlin)*, 1983, **60**, 92.

108. Y. Hori and S. Miyazawa, Argyria: electron-microscopy and X-ray microanalysis, *J. Electron Microsc.*, 1977, **26**, 193.

109. R. O. Becker, Silver in medicine, in *Precious Metal*, Proceedings of the 9th International Precious Metal Institute Conference, ed. E. D. Zysk and

J. A. Bonnuci, International Precious Metal Institute, Allentown, NJ, 1986, p. 351.

110. J. A. Spadaro, T. J. Berger, S. D. Barranco, S. E. Chaplin and R. O. Becker, Antibacterial effects of silver electrodes with a weak current, *Antimicrob. Agents Chemother.*, 1974, **6**, 637.

111. R. O. Becker and J. A. Spadaro, Treatment of chronic osteomyelitis with silver ions, *J. Bone Joint Surg.*, 1978, **60A**, 871.

112. L. G. Ovington, Nanocrystalline silver: where the old and familiar meets a new frontier, *Wounds*, 2001, **13**(Suppl. B), 6.

113. Smith & Nephew Healthcare, Acticoat® Scientific Background. 7. Silcryst nanocrystals.

114. R. Birringer, Nanocrystalline materials, *Mater. Sci. Eng.*, 1989, **A117**, 33.

115. D. W. Voigt and C. N. Paul, The use of Acticoat® as silver impregnated Telfa dressings in a regional burn center: the clinicians view, *Wounds*, 2001, **13**(Suppl. B2), 11.

116. D. G. Maki and P. A. Tambyah, Engineering out the risk of infection with urinary catheters, *Emerging Infect. Dis.*, 2001, **7**(Special issue), 1.

117. S. Saint, D. L. Veenstra, S. D. Sullivan, C. Chenoweth and A. M. Fendrick, The potential clinical and economic benefits of silver alloy urinary catheters in preventing urinary tract infection, *Arch. Intern. Med.*, 2000, **160**, 2670.

118. H. A. Moss and T. S. J. Elliott, The cost of infections related to central venous catheters designed for long-term use, *Br. J .Med. Econom.*, 1997, **11**, 1.

119. G. A. Farnum, W. R. Knapp and R. L. LaDuca, Silver nitrate molecular species and co-ordination polymers with divergent supramolecular morphology constructed from hydrogen-bonding capable dipyridyl ligands, *Polyhedron*, 2009, **28**, 291.

120. T. Sollemann, Silver, in *A Manual of Pharmacology and its Applications to Therapeutics and Toxicology*, Saunders, Philadelphia, 1942, p. 1102.

121. Bray Healthcare, *Toughened silver nitrate caustics in medicine: warts, verrucae, cautery, granuloma*, 1997, www.bray.co.uk..

122. N. S. Tomi, B. Kränke and W. Aberer, A silver man, *Lancet*, 2004, **363**, 532.

123. W. W. Wilkinson, Silver stains, *J. Am. Med. Assoc.*, 1932, **98**, 72.

124. Mankeiwicz, cited in *A Manual of Pharmacology and its Applications to Therapeutics and Toxicology*, ed. T. Sollemann, Saunders, Philadelphia, 6th edn, 1942, p. 1103.

125. C. S. F. Credé, *Die Verhütung der Augenentzündung der Neugeboren, der häufigsten und wichtigsten Ursache der Blindheit*, Hirschwald, Berlin, 1895.

126. Von Behring, 1887, cited in *A Manual of Pharmacology and its Applications to Therapeutics and Toxicology*, ed. T. Sollemann, Saunders, Philadelphia, 6th edn, 1942, p. 1102.

127. L. Lehrfeld, Ophthalamia neonatorum, *J. Am. Med. Assoc.*, 1935, **104**, 1468.

128. E. J. L. Lowbury, Infection associated with burns, *Postgrad. Med. J.*, 1972, **48**, 338.
129. E. J. L. Lowbury, Problems of resistance in open wounds and burns, in *The Rational Choice of Antibacterial Agents*, ed. R. P. Mouton, W. Brumfitt and J. M. T. Hamilton-Miller, Kluwer Harrap Handbooks, London, 1977, pp. 18–31.
130. C. L. Fox, Silver sulfadiazine, a new topical therapy for *Pseudomonas aeruginosa*, *AMA Arch. Surg.*, 1968, **96**, 184.
131. E. J. L. Lowbury, K. Bridges and D. M. Jackson, Topical prophylaxis with silver sulphadiazine and silver nitrate-chlorhexidine creams: emergence of sulphonamide-resistant Gram-negative bacteria, *Br. Med. J*, 1976, **I**, 493.
132. R. J. White and R. Cooper, Silver sulphadiazine: a review of the evidence, *Wounds UK*, 2005, **1**, 51.
133. C. L. Fox and H. M. Rose, Ionisation of sulphonamides, *Proc. Soc. Exp. Biol. Med.*, 1942, **50**, 142.
134. R. B. Lindberg, The use of topical sulphonamide in the control of burn wound sepsis, *J. Trauma*, 1966, **6**, 407.
135. C. L. Fox, Pharmacology and clinical use of silver sulfadiazine and related topical antimicrobial agents, *Pahlavi Med. J.*, 1977, **8**, 45.
136. F. L. Meleney, Silver sulphadiazine, Discussion, *Ann. Surg.*, 1952, **134**, 640.
137. C. Dollery, Silver sulphadiazine, in *Therapeutic Drugs*, Churchill Livingstone, Edinburgh, 1991, Vol. **2**.
138. A. J. Clark, General pharmacology, in *Handbuch der experimentallen Pharmacologie*, ed. A. Heffter, Springer, Berlin, 1937, Vol. **4**.
139. A. Fraser-Moodie, Sensitivity to silver in a patient treated with silver sulphadiazine (Flamazine), *Burns*, 1992, **18**, 74.
140. A. M. Clarke, Febrile reactions to silver sulphadiazine cream, *Med. J. Aust.*, 1981, **2**, 456.
141. D. S. Cook and M. F. Turner, Crystal and molecular structure of silver sulphadiazine (N$_1$-pyrimidin-2-ylsulphanilamide), *J. Chem. Soc., Perkin Trans. 2*, 1975, 1021.
142. B. J. Sandmann, R. U. Nesbitt and R. H. Sandmann, Characterisation of silver sulfadiazine and related compounds, *J. Pharmacol. Sci.*, 1974, **63**, 948.
143. C. L. Fox and S. M. Modak, Mechanism of silver sulphadiazine action on burn wound infections, *Antimicrob. Agents Chemother.*, 1974, **5**, 582.
144. T. B. Vree and Y. A. Hexter, Sulphadiazine, in *Clinical Pharmacokinetics of Sulphonamides and their Metabolites: An Encyclopaedia*, ed. H. Schönfeld, Karger, Basel, 1987, Antibiotics and Chemotherapy, Vol. **37**, p. 1 and p. 16.
145. X. W. Wang, N. Z. Wang, O. Z. Zhang, R. L. Zapata-Sirvant and J. W. Davies, Tissue deposition of silver following topical use of silver sulphadiazine in extensive burns, *Burns Incl. Therm. Inj.*, 1985, **11**, 197.

146. C. W. Cruse and S. Daniels, Minor burns: treatment using a new drug delivery system with silver sulphadiazine, *South. Med. J.*, 1989, **82**, 1135.
147. C. I. Price, J. W. Horton and C. R. Baxter, Topical lysosomal delivery of antibiotics in soft tissue infection, *J. Surg. Res.*, 1990, **49**, 174.
148. D. J. Stickler, Biomaterials to prevent nosocomial infections: silver is the gold standard, *Curr. Opin. Infect. Dis.*, 2000, **13**, 389.
149. National Center for Complimentary and Alternative Medicine (NCCAM), *Colloidal Silver Products*, Bethesda, MD, 2004, Consumer Advisory Bulletin.
150. Department of Health and Human Services, US Food and Drug Administration, Over-the-counter product containing colloidal silver ingredients or silver salts [proposed rule], *Fed. Regist.*, 1996, **61**, No. 200 (October 15, 1996), 53685.
151. Therapeutic Goods Administration (TGA) Australia, Regulation of colloidal silver and related products, 2002, www.tga.gov.au/docs/html/csilver.htm, accessed 12 November 2009.
152. Neisser, in *A Manual of Pharmacology and its Applications to Therapeutics and Toxicology*, ed. T. Sollemann, Saunders, Philadelphia, 6th edn, 1942, p. 1103.
153. Bechhold, in *A Manual of Pharmacology and its Applications to Therapeutics and Toxicology*, ed. T. Sollemann, Saunders, Philadelphia, 6th edn, 1942, p. 1103.
154. A. Bottner, Collargol, *Münch. Med. Wochenschr.*, 1921, **68**, 876.
155. P. Van Hassault, B. A. Gashe and J. Ahmad, Colloidal silver as an antimicrobial agent: fact or fiction, *J. Wound Care*, 2004, **13**, 154.
156. W. C. Bell and M. L. Myrick, Preparation and characterisation of nanoscale silver colloids by two novel synthetic routes, *J. Colloid Interface Sci.*, 2001, **242**, 300.

CHAPTER 3

Uptake and Metabolism of Silver in the Human Body

3.1 Introduction

Silver is ubiquitous in the human environment. It is not a trace metal in the human body and serves no recognised physiological function, but the majority of people exhibit low levels of silver in their blood plasma or body tissues even though they have not knowingly been exposed to silver or silver compounds occupationally, worn hygiene clothing or been treated with antibiotic dressings or medicinal products containing silver.[1] Silver is absorbed into the body:

- by ingestion in the diet or drinking water, oral hygiene products or therapies for gastrointestinal complaints;
- by topical administration (including exposure to the eye);
- by inhalation;
- through the implantation or clinical use of silvered devices and catheters;
- through exposure to silver-containing clothing and domestic products.

Silver absorption by any route is determined by the nature of the product or source of silver contamination, its capacity to release biologically active silver ion (Ag^+), and the route and duration of exposure. Absorption is reduced in the presence of anions, proteins and materials that chelate or otherwise bind the free silver ion in the form of insoluble precipitates.[2]

The greatest part of present knowledge on the uptake and metabolism of silver in the human body has been generated through occupational health studies of workers employed in silver mining and related industries, and clinical studies on the use of silver nitrate and silver sulfadiazine. As more accurate means of determining silver in blood, urine and tissues have become available, so information has improved on the relative safety of silver as an antibiotic and its clinical relevance as a bactericide and fungicide in medical devices.

Issues in Toxicology No. 6
Silver in Healthcare: Its Antimicrobial Efficacy and Safety in Use
By Alan B. G. Lansdown
© Alan B. G. Lansdown 2010
Published by the Royal Society of Chemistry, www.rsc.org

Argyria, argyrosis and argyraemia present cardinal and irrefutable evidence of silver absorption by all routes. The clinical and toxicological implications of argyria are discussed more fully in Chapter 8, but it is sufficient here to define the condition as a local or generalised discolouration of the tissues attributable to deposition of fine brown-black granules of metallic silver, silver sulfide or silver selenide.[3–5] It is a cosmetically undesirable feature of acute or sub-acute exposure to high levels of silver or occupational exposures over several years.[3] In contrast, argyria-like discolourations in superficial layers of the skin (attributable to topical application of silver nitrate or other soluble silver salts to the skin) tend to fade gradually as silver sulfide impregnated keratinocytes are progressively desquamated from the skin surface through natural wear and tear. Bleehan[3] did not measure total body silver in his patients but earlier estimates suggest that as little as 4–5 g of silver can produce a clinical picture of argyria.[6] Attempts to mitigate discolourations due to argyria have proved disappointing. At one time, topical argyrism was reduced by rubbing the skin with crystals of potassium iodide or painting the area with a 10% iodide solution and allowing several hours for effective action.[7] Earlier clinicians recommended application of 10% potassium cyanide or 10% mercuric chloride in 10% ammonium chloride, but these therapies were clinically unsafe and abandoned.[8] Dermabrasion is clinically unsatisfactory and was largely abandoned in the 1950's (B.A. Bouts, personal communication).

Argyrosis is more specifically used to refer to deposition of silver or silver sulfide in the cornea with or without impairment of vision.[9] Some silver is absorbed through ocular exposure as documented in older literature where silver nitrate was used to treat conjunctival infections or in corneal surgery.[10–12] Reported cases of silver nitrate abuse or accidental application have been associated more with local toxicity due to the corrosive and astringent properties of silver nitrate, but perocular absorption and argyria do occur.[10,13] Silver nitrate (20% solution) is no longer used in optic surgery these days for treatment of corneal abscesses and conjunctivitis on account of its irritancy and serious risk of chemical conjunctivitis and corneal opacity.[12,14]

Argyraemia specifically refers to silver in the circulation, mostly complexed with proteins and is employed commonly as a measure of silver absorption from all routes of exposure.[1,15–17] Raised blood silver levels may not be associated with overt symptoms of argyria or other signs of silver-related changes,[18] but are commonly used as indices of the magnitude of occupational or therapeutic silver and duration of exposure. Argyraemia is useful as a monitor of silver absorption through the use of medical devices coated or impregnated with silver that are implanted in blood vessels or which come into close contact with the circulation in normal clinical use. In these situations, metallic silver, silver alloys, silver oxide and even silver sulfadiazine are used to control device-related infections and biofilm formation, each of which is a potential cause of patient discomfort, malfunctioning of devices and fatality. It is common practice in the development of such devices to monitor silver ion release from product surfaces by immersion in simulated human serum or urine as in the static silver dissolution test used in evaluation of SILCRYST™ technology (Smith & Nephew) or in continuous flow experiments (Figure 3.1).[19] In each

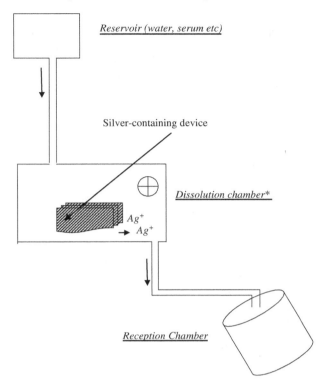

Figure 3.1   Schematic representation of simplified apparatus for static dissolution test. Silver ion released from the device is measured by mass spectrometry or other means.

case, a clinician becomes aware of the patterns of human exposure and can validate the quantity of silver ion released in relation to the severity of local infections and inherent toxic risk. Less well-defined are the risks of silver absorption through implantation of silver and silver–gold alloy acupuncture needles, in-dwelling silver coated or impregnated polyurethane catheters, jewellery and occupational exposures including photography (and X-radiography), dentistry, and the manufacture of silver nitrate, silver oxide and silver fulminate (for explosives).

Contrary to reports, that "... all forms of silver are cumulative once they enter body tissues and that very little is excreted ...",[20] and that "... absorbed silver is tenaciously retained ... in elastic tissues throughout the body ...",[21] there is very little substantive evidence that silver acts either as a cumulative poison in the human body like lead and mercury, or that it reaches toxic levels in any tissue. Silver does accumulate preferentially in the basement membrane region of the dermis, but no evidence has been seen to show that this is either life-threatening or a clinical manifestation of toxicity. Like many heavy metals, silver is absorbed through topical application, mucus membranes of the buccal cavity, gastrointestinal and respiratory tracts, and eye and is eliminated in bile or *via* the urinary tract.[1,15,22] A small proportion of silver absorbed is deposited in bone, probably as a long-lived hydroxyapatite complex.[23] A small proportion is voided in the hair,[24] but there is equivocal evidence that silver is eliminated from the systemic circulation through the skin.[4,25]

3.2   Oral Exposure

Silver is ingested and absorbed through the mucus membranes of the buccal cavity and proximal aspects of gastrointestinal tract from:

- Drinking water from contaminated sources or exposed to silver–copper filtration as a means of eliminating *Legionella* sp. and nosocomial infections;[26]
- Contaminated food;
- Occupational exposures;
- Silver coating of pills for therapeutic use (Figure 3.2);
- Silver nitrate or colloidal silver preparations used in oral hygiene and in treating gastrointestinal infections and peptic ulcers;
- Antismoking remedies;
- Accidental or suicidal consumption of silver nitrate solutions.

Fung and Bowen reviewed the risk–benefit profile for a range of silver-containing products for human medicine and concluded that as much as 10% of "silver salts" may be absorbed following ingestion and that silver is

Figure 3.2   Pills coated in metallic silver. *By courtesy of Dr M. Kessler, Pharmazie-Historisches Museum, Basel.*

deposited in many organs—notably skin, liver, spleen and adrenal glands, with lesser deposits in muscle and brain.[27] Whereas the amount of silver present in domestic drinking water, prescribed medicinal products and antismoking remedies may be controlled and approximate exposure levels appreciated, the amount of silver entering the body through occupational exposures, or consumption of contaminated food and silver in food chains is more difficult to assess.[28] The human risks associated with massive long-term emission of silver in factory wastes and sludges as in the San Francisco Bay area are real, but the magnitude of the problem of silver accumulating in sea food and human food chains is still not fully appreciated.[29] Accumulating evidence suggests that silver is toxic to fish and crustaceans.[30–32]

In clinical practice, silver has been largely phased out of oral, gastrointestinal and respiratory hygiene medicines in many countries on account of its irritancy and corrosive properties, but is still used in dental amalgams (with mercury and other metals) and is an active constituent of numerous colloidal silver preparations sold through clandestine markets under the guise of "alternative medicines" or "nutritional additives". These colloidal silver products carry unsubstantiated claims as efficacious remedies for gastrointestinal infections, peptic ulcers and conditions broadly associated with the so-called "silver deficiency syndrome",[33,34] but published studies illustrate the strong tendency for silver (as Ag^+) to be absorbed and risks of argyria.

Earliest evidence for absorption of silver from the gastrointestinal tract is provided by reference to silver nitrate in the form of pills (with kaolin or petrolatum) or dilute solutions which were prescribed for mouth infections, gastrointestinal infections and peptic ulcer.[35] Ingestion of silver nitrate for gastrointestinal complaints or dental hygiene is a well-documented cause of argyria and argyrosis, but its value in 21st century medicine is strongly contraindicated on account of the astringent and corrosive action of the preparations available and their capacity to cause severe inflammatory changes, stomach pain, convulsions, shock, coma, paralysis and even death.[36–38] Less severe reactions and risks of argyria were associated with the consumption of pills containing 10–20 mg (0.17–0.33 grains) of silver nitrate for gastric ulcer, but these were abandoned many years ago on account of their low clinical efficacy. Interestingly, whereas silver nitrate pills were mildly irritant, accidental swallowing of 2.5 g pieces of silver nitrate "pencil" (silver nitrate mitigated with potassium nitrate) was apparently without obvious effect.[35] When silver nitrate was applied acutely for reducing gingival infections and acute pharyngitis, the principal complication included blackening of buccal membranes, severe mouth pain and necrotising ulcerative gingivitis though, in severe cases, reduced blood pressure, spasms and paralysis were reported.[21,38,39] Although the actual amount of silver absorbed into the body following use of silver nitrate in oral hygiene is rarely reported, it should be expected that the corrosive damage evoked in the stomach or upper intestine tract is attributable to nitric acid, which also may enhance the permucosal absorption of the silver ion.

These days the risk of oral absorption of silver through occupational exposure is greatly reduced on account of far-reaching occupational health and safety requirements and the introduction of threshold limit values of $0.1\,mg\,m^{-3}$ for metallic silver dust and $0.01\,mg\,m^{-3}$ exposures to soluble silver compounds in the workplace.[40] An earlier report illustrates how workers occupationally exposed to $0.039–0.378\,mg$ silver per m^3 silver nitrate and silver oxide dust developed severe abdominal pain, which was relieved by administration of antacids.[41] Chronic exposure to silver nitrate or silver oxide dust complications may be a cause of progressive taste and smell disorders, hypogeusia and vertigo through silver absorption and deposition of silver sulfide or other complex particles in the region of sensory nerve endings.[42]

Although silver has been used as an antismoking remedy for more than 50 years, only recently have the metabolic and toxic implications of this route of silver exposure been examined.[35,43] At one time, painting the tongue with 10% silver nitrate or rinsing the mouth with 0.5% silver nitrate after meals proved a disincentive to smoking on account of the disagreeable taste created in the presence of cigarette smoke. More recently silver acetate pills or lozenges have been used.[43,44] Case histories of patients consuming large amounts of these remedies over long periods show that silver is absorbed and that argyria, high levels of blood silver and convulsive seizures are complications. So far, there is no substantive evidence that silver absorption through antismoking remedies or other route is a cause of neurological changes as inferred in this work.[45] East and his colleagues monitored silver uptake and retention patterns in a severely argyric 47-year-old lady using radioactive tracer (^{110m}Ag) with neutron activation analysis and whole body counting.[43,46] The patient had consumed excessive high doses of an antismoking remedy containing silver acetate for two years and her total body silver was estimated to be 6.4 g. The silver content of her forearm skin was $71.7\,\mu g\,g^{-1}$ wet weight. To appreciate silver absorption, East *et al.* pipetted 8 mL of labelled silver acetate solution onto her tongue (4.5 mg containing 4.43 μCi ^{110m}Ag) and estimated percentage total body silver retention after 1–30 weeks. After a rapid decrease in tracer level in first day, approximately 18% of the administered dose was retained for up to 30 weeks and urinary silver from days 1–7 remained fairly constant at 2.5% of the administered dose. Blood silver declined rapidly and after 2 h of dosing measured only $4.5 \times 10^{-4}\%$ dose mL^{-1}. This observation is entirely consistent with data published by the International Commission on Radiological Protection, which stated that absorbed silver has a half-life of 5 d in the body, but the half-life of silver in specific tissues such as bone, liver and kidney is longer, *i.e.* 30, 15 and 10 d respectively.[47] The low levels of silver retained in the body after 30 weeks were held to be clinically insignificant.

Conditions like "the Blue Man of Barnum and Baileys Circus", the silver man, the silver-plated boy and skin changes associated with the so-called "Great Silver Fraud", are the inevitable consequence of consuming colloidal silver products for gastrointestinal or respiratory infections over a long period.[48–50] All present with profound brown-black or slate grey discolourations of the skin and excessive pigmentation of eyes. The dubious clinical value of colloidal silver preparations in medicine was examined in detail by Hill and Pillsbury in 1939,[51] and more recently in Goodman and Gillman's treatise on the *Pharmacological Basis of Therapeutics.*[52] Both substantiate that silver is readily absorbed from oral colloidal preparations and that risks of argyria and argyrosis vastly outweigh the dubious clinical benefits offered. If predictions that up to 10% of the silver content of oral medications may be absorbed are true,[27] dosage with even a mild silver preparation containing 19–23% of microparticulate silver will greatly increase the body silver burden and possibly overwhelm the body's own defence systems (metal–protein binding proteins, precipitation with anions such as chloride, phosphate, sulfide, *etc.*). Mucosal absorption of colloidal silver particles is greatly enhanced by their micro-particulate size and higher rates of ionisation in body fluids and exudates. Recent quality control evaluations conducted by the US Food and Drug Administration concluded that the amount of silver in some products varies by 15.2–124% of the amount identified on product labels and that the actual amount entering the human body is virtually impossible to assess.[53] One anonymous communication entitled *"Colloidal Silver: the rediscovery of a super antibiotic"* states that when taken orally, the silver solution is "... readily absorbed into the bloodstream and accumulates in the tissues to provide protection against pathogenic organisms".

The oral toxicity of silver nitrate has been investigated in a range of laboratory animal species, but published information suggests that no single laboratory species can provide a surrogate model for the human individual.[27,54] Patterns of silver uptake, metabolic pathways and excretion patterns vary greatly between species and reported studies are inconsistent in design, detail and objectives to be of useful predictive value. Predictive safety evaluation of silver nitrate in dogs, cats, goats and rodents substantiated the gastrointestinal irritancy seen in humans and illustrated pathways by which absorbed silver is metabolised to soft tissues leading to changes resembling argyria and argyrosis.[55,56] In the cat and rat in particular, the eye seems to be a target organ for deposition of silver absorbed intestinally.

3.3 Dermal Contact and Percutaneous Absorption

The skin constitutes 12–15% of the total weight of the human body, but covers every external surface and provides the essential functions of regulating loss of excess water as "insensible perspiration" and acting as a barrier against the penetration of xenobiotic and potentially toxic materials from the environment.[57] Human skin varies greatly in its external appearance, thickness, hair cover, sweat gland distribution and colour according to the region of the body and the age, sex, genotype, state of health and geographic location of an individual. Kligman questioned what normal skin really is and emphasised the formidable problems confronting toxicologists in understanding the wide diversity of skin types in *Homo Sapiens* in predicting risks of percutaneous toxicity.[58] The vast majority of substances applied topically penetrate the skin to some extent and enter the systemic circulation, albeit that in the case of silver, the influx is exceedingly low. Using greatly improved technology these days, we have a far greater understanding of the absorption of metals like gold, silver, aluminium, manganese which at one time were thought to be inert in contact with the skin surface.[22]

The majority of products containing metallic silver or silver compounds come into contact with human skin at some time, and in view of the large number of silver-containing dressings designed for wound care, the skin might be considered a major route of body silver uptake. This appears not to be the case. Whereas metallic silver is generally inert, ionised silver does penetrate intact skin to a low extent by an active mechanism involving carrier proteins (like many other metallic elements). However, the propensity of Ag^+ to interact with and irreversibly bind proteins in superficial layers of the stratum corneum—notably those displaying sulfydryl, carbonyl and imidazole moieties in epidermal keratin—greatly reduces the concentration available for apsorption.[59] Additionally, Ag^+ precipitates readily as minimally soluble silver chloride with Cl^- in sweat and perspiration, as well as proteins in cutaneous secretions and surface debris.

Interaction between silver compounds and the skin is illustrated by experimental and clinical studies with silver nitrate, silver sulfadiazine and sustained silver-release wound dressings.[60–62] In each case the epidermal barrier function afforded by mature keratinocytes controlled silver penetration from topical applications.[35,63] Whereas the skin stained black-brown in the presence of highly ionised silver nitrate, discolourations were greatly reduced following silver sulfadiazine, possibly on account of the lower and more sustained release of Ag^+. In either case, the superficial deposits of silver sulfide, silver selenide or other insoluble complexes were rarely long-lasting and were progressively lost as keratinocytes desquamated and wounds healed in keeping with sequential events of the wound healing cascade.[61]

Early studies commended the value of silver nitrate as a topical antibiotic on account of its efficacious topical action and its ability to penetrate "quite deeply" into a wound.[64] Silver was thought to form double salts of silver albuminate and sodium chloride in the tissues, with the caustic action of the nitrate anion being blocked by sodium chloride douching.[35] Few studies have been published in more recent times documenting the actual amount of silver penetrating the skin following topical application of wound therapies and virtually nothing is presently known of the inherent risks following the use of silver in textiles and hygiene clothing.[65] Current views based on radioisotope studies using [111]Ag, suggest that less than 4% of the total silver released from silver nitrate is absorbed by normal human arm skin.[66] Guinea pig skin, which is thought by some to resemble human skin in the percutaneous absorption of metals, permitted percutaneous absorption of less than 1% of topically applied silver.[67] However, when 1.0% silver sulfadiazine (Flamazine) was used therapeutically and prophylactically in severely burned patients, uptake of silver was greatly increased.[1,68] Baxter reported that up to 10% of topically applied 1.0% silver sulfadiazine (Flamazine) is absorbed and that absorption is significantly higher in deep partial thickness burns on account of the greater vascularity of the wound bed.[69] Coombs monitored silver sulfadiazine absorption in 26 severely burned patients in Alfred Hospital in Melbourne during 1989–1991 and accurately related the amount of cream applied and blood and urine silver levels.[1,68] In patients with burns extending to >5% of the total body surface area (TBSA), silver levels increased in line with the severity of wounds and the amount of silver sulfadiazine applied. Patients with >20% TBSA burns exhibited plasma silver levels of greater than 200 µg L^{-1}; the highest blood silver of 310 µg L^{-1} was seen in a patient with 35% TBSA burns. It was not uncommon for blood silver to be increased 20 fold to >40 µg L^{-1} within 6 h of administering an initial silver sulfadiazine therapy. In patients with less extensive wounds (>10% TBSA), plasma silver showed a rapid rise reaching a plateau in 4–7 d. Only when the silver sulfadiazine therapy was withdrawn did the argyraemia decline. As a placebo, Coombs applied 100 g of silver sulfadiazine to the unburned skin of two volunteers over two weeks to monitor uptake by undamaged skin and noted a complete lack of absorption using highly sensitive graphite furnace with atomic absorption spectrometry for tissue analysis.[1,68] These critical observations contrast to the work of Charles Fox, who in assessing the efficacy of his new product, applied 500–1000 g of 30 mM silver sulfadiazine ointment to 1 m^2 of burned skin and noted that 10–30% of the total dose of the sulfadiazine was excreted in 24 h.[70] In his experimental studies, Fox measured percutaneous absorption of 100 mg of silver sulfadiazine as a 1% suspension or 0.5% ointment in experimentally burned mice and reported urinary excretion rates of less than 10 mg d^{-1} for three days.

Silver nitrate is appreciably more astringent and corrosive to the human skin than silver sulfadiazine and, as Toughened Silver Nitrate (with potassium nitrate) (Bray), is used clinically in the ablation of warts, verrucae and indolent granulomas.[71] The manufacturers claim that their product is easy to apply to a target site and that the 40 or 95% solutions "soak into the unwanted tissue" but leave intact skin unharmed, other than causing a temporary black-brown discolouration. Humphreys and Routledge (1998) reviewed several cases where silver nitrate had been used to remove umbilical granulations in young children.[54] Silver nitrate sticks (75% silver nitrate) were shown to cause severe local blistering and erythema in some patients, but absorbed silver in children was a cause of electrolyte imbalances. Children with severe burns (>50% of total body surface area) treated topically with 0.5% silver nitrate solutions to combat life-threatening infections were also shown to be at risk of methaemaglobinaemia with cyanosis, tachypnoea and drowsyness.[72–75] The characteristic pigment of methaemaglobinaemia results from the oxidation of ferric ion to the ferric state, thereby impairing the oxygen-carrying capacity of the blood.[76] Molecular changes in the haemoglobin molecule are such that oxygen uptake is not a reversible process and tissues become subject to hypoxia and related changes. At blood levels of 50%, methaemaglobinaemia is potentially fatal. (Methaemaglobinaemia is evoked by a number of toxins absorbed into the blood stream notably nitrites and aniline analogues.) When silver nitrate is used therapeutically as an antibiotic, bacteria commonly present in the body's microflora including *Escherichia coli*, *Staphylococcus aureus*, *Pseudomonas aeruginosa* and *Proteus* sp. reduce nitrate to the more toxic nitrite, which is absorbed systemically. Navajo children are particularly vulnerable to congenital methaemaglobinaemia and are at greater risk following silver nitrate therapy.[74] Methylene blue solution (3,7-bis-[trimethylamino]phenazothonium chloride, trihydrate) (1%) administered intravenously at 1 mg kg^{-1} body weight is used therapeutically to alleviate methaemaglobinaemia in vulnerable young children of up to six months of age.[77] The reagent is able to reduce the methaemoglobin to haemoglobin, but is contraindicated in high doses as it has the ability itself to evoke methaemoglobin formation. Methylene blue solution is a mild antiseptic agent and is reported to inhibit amine oxidase in tissues.[77]

Topical application of 0.5% silver nitrate is still standard therapy in burn wound clinics where infections of *Pseudomonas aeruginosa* and *Staphylococcus aureus* are life-threatening.[61,78] Apart from the cosmetically undesirable staining of undamaged tissue, silver ion does not penetrate wounds to any great extent and that reaching the systemic circulation is deposited in bone, skin and most major soft tissues.[79,80] A case is recorded of a 60-year-old man who received silver nitrate therapy for burns 8 h daily for 30 days, but who died from a brain tumour. Post-mortem analysis of soft tissues showed increased levels of silver in his kidney, heart and bone but silver concentrations of 2800 µg g^{-1} in his skin were at least tenfold higher.[79]

Whereas argyria and deposition of silver residues in the skin, eye and soft tissues are not in themselves life-threatening, the capacity of silver ion in blood to disturb electrolyte balances, particularly hypochloraemia, hyponatraemia and occasionally hypokalaemia, is of concern.[29,74] A clinical study of ten children treated with 0.5% silver nitrate for severe burns recorded an initial inflammatory response to the silver nitrate followed by a pronounced decrease in body sodium and potassium. These effects were more than fourfold higher than when non-silver therapies were used and point to the importance of measuring electrolyte balances when silver nitrate therapy is used for burn wounds in children and vulnerable adults.[81]

It is common practice in the development and efficacy evaluation of sustained silver-release wound dressings to concentrate on the ability of the products to control infections and to advance healing without measurement of percutaneous absorption of silver ion.[82–86] Available evidence suggests that blood silver levels are only marginally raised or not significantly different from non-silver therapies.[61,87] Thus, clinical evaluation of a high silver dressing (Contreet Foam, Coloplast, Denmark) comprising a soft hydrophilic polyurethane foam containing silver as an integral part of the matrix, demonstrated that blood silver concentrations increased from less than 4 nMol L^{-1} to a maximum of 80 nMol L^{-1} over 28 d therapy.[60] They declined in the period following removal of dressings but tended to be higher in patients with larger leg ulcers, though there was no consistent relationship between the blood silver, ulcer size and number of dressing changes. A narrow band of blackish residues was reported in a small number of peri-ulcer areas at dressing changes, but this staining was removed by washing in water. Other studies of sustained silver release wound dressings have confirmed that amount of silver absorbed systemically is of marginal significance, irrespective of the silver content or release patterns of Ag^{+}[61]. However, as in Karlsmark's study,[60] blood silver levels in a small number of patients treated with Contreet Foam were higher than normal baseline levels at the commencement of the study, possibly through occupational silver exposure or consumption of silver in food and water.

Sibbald *et al.* noted that, in-patients with leg ulcers with a mean surface area of 4.8 cm treated with a high silver-containing rapid release dressing (Acticoat®, Smith & Nephew), blood silver levels after 8–86 d were 0.15 mg above median baseline levels at the commencement of the study.[88] Acticoat® is defined as a "silver-coated absorbent dressing" containing a mean of 2.917 mg silver per square inch (nanocrystalline silver) and which, in simulated wound fluid, releases 70 μg mL^{-1} silver ion at equilibrium after 8-10 h.[82] A prospective "open-label" study designed more specifically to evaluate absorption of silver from Acticoat® dressings (12% TBSA) indicated that the mean blood silver level attained after 9 d therapy was 56.8 μg L^{-1}, but this declined to 0.8 μg L^{-1} by six months following treatment.[89] No haematological changes were observed in response to absorbed silver in any of these observations.

The percutaneous absorption of silver ion following occupational exposure to metallic silver fumes or dusts or silver nitrate droplets in air is exceedingly difficult to estimate in view of the long periods of exposure involved. Many authors have reviewed cases of occupational silver exposure and have specifically examined symptoms and pathological features of argyria, but in each case the actual route of silver absorption is unclear.[3,4,90,91] Wolbling evaluated the health status of 50 men exposed to silver nitrate and silver chloride in a German silver processing plant with environmental silver levels of 0.001–0.54 mg m^{-3} for 3–20 years and reported mean skin silver concentrations of 0.115 mg g^{-1} compared to 0.02 mg g^{-1} in non-exposed control individuals.[90]

3.4 Inhalation and Absorption of Silver through the Respiratory Tract

No silver-containing products are currently marketed in North America, the European Union or Australia these days for the treatment of respiratory complaints. The majority of the information available regarding the uptake of silver *via* the respiratory route derives from inhalation of metallic silver dust, silver oxide and silver nitrate in occupational health studies.[90] Incomplete and widely scattered studies relating to addictive or ill-defined cases of the use of colloidal silver in anti-infective sprays, nose drops and breath fresheners provide some additional information.[5,92,93] In each case, patients have presented with manifestations of local or profound generalised argyria. Detailed study of the respiratory uptake of silver absorption through the respiratory membranes has not been seen, but it is reasonable to expect that metallic silver particles, silver oxide or other silver compounds ionising in mucoid secretions and in the thin fluid film on alveolar surfaces will be readily absorbed into the circulation to be distributed to other parts of the body. Absorptive mechanisms are not fully understood but probably involve a passive diffusion through the exceedingly thin membranous linings of the alveoli or a carrier-mediated process as seen in the gastrointestinal tract and in the skin. Alternatively, silver dust may be taken up through the pinocytic action of alveolar macrophages (colloquially identified as "dust cells) found in interstitial spaces, alveolar cavities and in the region of the blood–air barrier of the interalveolar septae.

Chronic inhalation of silver or silver nitrate dust in occupational settings is commonly associated with progressive discolouration of the skin, hair, eyes and fingernails which may be preceded by a mild to severe chronic bronchitis with inflammation of membranes of the nasal and upper respiratory tract.[13,90,94,95] In the early days prior to introduction of health and safety at work regulations, acid was poured onto silver ores in open vessels and evaporated in the absence of ventilation. Crystals of silver nitrate were dried in ovens and finely sieved thereby generating fine dust. On occasions, patients with symptoms of bronchitis have revealed squamous metaplasia and pigmentations resembling anthracosis and siderosis.[28] Occupational health studies of 49 employees exposed to silver chloride or silver nitrate $(0.001–0.540\,mg\,m^{-3})$ in a silver processing plant for 3–20 years showed that exposures to soluble silver compounds presented appreciably greater risk of argyria and argyrosis, but that silver concentrations in skin biopsies did not correlate well with ocular deposits or the duration of exposure.[90] Rosenman et al. conducted a comprehensive study of occupational exposure to silver, silver oxide and silver nitrate in an industrial plant engaged in the production of silver and other precious metals.[96] At environmental exposures of $0.04–0.35\,mm\,m^{-3}$, 15 out of 27 workers (of a workforce of 220) experienced irritation in their eyes, nose and throat and 30% of the remainder reported coughing, wheezing, respiratory distress and occasional nose-bleeding during their time at the workplace. The incidence of these symptoms was related to the job status of the workers and the location of their work (especially in melting, refining, crystal and powder production). Ophthalmic changes (including deposition of silver precipitates in the cornea and conjunctiva) were greatest in workers exposed to silver for more than 10 years (63%). Wide variations in blood and urinary silver levels of $0.5–62.0\,\mu g\,L^{-1}$ and $0.05–52\,\mu g\,L^{-1}$, respectively, were recorded but 26 patients exhibited higher than expected blood and/or urine silver levels. Increased urinary silver excretion was associated with changes in renal function and these were more prevalent in workers exposed to silver for longer periods. Fourfold increases in the renal enzyme N-acetyl-β-D-glucosaminidase (NAG) as reported in four workers with high blood silver may have clinical significance as a marker for the nephrotoxicity of silver, but the authors could not discount the possibility that mild functional changes in the tissue were in part attributable to contamination with other nephrotoxic metals such as cadmium.

DiVincenzo et al. monitored 37 workers engaged in the smelting and refining of silver with view to assessing the absorption and elimination of inhaled silver and excretion patterns.[16] These workers exhibited mean silver concentrations in the blood, urine and faeces of $11\,\mu g\,L^{-1}$, $<5\,\mu g\,L^{-1}$ and $15\,\mu g\,g^{-1}$, respectively, but silver in hair was much higher at $130\pm160\,\mu g\,g^{-1}$ compared to $0.57\pm0.56\,\mu g\,g^{-1}$ in the control group. The authors cautioned on interpreting hair silver as a true reflection of the amount of silver inhaled since ionised silver in air droplets binds strongly to hair shaft keratin and may lead to misleadingly high values. In their estimation, exposure to $0.1\,mg\,m^{-3}$ atmospheric silver (TLV) can be expected to lead to faecal excretion of about $1\,mg\,d^{-1}$.[16] DiVincenzo *et al.* calculated the body burden of silver for these 37 workers and concluded that "the incremental body burden for silver workers was $14\,\mu g\,kg^{-1}$ of body weight (non-silver exposed workers, $2\,\mu g\,kg^{-1}$), with risks of argyria being negligible".[16]

The American Conference of Governmental Industrial Hygienists (ACGIH) published threshold limit values (TLV) for atmospheric contamination of $0.1\,mg\,m^{-3}$ for metallic silver and $0.1\,mg\,m^{-3}$ for soluble silver compounds,[40] whereas the US Occupational Safety and Health Administration and Mine Safety and Health Administration[97] proposed that maximal exposure limits to all forms of silver should be $0.01\,mg\,m^{-3}$. Prolonged exposures to $1\,mg\,m^{-3}$ of silver can cause generalised argyria but inhalation of $0.1\,mg\,m^{-3}$ is a potential cause of discolouration of the nose, throat and eyes.[98] Limited information is available to show elimination patterns of silver deposited in the lungs following inhalation. In the case of a 29-year-old man inadvertently inhaling an unknown amount of radiolabelled silver (^{110m}Ag), the half-life of silver in his lungs was estimated to be 24 h but at least 50% of the presumed inhaled dose persisted in his body (mostly in his liver) for 16 h.[99] Faecal silver excretion persisted for at least 300 d in this patient. The US Environmental Protection Agency cited evidence from experimental studies in the rat which showed that inhaled silver is eliminated from the lungs within 0.3–1.7 d in a biphasic process.[100]

The extent to which silver nitrate is used therapeutically these days for removing nasal polyps or other lesions is not known, but the few published studies indicate corrosive damage with local or more generalised symptoms of argyria. Thus, a 72-year-old lady treated for a painful nasal polyp exhibited silver staining resembling a "tattoo" four years after therapy[101] and a 67-year-old lady with a history of epistaxis in both nostrils, treated with silver nitrate to cauterise abscesses and a Bartholin's cyst, developed to severe local inflammation and corrosive changes on her face due to application of silver nitrate tipped sticks.[102] It is unclear from either case how much silver was inhaled from the instillates or whether damage to the respiratory tract was experienced.

Applications of unregulated colloidal silver products for respiratory tract infections, rhinitis, *etc.* are more worrying causes of generalised argyria and respiratory tract changes. The amount of silver or Ag^+ inhaled is rarely known and overt changes may not be recognisable for several months or even years after therapy has commenced. Colloidal silver proteins are prepared by mixing silver nitrate with sodium hydroxide and gelatine may contain 19–23% silver with low ionising potential (mild silver protein) or 7.8–8.5% highly ionised silver (strong silver protein).[27] Whilst they are appreciably less corrosive than silver nitrate, argyria is a common side effect.[48,103–105] In reported cases of argyria attributable to inhalation of colloidal silver, patterns of silver absorption and excretion are rarely known and the actual blood silver consistent with overt discolouration of skin or eyes is not known. On rare occasions, pathological reports confirm the cellular localisation of silver, silver sulfide or silver selenide and local tissues responses.[3-5]

3.5 Miscellaneous Routes of Silver Exposure and Systemic Absorption

Silver is employed in catheters for renal drainage, central vascular insertion, intraventricular drainage in patients with hydrocephaly and cerebrospinal fluid disorders, and in-dwelling intraperitoneal use. In addition silver, silver oxide, silver sulfadiazine or other Ag^+-releasing products are incorporated or coated onto orthopaedic devices (prostheses, bone cements and external fixation pins), dental amalgams, cardiovascular prostheses, heart valve sewing cuffs, stents, surgical instruments (sutures, staples) and acupuncture needles. In each case, a sustained release of biologically active Ag^+ is necessary to achieve anti-microbial efficacy, but an unknown amount is released into the systemic circulation where it is strongly bound by serum proteins, notably albumins and macroglobulins. Schierholtz *et al.* considered that most of this irreversibly bound silver has no toxicological, physiological or antimicrobial significance,[59] and that technical achievements in antibiotic silver-coating or impregnation of medical devices can only be effective when the concentration of "free" Ag^+ is increased and where contact with serum proteins and inorganic anions is minimised. They also concluded that a lot of the silver released from silvering may be precipitated as virtually insoluble silver chloride and that it is not feasible to achieve concentrations of blood silver exceeding $0.15\,mg\,L^{-1}$ in this situation.

Hydrogel technology employed in some intraurethral catheters acts by absorbing moisture and pathogenic bacteria into an absorbent sheath impregnated with an ionisable silver compound. The absorbed moisture leads to ionisation of the silver source and bactericidal action, with minimal amounts of silver being released in the close vicinity of the urethral membranes.[106] However, there seems to be a general view amongst clinicians that the amount of silver ion available for absorption is low irrespective of the design of intraurethral catheters or the form in which silver is incorporated, hence the toxicological significance is negligible.[59,107] We know little about how much silver is absorbed into a patient's circulation from the wide range of central venous, intraperitoneal or intracerebral catheters even though *in vitro* evidence is published to demonstrate patterns of Ag^+ release under simulated conditions of use.[108-110] No cases of argyria or argyrosis have been reported as far as I am aware, but it can be assumed that any silver absorbed systemically from these devices will be eliminated as protein complexes *via* the liver or kidney. Total systemic silver concentrations can be expected to reflect ionisation patterns of silver coatings or impregnates less that which becomes irreversibly bound or deposited in skin, bone or other tissues.[59,111]

Samuel and Guggenbichler evaluated a new catheter impregnated with nanocrystalline silver particles in a hydrophilic matrix in an *in vitro* model using simulated human urine.[112] The technology for catheter construction developed at the University of Erlangen, Germany, uses a polyurethane silicone catheter impregnated with nanoparticles of silver (309 nm diameter) comprising 0.8–1.5% of its total weight and with a surface area of metallic silver of $2500\,cm^2\,g^{-1}$. Concentrations of ionised silver released following continuous perfusion were estimated to be $25\,ng\,cm^2\,day^{-1}$ initially but decreasing over 5 d to $50\,ng\,cm^2\,day^{-1}$ and reaching a steady state by 10 d of $35\,ng\,cm^2\,day^{-1}$. This compared with total release of $1–2\,\mu g\,mL^{-1}$ of silver from an antibiotic Foley catheter.[113]

Clinical evidence for silver absorption through the urinogenital tract is illustrated by a bizarre case of a lady who was treated intravaginally with 2 mL of 7% silver nitrate in attempt to induce abortion.[114] The patient died within 3 h with acute abdominal pain, serious corrosive internal injuries and nausea, but silver was identified in her blood and urine indicating systemic absorption.

Other rare cases reported in earlier literature show that 10–20% solutions and pastes of silver nitrate have been used to induce abortion and sterility.[115,116] Apart from local irritancy, severe uterine contractions and vaso-vaginal reactions, silver was absorbed and patients exhibited dose-related hypokalaemia, vascular changes and neurological disorders. A 69-year-old Japanese patient presenting with pyuria and renal dysfunction underwent retrograde instillation of silver nitrate to stem renal haemorrhage; she developed argyria of the upper urinary tract and abnormality in her renal function tests, but the pathogenicity is not known.[117]

A dramatic example of silver absorption from a cardiovascular device is documented in relation to the Silzone mechanical heart valve (St Jude Medical Inc).[118] This device was withdrawn from clinical use for safety reasons unrelated to silver but, in a litigation case, the cause of death of a 44-year-old lady was identified as a large blood clot forming on the silver-coated surface of the synthetic valve.[119] Blood silver concentrations were not reported in this or other clinical studies with the St Jude Silzone heart valve,[120] but preclinical studies have demonstrated that surgical implantation of the valve in sheep hearts led to a modest transitory rise in argyraemia [40 parts per billion (ppb)] within 10 d of implantation, returning to preoperative levels within 30 d.[121] Silver accumulation in liver ($16.75\,\mu g\,g^{-1}$ dry weight), kidney ($8.86\,\mu g\,g^{-1}$), brain, spleen and lung ($<5\,\mu g\,g^{-1}$) confirmed that observed in earlier studies.[122]

Silver-impregnated bone cements are an occasional cause of clinical problems.[123–124] A 76-year-old lady implanted with a Christiansen total hip replacement was admitted to hospital following neurological complaints and found to have high silver content in her hip joint fluid ($103\,\mu g\,L^{-1}$) and other tissues five years following surgery. Her blood silver was 1000-fold higher than normal but declined from $6.3\,\mu g\,L^{-1}$ in the 24 months following removal of the prosthesis and silver-containing bone cement to $2.2\,\mu g\,L^{-1}$, suggesting a modest retention of silver in the body. The neurological changes regressed.

Acupuncture needles containing silver and occasionally gold are an essential part of the ancient Hari therapy practised in Japan for relief of muscular pains, headache and other forms of discomfort. A 63-year-old lady implanted with traditional silvered (69% silver) needles regularly over more than 14 years for epilepsy related to hydrocephalus developed focal argyria-like discolourations on her extremities.[125] The bluish discolouration varied from dark in the centre to pale at the periphery of the macules and X-radiography revealed multiple needle-like "shadows" in the skin, but their pathology was typical of argyria. A second interesting case of argyria attributable to acupuncture refers to a 57-year-old housewife who commenced Hari therapy at the age of 33 and, over the space of 13 years, implanted 2500 needles into most parts of her body to induce sterility and relieve her general fatigue.[126] The blue-black macules of argyria appeared ten years after commencing therapy and gradually increased in number and intensity of pigmentation. Blood levels of silver in this patient were estimated at $16.7 \, \mu g \, L^{-1}$, but the silver content of the needles is not known. Electron dense granules of 40–500 nm diameter were observed in the spectrographs indicative of silver absorption. The argyraemic concentration in the first case is not known, but earlier work suggested that the minimum body silver levels consistent with evidence of argyria are at least 4–5 g.[6,127,128]

Silver absorption from silver-coated needles, catheters and stents implanted into major blood vessels is not well documented. Silver ion released by any device implanted intraparenterally will bind with albumins and macroglobulins, and be taken up by erythrocytes and reticulo-endothelial cells in the circulation. Rare cases are reported of suicidal or accidental injection of silver nitrate, with fatal consequences.[54] However, an interesting case of accidental intravenous injection of radioactive silver nitrate (0.1 mg) in a cancer patient provides useful information regarding uptake of silver and its rapid elimination in faeces and urine. Silver absorbed into the blood during the first three days after injection was primarily associated with erythrocytes but subsequently 77% became bound to globulins, 15% to albumins and 8% to fibrinogens.[129] Silver was rapidly metabolised from the blood and, within seven minutes of injection, 70% of the injected dose was metabolised to the liver and other tissues; within 2 h, 90% had been lost. Estimates suggest that approximately 2% of the initial radioactivity was retained in the body for at least 20 d.

3.6   Silver in Textiles

Early studies showed that silver in braided nylon fabric provides commendable antibacterial activity in tissue cultures,[130–133] but minimal attention was given to the possibility that Ag^+ released would interact with the tissues of the skin or would be absorbed percutaneously. Presently, development of silver in textiles and in hygiene clothing as a means of alleviating odour, socially undesirable skin infections and pathogenic conditions (*e.g.* MRSA or athlete's foot, *Tinea pedis*) is still at an early stage. Although several products presently marketed contain high levels of metallic silver in the form of threads or minute particles, available evidence suggests that release of Ag^+ into the surface moisture of the skin is very low and that free ion available for absorption through intact or injured human skin is of minimal significance.[65,134,135] Kumar and Munstedt commented that Ag^+ release from a silver-loaded polyamide composite material soaked in water is time-dependant and increases according to the "physical state" of the material and its silver content.[136] Scanning electron microscopy has shown that composites containing silver nanoparticles with fine dispersion and relatively high surface area ($0.78 \, g \, cm^{-2}$) exhibited considerably higher Ag^+ release than aggregates of larger particles.

Manufacturers balance the silver ion release from a fabric in keeping with concepts of the oligodynamic action in which silver-sensitive organisms concentrate toxic levels of silver ion from dilute solutions.[137] Products are designed to have a long-lasting antibiotic action. Minute amounts of silver liver released from clothing are expected to be sufficient to control bacterial and fungal infections on the skin surface without risk of percutaneous absorption. Adsorption of silver to the surface of mature keratin of hair follicles is hypothetically a further means of de-activating silver ion.[24] Sea Cell® Active fibres are produced by loading pure Sea Cell® algal fibres (Lyocell Process) with silver through immersion in dilute silver nitrate solution containing $6900 \, mg \, kg^{-1}$,[138] but only minimal amounts of silver are released from raw fabric containing 20% of the silver-treated fibre. Sea Cell® Active fibre did not influence the activity of the inflammatory cytokines IL-1α, IL-6, IL-8 or GM-CSF in cultured skin fibroblasts and is expected to be entirely compatible with healthy human skin. Padycare® textiles and similar silver-containing fabrics comprising woven silver filaments with a total silver content of 20% designed for the treatment of atopic dermatitis[139] show commendable antibiotic efficacy *in vitro*, but it is presently unclear whether or to what extent silver released is absorbed by diseased skin. It is expected that silver in textiles designed for use in tropical climates where excessive sweating leads to increased hydration of the skin will be associated with higher levels of ionisation and percutaneous absorption.[140] Further research is awaited.

Silver-loaded fabrics such as polyethylene terephthalate have wide ranging applications in medical devices for intra-parenteral use. The St Jude mechanical heart valve discussed above for prevention of prosthetic valve endocarditis is an illustration of polyethylene terephthalate cloth incorporating ion beam-assisted deposition of silver (IBAD).[122] The product showed commendable antibacterial action *in vitro* and when implanted into sheep hearts led to only marginally raised blood silver levels (<20 ppb), even though it proved unsuitable for human use.[118,119]

3.7 Metabolism of Silver in the Human Body

Contrary to statements that all forms of silver are cumulative once they enter body tissues and that very little is excreted,[20,21] silver *is* actively metabolised in the human body and a large part eliminated eventually *via* the liver, urine and hair.[1,15,16,68] Metallic silver is biologically inert, but it slowly ionises in the presence of moisture and body fluids and secretions to release biologically active Ag^+ (or other unidentified ion) which is absorbed to a variable extent through the skin and mucus membranes of the respiratory, gastrointestinal and urinogenital tracts.[1,2]

The most useful and reliable information concerning the metabolism of silver in the human body is provided by clinical and experimental studies relating to the development and safety evaluation of 1% silver sulfadiazine (Flamazine).[63,68,141] When Charles Fox first synthesised silver sulfadiazine in 1968, he noticed that "the velocity of absorption of sulfadiazine" from the insoluble complex silver salt depends upon the contact of the ointment or other suspending medium with the body and wound fluids.[70] Current views are that silver is eliminated from the human body principally in bile, with some in urine and minimal amounts in hair and nail. Kanabrocki measured silver in the thumb nails of 60 normal ladies and reported low levels of $0.74 \pm 0.75\,\mu\mathrm{g\,g}^{-1}$.[142] Silver levels in nail noted by East *et al.* were slightly lower at $0.45 \pm 0.04\,\mu\mathrm{g\,g}^{-1}$ wet weight but, in their patient, long-term argyria was associated with skin silver of $71.7 \pm 3.7\,\mu\mathrm{g\,g}^{-1}$ wet weight.[43] The rate of silver accumulation and mobilisation in skin, hair and nail is not known from these studies or earlier work by Tipton for the International Commission for Radiological Protection.[143]

Absorbed silver sulfadiazine dissociates rapidly in the circulation with as much as 50% of the sulfadiazine moiety being excreted as unchanged drug, and the remainder N_4-acetylated or subject to 5-hydroxylation.[144] Silver is absorbed to form a reservoir of silver ion in the skin for prolonged antibiotic action.[145] From this dermal reservoir, protein-bound silver is transported to bone and most soft tissues.[15,79,146,147] Silver uptake by teeth and bone is low but tends to be long-lasting or permanent. Case reports have occasionally stated that silver is deposited in brain and neurological tissues and that it is a cause of certain neurological changes,[123–124,148] but critical evaluation of these and other studies indicate that silver is not absorbed into neurological tissues but becomes bound in lysosomal vacuoles of the blood–brain barrier and in the blood–cerebrospinal fluid (CSF) barrier.[45] Excessive amounts of silver absorbed from the respiratory or gastrointestinal tracts are commonly deposited in the cornea and conjunctiva of the eye (argyrosis) and in the dermal regions of the skin (argyria), particularly in areas exposed to solar radiation (see Chapter 9). These deposits tend to be long-lasting but may fade as a response to normal wear-and-tear and the action of dermal macrophages. In a similar way, dermal deposits of silver sulfide or silver selenide in severe cases of generalised argyria are normally long-lasting. Good clinical evidence supported by accurate silver estimations indicates that silver sulfadiazine absorption through injured skin is proportional to the severity and depth of wounds/burns, their vascularity, the duration of therapy and presumably the amount of silver sulfadiazine applied as discussed above.[15,17,68] Urinary silver analysis provides a convenient means of assessing silver exposure by all routes and may provided a useful guide to the silver content of the body.

Absorbed silver may persist in the body and concentrate in the liver where it evokes transitory changes in drug-metabolising enzymes including γ-glutamyl transpeptidase, aspartate aminotransferase and alanine aminotransferase.[68] Toxic changes in the liver are rare in burn wound cases treated with silver sulfadiazine or sustained silver release wound dressings, but a case was reported where a nanocrystalline silver dressing was associated with mild manifestations of argyria and elevated serum levels of glutamic–oxaloacetic transaminase

(GOT), glutamic–pyruvic transaminase (GPT) and γ-glutamyl-transferase (γ-GT).[149] The full hepatic pathology is not known, but the blood and urine silver levels in this patient were greatly elevated and no evidence found indicating liver damage. A post-mortem examination of another patient with severe burns treated with silver sulfadiazine revealed liver silver of $14\,\mu\mathrm{g\,g}^{-1}$, with extensive deposition of the electron dense silver particles enclosed within lysosomal vesicles. No obvious correlation was seen in this patient between blood silver and liver dysfunction. The kidney is also a target organ for deposition of absorbed silver, but concentrations reported consistently show renal silver to be appreciably lower than in the liver. Coombs *et al.* concluded that, in view of inconsistencies in urinary silver excretion at blood silver levels of more than $100\,\mu\mathrm{g\,L}^{-1}$, urinary silver excretion is not a reliable guide to blood silver concentrations.[68]

Intracellular handling of silver ion is illustrated by cytochemical changes in the wound margin following application of 1.0% aqueous silver nitrate or 1.0% silver sulfadiazine cream.[63,150] In each case, silver absorbed through epidermal cell epithelia induced synthesis of the cysteine-rich proteins metallothionein I and II. Metallothioneins are complex molecules synthesised by a wide range of cell types in the human body to provide an essential function as binding and carrier proteins for essential trace elements, but they act also as cytoprotectants by irreversibly binding toxic metals and excess quantities of trace metals. Only when this protective mechanism is saturated, is toxicity manifest.[151] They are induced by a number of trace and xenobiotic metals (*e.g.* Zn, Mg, Cu, Se, Ag, Cd).[152–154] Using immunocytochemistry and very pure and specific antibodies (Daco, Denmark), topically applied silver was shown to induce and bind metallothionein I and II in the cytosol of metabolically active cells in the wound margin.[63] Increased metallothionein in metabolically active cells favours the uptake of trace metals including zinc and copper, which in turn promote RNA and DNA synthetases leading to cell proliferation and maturation.[155] There is clinical and experimental evidence to show that topical application of dilute silver nitrate, silver sulfadiazine and the various sustained silver-release wound dressings to acute and chronic skin wounds promotes healing.[156,157] In other respects, silver from silver nitrate and silver sulfadiazine binds albumins and macroglobulins in the wound bed, possibly as a preliminary to percutaneous absorption.[61] Other metabolic changes seen in the skin following topical silver application include induction of epidermal growth factor (EGF), a critical factor in the wound healing cascade.[158]

References

1. A. T. Wan, R. A. J. Conyers, C. J. Coombs and J. P. Masterton, Determination of silver in blood, urine and tissues of volunteers and burn patients, *Clin. Chem.*, 1991, **37**, 1683.
2. R. E. Burrell, A scientific perspective in the use of topical silver preparations, *Ostomy Wound Manage.*, 2003, **49** (Suppl), 19.

3. S. S. Bleehan, D. J. Gould, C. I. Harrington, T. E. Durrant, D. N. Slater and J. C. Underwood, Occupational argyria: light and electron microscopic studies and X-ray microanalysis, *Br. J. Dermatol.*, 1990, **104**, 19.

4. W. R. Buckley, C. F. Oster and D. W. Fassett, Localised argyria. II. The chemical nature of the silver-containing granules, *Arch. Dermatol.*, 1978, **92**, 697.

5. R. J. Pariser, Generalised argyria: clinicopathologic features and histochemical studies, *Arch. Dermatol.*, 1978, **114**, 373.

6. J. Siemund and A. Stolp, Argyrose, *Z. Haut. Geschlechtskr.*, 1968, **43**, 71.

7. W. W. Wilkinson, Silver stains, *J. Am. Med. Assoc.*, 1932, **98**, 72.

8. E. Mankiewicz, 1916, cited by T. Sollemann, in *A Manual of Pharmacology and its Applications to Therapeutics and Toxicology*, Saunders, Philadelphia, 6th edn, 1942, p. 1001.

9. A. P. Moss, A. Sugar, N. A. Hargett, A. Atkin, M. Wolkstein and K. D. Rosenman, The ocular manifestations and functional effects of occupational argyrosis, *Arch. Ophthalmol.*, 1979, **97**, 906.

10. W. M. Grant, *The Toxicology of the Eye*, Chas. Thomas, Springfield, IL, 1986, p. 818.

11. P. A. Laughrea, J. J. Arentsen and P. R. Laibson, Iatrogenic ocular silver nitrate burn, *Cornea*, 1985, **4**, 47.

12. R. M. Stein, W. M. Bourne and T. J. Liesegang, Silver nitrate injury to the cornea, *Can. J. Ophthalmol.*, 1987, **22**, 279.

13. E. Browning, *The Toxicology of Industrial Metals*, Butterworths, London, 1969, pp. 296–301.

14. E. B. Shaw, Questions need for prophylaxis with silver nitrate, *Pediatrics*, 1977, **59**, 792.

15. M. G. Boosalis, J. T. McCall, D. H. Ahrenholz, L. D. Solem and C. J. McClain, Serum and urinary silver levels in thermal injury patients, *Surgery*, 1987, **101**, 40.

16. G. D. DiVincenzo, C. J. Gordiano and L. S. Schriever, Biologic monitoring of workers exposed to silver, *Int. Arch. Occup. Environ. Health*, 1985, **56**, 207.

17. S. Maitre, K. Jaber, J. L. Perot, C. Guy and F. Cambazard, Increased serum and urinary levels of silver during treatment with topical silver sulphadiazine, *Ann. Dermatol. Venereol.*, 2002, **129**, 217.

18. N. Williams and I. Gardener, Absence of symptoms in silver refiners with raised blood silver levels, *Occup. Med. (London)*, 1995, **45**, 205.

19. Smith & Nephew, Acticoat® (Silcryst™) Scientific Background, Scientific Data Sheet, Technology.

20. M. Sittig, *Handbook of Toxic and Hazardous Chemicals and Carcinogens*, Noyes Data Corporation, Park Ridge, NY, 1985, p. 789.

21. G. D. Clayton and F. E. Clayton, *Patty's Industrial Hygiene and Toxicology*, Wiley, New York, 3rd edn, 1981, Vol. **2**, pp. 1881–1894.

22. J. J. Hostýnek, R. S. Hinz, C. R. Lorence, M. Price and R. Guy, Metals and the skin, *CRC Crit. Rev. Toxicol.*, 1993, **23**, 171.

23. A. B. G. Lansdown, Cartilage and bone as target materials for toxic materials, in *General and Applied Toxicology*, ed. B. Ballantyne, T. C. Marrs and T. Sylversen, John Wiley, Chichester, 3rd edn, 2009, Vol. **3**, Ch. 62, pp. 1491–1523.
24. S. N. Kales and D. C. Cristiani, Hair and metal toxicity, in *Hair in Toxicology*, ed. D. Tobin, Royal Society of Chemistry, Cambridge, 2005, pp. 125–158.
25. W. R. Buckley and C. J. Terhaar, The skin as an excretory organ in argyria, *Trans. St John's Hosp. Dermatol. Soc.*, 1973, **59**, 39.
26. A. Hambidge, Reviewing the efficacy of alternative water treatment techniques, *Health Estate*, 2001, **55**, 23.
27. M. C. Fung and D. L. Bowen, Silver products for medical indications: risk benefit assessment, *J. Toxicol. Clin. Toxicol.*, 1996, **34**, 119.
28. US Environmental Protection Agency, *Ambient Water Quality Criteria for Silver*, USEPA, Washington DC, 1980, EPA Report Number 440/5-80-071.
29. A. R. Flegal, C. L. Brown, S. Squire, J. R. M. Ross, G. M. Scelfo and S. Hibden, Spatial and temporal variations in the silver contamination and toxicity in San Francisco Bay, *Environ. Res.*, 2007, **105**, 34.
30. N. G. Rose-James and R. C. Playle, Protection by two complexing agents, thiosulphate and dissolved organic matter, against the physiological effects of silver nitrate to rainbow trout (*Ornithorhynchus mykiss*) in ion-poor water, *Aquat. Toxicol.*, 2000, **51**, 1.
31. N. A. Webb, J. R. Shaw, J. Morgan, C. Hogstrand and C. H. Wood, Acute and chronic physiological effects of silver exposure on three marine teleosts, *Aquat. Toxicol.*, 2001, **54**, 161.
32. W. L. Roper, *Toxicological Profile for Silver*, Agency for Toxic Substances and Disease Registry, US Public Health Service, Atlanta, GA, 1990.
33. A. B. G. Lansdown, Controversies over colloidal silver, *J. Wound Care*, 2003, **12**, 120.
34. National Center for Complementary and Alternative Medicine, *Colloidal Silver Products*, National Institutes of Health, Bethesda, MD, 2004, Consumer Advisory Bulletin, p. 1.
35. T. Sollemann, Silver, in *A Manual of Pharmacology and its Applications to Therapeutics and Toxicology*, Saunders, Philadelphia, 1942, pp. 1102–1109.
36. R. E. Gosselin, R. P. Smith and H. C. Hodge, in *Clinical Toxicology of Commercial Products*, Williams and Wilkins, Baltimore, 5th edn, 1984, pp. 11–145.
37. J. Doull, C. D. Klaasen and M. D. Amdur, in *Casarett and Doull's Toxicology*, MacMillan, New York, 3rd edn, 1986, p. 625.
38. B. Venugopal and T. D. Lucky, in *Metal Toxicity in Mammals-2: Chemical Toxicity of Metals and Metalloids*, Plenum Press, New York, 1978, pp. 32–36.
39. R. H. Dreisbach, in *Handbook of Poisoning*, ed. R. H. Dreisbach and W. O. Robertson, Appleton & Lange, Norwalk, CT, 12th edn, 1987, p. 375.

40. P. L. Drake and K. J. Hazelwood, Exposure-related health effects of silver and silver compounds: a review, *Ann. Occup. Hyg.*, 2005, **49**, 575.
41. K. D. Rosenman, A. Moss and S. Kon Argyria: clinical implications of exposure to silver nitrate and silver oxide, *J. Occup. Health*, 1979, **21**, 430.
42. M. Westhofen and H. Shafer, Generalised argyrosis in man: neurological ultra-structural and X-ray microanalytical findings, *Arch. Otolaryngol.*, 1986, **243**, 260.
43. B. W. East, K. Boddy, E. D. Williams, D. MacIntyre and A. A. L. C. McLay, Silver retention, total body silver and tissue silver concentrations in argyria associated with exposure to an anti-smoking remedy containing silver nitrate, *Clin. Exp. Dermatol.*, 1980, **5**, 305.
44. Y. Ohbo, H. Fukuzako, K. Tacheuchi and M. Takigawa, Argyria and convulsive seizures caused by ingestion of silver in a patient with schizophrenia, *Psychiatry Clin. Neurosci.*, 1996, **50**, 89.
45. A. B. G. Lansdown, Critical observations on the neurotoxicology of silver, *CRC Crit. Rev. Toxicol.*, 2007, **37**, 237.
46. K. Boddy, A high sensitivity shadow-shield-whole-body monitor with scanning bed and tilting chair geometrics incorporated in a mobile laboratory, *Br. J. Radiol.*, 1967, **40**, 631.
47. International Commission on Radiological Protection, *Recommendations of the International Commission on Radiological Protection. Report of Committee II on permissible dose for internal radiation*, Pergamon Press, Oxford, 1959, ICRP Publication 2.
48. N. S. Tomi, B. Kränke and W. Aberer, A silver man, *Lancet*, 2004, **363**, 532.
49. A. O. Gettler, C. P. Rhoads and A. Weiss, A contribution to the pathology of generalised argyria with a discussion on the fate of silver in the human body, *Am. J. Pathol.*, 1927, **3**, 631.
50. B. A. Bouts, Images in clinical medicine; argyria, *N. Engl. J. Med.*, 1999, **340**, 1554.
51. W. R. Hill and D. M. Pillsbury, *Argyria: The Pharmacology of Silver*, Williams and Wilkins, Baltimore, MD, 1939.
52. L. S. Goodman and A. Gilman, in *Pharmacological Basis of Therapeutics*, MacMillan, New York, 5th edn, 1975, pp. 930–931.
53. US Food and Drug Administration, Colloidal silver, *Health and Fraud Bulletin*, 1994, No. 19.
54. S. D. Humphreys and P. A. Routledge, The toxicology of silver nitrate, *Adverse Drug React. Toxicol. Rev.*, 1998, **17**, 115.
55. M. L. Clarke, D. G. Harvey and D. J. Humphreys, in *Veterinary Toxicology*, Baillière Tindall, London, 2nd edn, 1981, p. 73.
56. US Environmental Protection Agency, *Registration Eligibility Decision Document, Silver*, Office of Pesticide Programs, Washington, DC, 1992, Ref. 738-R-94-021.
57. B. Forslind, S. Engström, J. Engblom and L. Norlén, A novel approach to the understanding of the human skin barrier function, *J. Dermatol. Sci.*, 1997, **14**, 115.

58. A. M. Kligman, Cutaneous toxicology: an overview from the underside, *Curr. Probl. Dermatol.*, 1978, **7**, 1.
59. J. M. Schierholz, J. Beuth, G. Pulverer and D.-P. König, Silver-containing polymers, *Antimicrob. Agents Chemother.*, 1999, **43**, 2819.
60. T. Karlsmark, R. H. Agerslev, S. H. Bendz, J. R. Larsen, J. Roed-Petersen and K. E. Andersen, Clinical performance of a new silver dressing, Contreet Foam, for chronic exuding venous leg ulcers, *J. Wound Care*, 2003, **12**, 351.
61. A. B. G. Lansdown, A. Williams, S. Chandler and S. Benfield, Silver absorption, and antibacterial efficacy of silver dressings, *J. Wound Care*, 2005, **14**, 131.
62. A. B. G. Lansdown, K. Jensen and M. Q. Jensen, Contreet foam and Contreet hydrocolloid: an insight into two new silver-containing dressings, *J. Wound Care*, 2003, **12**, 205.
63. A. B. G. Lansdown, B. Sampson, P. Laupattarakasem and A. Vuttivirojana, Silver aids healing in sterile skin wounds: experimental studies in the laboratory rat, *Br. J. Dermatol.*, 1997, **137**, 728.
64. W. Lublinsku, Silbernitrat oder Silbereiweiss, *Berl. Klin. Wochenschr.*, 1914, **51**, 1643.
65. A. B. G. Lansdown, Silver in health care: antimicrobial effects and safety in use, *Curr. Probl. Dermatol.*, 2006, **33**, 17.
66. O. Nørgaard, Investigations with radioactive ^{111}Ag into the resorption of silver through human skin, *Acta Dermatol. Venereol.*, 1954, **34**, 415.
67. E. Skog and J. E. Wahlberg, A comparative investigation of the percutaneous absorption of metal compounds in the guinea pig by means of radioactive isotopes ^{51}Cr, ^{58}Co, ^{65}Zn, ^{110m}Ag, ^{115m}Cd, ^{203}Hg, *J. Invest. Dermatol.*, 1964, **43**, 187.
68. R. C. Coombs, A. T. Wan, J. P. Masterton, R. A. J. Conyers, J. Pedersen and Y. T. Chia, Do burn wound patients have a silver lining?, *Burns*, 1992, **18**, 179.
69. C. R. Baxter, Topical use of 1% silver sulphadiazine, in *Contemporary Burn Management*, eds. H. C. Polk and H. H. Stone, Little Brown, Boston, 1971, pp. 217–225.
70. C. L. Fox, Silver sulphadiazine: a new topical therapy for *Pseudomonas* infection in burns, *Arch. Surg.*, 1968, **96**, 184.
71. Bray Healthcare, *Toughened silver nitrate caustics in medicine: warts, verrucae, cautery, granuloma*, 1997, www.bray.co.uk..
72. B. Strauch, W. Buch, W. Grey and D. Laub, Successful treatment of methaemaglobinaemia secondary to silver nitrate therapy, *N. Engl. J. Med.*, 1969, **281**, 257–258.
73. J. L. Ternberg and E. Luce, Methaemoglobinaemia: a complication of the silver nitrate treatment of burns, *Surgery*, 1968, **63**, 328.
74. A. H. Cushing and S. Smith, Methaemoglobinaemia with silver nitrate therapy of a burn: report of a case, *J. Pediat.*, 1969, **74**, 613.
75. M. E. Geffner, D. R. Powers and W. T. Choctaw, Acquired methaemaglobinaemia, *West. J. Med.*, 1981, **134**, 7.

76. T. C. Marrs and S. Warren, Haematology and toxicology, in *General and Applied Toxicology*, eds. B. Ballantyne, T. C. Marrs and T. Sylversen, Macmillan Reference, London, 2nd edn, 1999, Vol. **1**, pp. 383–399.
77. Medsafe, *Information for Health Professions: Methylene Blue Solution for Injection*, AFT Pharmaceuticals Ltd, Takapuna, New Zealand, 2007.
78. E. J. L. Lowbury, Problems of resistance in open wounds and burns, in *The Rational Choice of Antibacterial Agents*, eds. R. P. Mouton, W. Brumfitt and J. M. T. Hamilton-Miller, Kluwer Harrap, London, 1977, pp. 18–31.
79. K. F. Bader, Organ deposition of silver following silver nitrate therapy of burns, *Plast. Surg.*, 1966, **37**, 550.
80. J. P. Marshall and R. P. Sneider, Systemic argyria secondary to topical silver nitrate, *Arch. Dermatol.*, 1977, **113**, 1077.
81. C. C. Bondoc, P. J. Morris, T. Wee and J. F. Burke, Metabolic effects of 0.5% silver nitrate therapy for extensive burns in children, *Surg. Forum*, 1966, **17**, 477.
82. J. B. Wright, D. L. Hansen and R. E. Burrell, The comparative efficacy of two antimicrobial barrier dressings: *in vitro* examination of two controlled release of silver dressings, *Wounds*, 1998, **10**, 179.
83. M. E. Innes, N. Umraw, J. S. Fish, M. Gomez and R. C. Cartotto, The use of silver coated dressings on donor sites: a prospective, controlled matched pair study, *Burns*, 2001, **27**, 621.
84. G. R. Sibbald, P. Coutts, A. C. Brown and S. Coehlo, Acticoat®, a new ionised silver-coated dressing: its effect on bacterial load and healing rates, presented at First World Wound Healing Congress, Victoria, Australia, 2000.
85. S. Enoch and K. G. Harding, Wound bed preparation: the science behind barriers to healing, *Wounds*, 2003, **15**, 213.
86. G. S. Schultz, R. G. Sibbald and V. Falanga, Wound bed preparation, a systematic approach to wound management, *Wound Rep. Regen.*, 2003, **11**, 1.
87. L. Bolton, Are silver products safe and effective for chronic wound management? *J. Wound Ostomy Continence Nurs.*, 2006, **33**, 469.
88. R. G. Sibbald, A. C. Browne, P. Coutts and D. Queen, Screening evaluation of an ionized nanocrystalline silver dressing on chronic wound care, *Ostomy Wound Manage.*, 2001, **47**, 38.
89. E. Vlachau, E. Chipp, E. Shale, Y. T. Wilson, R. Papini and N. S. Moiemen, The safety of nanocrystalline silver dressings on burns: a study of systematic silver absorption, *Burns*, 2007, **33**, 979.
90. R. H. Wolbling, R. Milbradt, E. Schopenhauer-Germann, G. Euler and K. H. Konig, Argyrosis in workers in the silver-processing industry: dermatological investigations and quantitative measurements with atomic absorption spectrometry, *Arbietsmed. Sozialmed. Praventimed.*, 1988, **23**, 293.
91. W. R. Buckley and C. C. Terhaar, The skin as an excretory organ in argyria, *Trans. St John's Hosp. Dermatol. Soc.*, 1973, **59**, 39.

92. M. C. Fung, M. Weintraub and D. L. Bowen, Colloidal silver proteins marketed as health supplements, *J. Am. Med. Assoc.*, 1995, **274**, 1196.

93. W. H. Spencer, L. K. Garron, F. Contraras, T. L. Hayes and C. Lai, Endogenous ocular and systemic silver deposition, *Trans. Ophthalmol. Soc. U.K.*, 1980, **100**, 171.

94. J. M. Harker and D. Hunter, Occupational argyria, *Br. J. Dermatol.*, 1935, **47**, 441.

95. A. Montandon, Argyrose des voies respiratoires, *Arch. Mal. Prof.*, 1959, **20**, 419.

96. K. D. Rosenman, N. Seixas and I. Jacobs, Potential nephrotoxic effects of exposure to silver, *Br. J. Ind. Med.*, 1987, **44**, 267.

97. US Occupational Safety and Health Administration and Mine Safety and Health Administration, cited by P. L. Drake and R. J. Hazelwood (Ref. 40).

98. American Conference of Governmental Industrial Hygienists, *Documentation of Threshold Limit Values and Biological Exposure Limits*, Cincinnati, OH, 6th edn, 1991, Vol. **1–3**, p. 1397.

99. D. Newton and A. Holmes, A case of accidental inhalation of ^{65}Zn zinc and ^{110m}Ag silver, *Pediatr. Res.*, 1966, **29**, 403.

100. US Environmental Protection Agency, *Re-registration Eligibility Decision Document-Silver*, Office of Pesticide Programs, Washington, DC, 1992, Reg. No. 738-R-094-021.

101. F. Mayall and D. Wild, A silver tattoo of the nasal mucosa after silver nitrate cautery, *J. Laryngol. Otol.*, 1966, **110**, 609.

102. P. Murthy and M. R. Laing, An unusual severe adverse reaction to silver nitrate cautery for epistaxis in an immuno-compromised patient, *Rhinology*, 1996, **34**, 186.

103. Y. Shimamoto and H. Shimamoto, Systemic argyria secondary to a breath freshener "Jintan Silver Pills", *Hiroshima J. Med. Sci.*, 1987, **36**, 245.

104. S. Landas, S. M. Bonsib, R. Ellerbroek and J. Fischer, Argyria: micro-analytic morphologic correlation using paraffin-embedded tissue, *Ultrastruct. Pathol.*, 1986, **10**, 129.

105. A. C. Timmins and A. R. Morgan, Argyria or argyrosis, *Anaesthesia*, 1988, **43**, 755.

106. T. Cymet, Do silver alloy catheters increase the risk of systemic argyria? *Arch. Int. Med*, 1998, **161**, 1014.

107. P. Thibon, X. Le Coutour, R. Leroyer and J. Fabry, Randomised multi-centre trial of the effects of a catheter coated with hydrogel and silver salts on the incidence of hospital acquired urinary tract infections, *J. Hosp. Infect.*, 2000, **45**, 117.

108. K. K. Lai and S. A. Fontecchio, Use of silver-hydrogel urinary catheters on the incidence of catheter-associated infections in hospitalised patients, *Am. J. Infect. Control*, 2002, **30**, 221.

109. E. J. Tobin and R. Bambauer, Silver coating dialysis catheters to reduce bacterial colonisation and infection, *Ther. Apher. Dial.*, 2003, **7**, 504.

110. T. S. J. Elliott, Role of antimicrobial central venous catheters for the prevention of associated infections, *J. Antimicrob. Chemother.*, 1999, **43**, 441.

111. J. M. Schierholz, L. J. Lucas, A. Rumpt and G. Pulverer, Efficacy of silver-coated medical devices, *J. Hosp. Infect.*, 1998, **40**, 257.

112. U. Samuel and J. P. Guggenbichler, Prevention of catheter-related infections: the potential of a new nano-impregnated catheter, *Int. J. Antimicrob. Agents*, 2004, **23** (Suppl. 1), S75.

113. D. Stickler, N. S. Morris and T. J. Williams, An assessment of the ability of a silver-releasing device to prevent bacterial contamination of urethral catheter drainage system, *Br. J. Urol.*, 1996, **78**, 839.

114. G. Reinhardt, G. von Mallinck, H. Kittel and O. Opitz, Acute fatal poisoning with silver nitrate following an abortion attempt, *Arch. Kriminol.*, 1971, **148**, 69.

115. C. A. D. Ringrose, Office tubal sterilization, *Obstet. Gynecol.*, 1973, **42**, 151.

116. N. H. Dubin, T. H. Parmley, R. T. Coc and T. M. King, Effect of silver nitrate on pregnancy termination in cynomolgus monkeys, *Fertil. Steril.*, 1981, **36**, 106.

117. Y. Kojima, K. Uchida and H. Takiuchi, Argyrosis of the urinary tract after silver nitrate instillation: report of a case, *Hinyokika Kiyo*, 1993, **39**, 41.

118. Medicines and Healthcare Products Regulatory Agency, *AN 1999(06) Thromboembolic Complications Involving Silzone Mechanical Heart Valve*, MHRA, London, 1999, ref. 04/01/98122131, www.mhra.gov.uk/Publications/Safetywarnings/MedicalDeviceAlerts/Advicenotices/CON008871, accessed 19 November 2009.

119. O. Dyer, US pacemaker manufacturer faces British lawsuit, *Br. Med. J.*, 2003, **327**, 466.

120. P. Tozzi, A. Al-Darweesh, P. Vogt and F. Stumpe, Silver-coated prosthetic heart valve: a double bladed weapon, *Eur. J. Cardiothorac. Surg.*, 2001, **19**, 729.

121. D. Langaki, M. F. Ogle, D. Cameron, R. A. Lirtzmann, R. F. Schroeder and M. W. Mirsch, Evaluation of a novel bioprosthetic heart valve incorporating anti-calcification and antimicrobial technology in a sheep model, *J. Heart Valve Dis.*, 1998, **7**, 633.

122. K. S. Tweeden, J. D. Douglas, A. J. Razzouk, W. Holmberg and S. J. Kelly, Biocompatability of silver-modified polyester to antimicrobial protection heart valves, *J. Heart Valve Dis.*, 1997, **6**, 553.

123. E. Sudemann, H. Vik, M. Rait, K. Todnem, K. J. Andersen, K. Juhlsham, O. Friesland and J. Rungby, Systemic and local silver accumulation after total hip replacement using silver-impregnated bone cement, *Med. Prog. Technol.*, 1994, **20**, 179.

124. H. Vik, K. J. Andersen, J. Juhlsham and K. Todnem, Neuropathy caused by silver absorption from arthroplasty cement, *Lancet*, 1985, **1** (8433), 872.

125. Y. Tanita, T. Kato, K. Hanada and H. Tagami, Blue macules of localised argyria caused by implanted acupuncture needles, *Arch. Dermatol.*, 1985, **121**, 1550.

126. S. Sato, H. Sueki and A. Nishijima, Two unusual cases of argyria: the application of an improved tissue processing method for X-ray microanalysis of selenium and sulphur in silver-laden granules, *Br. J. Dermatol.*, 1999, **140**, 158.

127. A. B. Molokhia, B. Portnoy and A. Dyer, Neutron activation analysis of trace elements in the skin, *Br. J. Dermatol.*, 1969, **101**, 567.

128. H. Citober, H. Frischauf and I. Leodolter, Quantitative Untersuchungen bei universeller Argyrose mittels Neutronenaktivierungsanalyse, *Virchows Arch. Pathol. Anat.*, 1970, **350**, 44.

129. A. A. Polachek, C. B. Cope, R. F. Willard and T. Enns, Metabolism of radioactive silver in a patient with carcinoid, *J. Lab. Clin. Med.*, 1996, **56**, 499.

130. E. A. Deitch, A. A. Marino, T. E. Gillespie and J. A. Albright, Silvernylon: a new antibacterial agent, *Antimicrob. Agents Chemother.*, 1983, **23**, 356.

131. P. C. MacKeen, S. Person, S. C. Warner, W. Snipes and E. E. Stevens, Silver-coated nylon fibre as an antibacterial agent, *Antimicrob. Agents Chemother.*, 1987, **31**, 93.

132. C. C. Chu, W. C. Tsai, J. Y. Yao and S. S. Chiu, Newly made antibacterial braided nylon sutures. *In vitro* qualitative and quantitative and *in vivo* preliminary biocompatibility study, *J. Biomed. Mater. Res.*, 1987, **21**, 1281.

133. W. C. Tsai, C. C. Chu, S. S. Chiu and J. Y. Yao, *In vitro* quantitative study of newly made antibacterial braided sutures, *Surg. Gynecol. Obstet.*, 1987, **165**, 207.

134. A. B. G. Lansdown, Risk assessment of silver in antimicrobial textiles, presented to 2nd European Conference on Textiles and the Skin, Stuttgart, Germany, 2006, NR.6-1.

135. D. Höfer, Antimicrobial textiles: evaluation of their effectiveness and safety, *Curr. Probl. Dermatol.*, 2006, **33**, 35.

136. R. Kumar and H. Münstedt, Silver ion release from antimicrobial polyamide/silver composites, *Biomaterials*, 2005, **26**, 2081.

137. K. W. von Nägeli, Leben die oligoynamischen Erschreinungen an lebenden Zellen, *Denkschr. Schweiz. Naturforsch. Ges.*, 1893, **33**, 174.

138. S. Zikeli, Production process of a new cellulosic fiber with antimicrobial properties, *Curr. Probl. Dermatol.*, 2006, **33**, 110.

139. A. Gauger, Silver-coated textiles in the therapy of atopic eczema, *Curr. Probl. Dermatol.*, 2006, **33**, 152.

140. D. E. Wurster, Some physico-chemical factors influencing percutaneous absorption from dermatologicals, *Curr. Probl. Dermatol.*, 1978, **7**, 156.

141. R. J. White and R. Cooper, Silver sulphadiazine: a review of the evidence, *Wounds UK*, 2005, **1**, 51.

142. E. L. Kanabrocki, *Analysis of Trace Elements in Human Tissues*, World Health Organisation/International Atomic Energy Authority, Joint Research Programme on Trace Elements in Cardiovascular Diseases, 1973, Technical Report Series, No. 157, p. 5.

143. I. H. Tipton, *Report of the Task Group on Reference Man*, International Commission on Radiological Protection, Pergamon Press, Oxford, 1975, ICRP Publications Series No. 23.

144. T. B. Vree and Y. A. Hexter, Sulphadiazine, in *Clinical Pharmacokinetics of sulphonamides and their Metabolites. An Encyclopaedia, Antibiotics and Chemotherapy*, ed. H. Schönfeld, Karger, Basel, 1987, pp. 1–9, 16–24.

145. C. Dollery, Silver sulphadiazine, in *Therapeutic Drugs*, Churchill Livingstone, Edinburgh, 1991, Vol. **2**.

146. R. Lazarre, P. A. Watson and D. D. Winter, Distribution and excretion of silver sulphadiazine applied to scalds in the pig, *Burns*, 1974, **1**, 57.

147. S. Sano, R. Fujimori, M. Takashima and Y. Itokawa, Absorption excretion and tissue distribution of silver sulphadiazine, *Burns Incl. Therm. Inj.*, 1982, **8**, 278.

148. J. Rungby and G. Danscher, Localisation of excess silver in the brain and spinal cord of silver-exposed rats, *Acta Neuropathol. Berlin*, 1983, **60**, 92.

149. M. Trop, M. Novak, B. Hellbom, W. Kroll and W. Gösseir, Silver toxicity with use of Acticoat® in burn wound treatment, presented at European Burns Association conference, Bergen, Norway, 2003, Poster Abs. 14.

150. A. B. G. Lansdown, Metallothioneins: potential therapeutic aids for wound healing in the skin, *Wound Rep. Regen.*, 2002, **10**, 130–132.

151. A. B. G. Lansdown, B. Sampson and A. Rowe, Experimental observations in the rat on the influence of cadmium on skin wound repair, *Int. J. Exp. Pathol.*, 2001, **82**, 35.

152. J. J. Van den Oord and M. De Ley, Distribution of metallothionein in normal and pathological human skin, *Arch. Dermatol. Res.*, 1994, **268**, 62.

153. A. T. Miles, G. M. Hawksworth, J. H. Beattie and V. Rodilla, Induction, regulation and biological significance of metallothioneins, *Crit. Rev. Biochem. Mol. Biol.*, 2000, **35**, 35.

154. M. A. Dunn, T. L. Blalock and R. J. Cousins, Metallothioneins: a review, *Proc. Soc. Exp. Biol. Med.*, 1987, **185**, 107.

155. A. B. G. Lansdown, U. Mirastschijski, N. Stubbs, E. Scanlon and M. S. Ågren, Zinc in wound healing: theoretical, experimental and clinical aspects, *Wound Rep. Regen.*, 2007, **15**, 2.

156. D. Kjolseth, J. M. Frank, J. H. Barker, G. L. Anderson, A. I. Rosenthal, R. D. Ackland, D. Schuschke, F. R. Campbell, G. R. Tobin and L. J. Weiner, Comparison of the effects of commonly used wound agents on epithelialisation and neovascularisation, *J. Am. Coll. Surg.*, 1994, **179**, 305.

157. E. A. Deitch, K. Sittig, D. Heimbach, M. Jordan, W. Cruse, A. Khan and B. Achauer, Results of a multicenter outpatient burn study on the safety and efficiency of DIMAC-SSD, a new delivery system for silver sulpha-diazine, *J. Trauma*, 1989, **29**, 430.

158. G. L. Brown, L. B. Nanney, J. Griffen, A. B. Cramer, J. M. Yancey, L. J. Curtsinger, L. Holzin, G. S. Schultz, M. H. Jurkiewicz and J. B. Lynch, Enhancement of wound healing by topical treatment with epidermal growth factor, *N. Engl. J. Med.*, 1989, **321**, 76.

CHAPTER 4
Silver as an Antibiotic: Problems of Resistance

4.1 Historical Aspects

Silver became recognised for its ability to purify drinking water at least 2000 years ago, but its true antibiotic properties have only become evident in the past hundred years or so following the golden age of microbiology when Robert Koch and Louis Pasteur conducted their fundamental research on the concepts of infection and the transmission of disease.[1] Publication of the famous Koch's Postulates in 1882 (Table 4.1) set a new thinking on the pathogenesis of the many diseases prevalent at the time and the isolation of offending organisms, and perhaps heralding the development of antibacterial chemotherapy. Robert Koch established his principles of disease first when he diagnosed anthrax as a fatal infection in human patients and proved that it was caused by an "organic" pathogen.[2]

Subsequent research by Pasteur and Koch in 1876–1877 on anthrax and transfer of *Bacillus anthracis* is regarded as the virtual starting point of pathogenic bacteriology and presumably set the stage for the development of antibiotic therapies, vaccines and immunotherapy.[3,4]

It is unclear when the true antibiotic properties of metallic silver were first recognised, but reviews of silver in healthcare commonly refer to silver vessels being used to transport drinking water for the monarchs of ancient Babylon, Rome and the Persian Empires, and quote anecdotal evidence that silver spoons were used in an attempt to ward off neonatal and infantile infections. Records from Medieval times suggest that the value of silver in surgery as a means of alleviating disease was appreciated; one 14th century manuscript accredited to Petrus Hispanus prescribed a remedy comprising:

"... *pure silver, cadmium, unripe chelledonia, aloe, litharge, white and red sandalwood, oriental saffron, each four drams; make a powder and mix with juice of coriander, endive and rosewater. The material should be placed in a*

Issues in Toxicology No. 6
Silver in Healthcare: Its Antimicrobial Efficacy and Safety in Use
By Alan B. G. Lansdown
© Alan B. G. Lansdown 2010
Published by the Royal Society of Chemistry, www.rsc.org

Table 4.1 Koch's postulates[2] (Die Aetiologie der Tuberculose, 1882)

The four conditions which need to be satisfied to prove that a particular organism is the cause of a disease are:

- The microbe is present in all patients with the disease.
- It can be isolated from a case.
- It can be grown in pure cultures *in vitro*.
- It produces symptoms of the disease upon inoculation of a late pure culture into a susceptible animal.

glass flask and left to warm in the sun. Then one drop is placed in the eye, by which means many are freed of disease as 'if by a miracle''.[5]

Ambose Paré (1510–1590) the paediatric surgeon who attended to the gunshot wounds of Henri II of France, pioneered silver clips and instruments in craniofacial surgery and other life-threatening conditions.[6] Later surgeons such as John Woodall (1617), surgeon general of the East India Company, also claimed that silver clips, silver instruments, silver nitrate and silver foil reduced the incidence of infections.[7] William Halsted in 1895 appreciated the value of silver in surgical instruments and silver sutures for hernias, but recorded that wounds covered with silver foil were less susceptible to post-operative infection and healed more rapidly.[8]

John Woodall is also accredited with devising a therapy known as "The Surgeon's Mate" comprising one part silver and three parts nitric acid, which was heralded as an efficacious treatment for leg ulcers, head injuries and facial wounds. Paul Ehrlich, who may be regarded as the father of chemotherapy, recognised that chemicals could be found with "*selective and direct action on parasites*" and that strains of parasite could emerge with specific drug resistance.[9,10] In a search for new chemotherapeutic agents, Ehrlich tested 606 compounds before discovering the arsenical drug, Salvarsan. Following initial successes, he followed this with silver-salvarsan and neo-silver-salvarsan for the treatment of venereal diseases which were prevalent at the time.

At the turn of the 20th century, soluble salts of metals including lead, arsenic, copper, bismuth, antimony, mercury and silver were commonly used to control bacterial and fungal infections.[11] Mercury and silver ions exhibited most efficacious action and provided greatest antibacterial action at concentrations of 1 part per million (ppm).[12] The expression "oligodynamic" originally coined by the German botanist Karl von Nägeli to describe the ability of micro-organisms to selectively absorb metal ions from dilute solutions,[13] was used by pharmacologists at the time to denote high antibiotic efficacy of low concentrations of these metal ions. They assumed that "*the presence of comparatively few ions exerted a remarkable effect on the cell*".[12] Clark estimated that lethal concentrations of 10^5 to 10^7 silver ions in bacteria, trypanosomes and yeast cells were of the same order of magnitude as the number of enzyme proteins per cell.[14] He was first to recognise that the bactericidal concentrations of silver solutions are markedly affected by the size of the inoculum and the proteins in the medium, or as discussed later, the amount of free Ag^+ available within a system.[15]

Even in 1942, Sollemann recognised that the prophylactic value of 0.5–1% silver nitrate in reducing the incidence of ophthalmological neonatorum was not as well-established as was commonly assumed.[11] Dr Credé's claims that vaginal douching with 2% silver nitrate dramatically reduced the incidence of infection from 10.8 to <0.2% in his clinic[16] have been seriously questioned by later workers, who felt that his results were largely attributable to 2% phenol douches and not silver nitrate *per se*. Other early therapies involved silver nitrate fused into sticks and pencils as "lunar caustic" in the removal of excessive granulations, warts and verrucae but at concentrations of 20–95%; these preparations are particularly caustic and used with extreme care.[11,17]

Colloidal silver preparations identified as "strong silver proteins" and "mild silver proteins" were developed in the early 20th century in an attempt to overcome the astringency, irritancy and corrosive effects of silver nitrate.[11] Dr Henry Crooks may have pioneered this research and is supposed to have claimed that:

"certain metals in a colloidal state exhibit profound germicidal action, but are quite harmless to human beings. There is no microbe known that is not killed by colloidal silver in laboratory tests within six minutes".[18]

The credibility of colloidal silver proteins as efficacious antibiotics in pre-1950s medicine was never fully established and their clinical acceptability was not recognised on account of their local toxicity and propensity to evoke generalised argyria. Davis considered them to be relatively ineffective on account of their slow ionisation rate.[12]

The period from about 1950 until the present time saw many dramatic advances in the antibiotic development of silver. During a period of unprecedented advances in biotechnology and materials science, the molecular and genetic basis for bacterial sensitivity and resistance to chemotherapy became evident. Microbiologists achieved an understanding of the far-reaching problems of environmental, nosocomial (hospital-related) infections and opportunistic pathogens.

The history of silver as an antibiotic shows many high spots, notably:

- The formulation of silver sulfadiazine as a prophylactic for life-threatening *Pseudomonas aeruginosa* infections in burn wounds;[19]
- Development of a range of wound dressings, (the so-called sustained silver-release dressings) tailored to control infections in indolent and painful diabetic and chronic wounds;[20–24]
- Nanotechnology with the ability to create minute particles of silver with appreciably higher ionisation rate and correspondingly higher antibiotic potency for incorporation in wound dressings, polymers for biomedical devices and prostheses;[25,26]
- Development of silver in textiles;[27,28]
- Silver as a means of controlling water born infections in water systems.[29–31]

Selection and validation of silver antibiotics including silver nitrate and silver sulfadiazine as therapeutics and prophylactics in open wounds and burns owes much to the extensive clinical and laboratory studies of Edward Lowbury at the Birmingham Accident Hospital. Lowbury was aware of the increasing concern over the rise in bacterial resistance to many of the antibiotics available in the 1970s and was aware of the possibility of bacterial resistance to silver.[32-34] Emergence of bacterial resistance is still not fully understood, but in the opinion of some, antimicrobial chemotherapy would be played out by the end of the century due to the upsurge of multi drug-resistant bacteria. This has not occurred; chemoprophylaxis and antibiotic therapies still provide an essential feature of current practice in burn and wound clinics.[35-37] There is no evidence that bacteria develop resistance to silver ion in wound care or in the use of silver-treated medical devices, and there is no evidence that continued exposure to silver in any ionisable form predisposes to emergence of silver-resistant strains.[38]

4.2 Mechanisms of Antibiotic Action

Clinical studies designed to evaluate the suitability of new devices or products containing silver, commonly focus upon the efficacy of the Ag^+ release in inhibiting nosocomial and opportunist pathogens and neutralising their toxins. Rarely is the discussion broadened to predict the possible mechanisms of antibiotic action of the silver released and why, in some situations, silver is not fully effective in clearing infections.[39] Silver effectively controls most Gram-positive and Gram-negative bacteria and fungal (*e.g. Candida albicans*) infections in wounds,[32] and inhibits some infections associated with other medical devices (notably the catheter-related intra-urethral tract infections),[40-43] but limited information is available concerning inter-species variations in bacterial or fungal sensitivity to silver, or why some strains are seemingly resistant to it.[38]

Ag^+-released from wound dressings, medical devices, textiles and domestic products is expected to show a three step antibiotic sequence involving:

1. Attachment of silver cation to exposed anionic groups (notably -SH) on the cell membrane/envelope or enzymes with a special role in maintaining cell wall structure;[44]
2. Denaturation of the cell membrane → loss of structural integrity and functional capacity → leakage of essential nutrients, electrolytes and metabolites;[45-46]
3. Internalisation of silver by the bacterium/fungus → silver bound to proteins, inhibits enzyme synthesis and respiratory activity, impaired DNA/RNA synthesis, loss of homeostatic state (Figure 4.1).[47-49]

Cell death is the ultimate event. Laboratory studies using electron microscopy and X-ray spectrophotometric analysis show that the antibiotic action of silver is complex and that target organisms differ in their sensitivity to Ag^+.

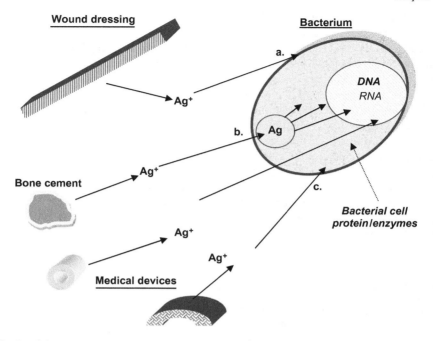

Figure 4.1   Antimicrobial action of silver (Ag$^+$) illustrating three major pathways: (a) attachment to cell membrane; (b) absorption or transmembrane diffusion into the cell (pinocytic vesicle); (c) coagulation and denaturation of bacterial enzymes and proteins.

Importantly, resistance to silver is determined genetically, and determinants for resistance can be transferred to another organism by conjugation.[50–52]

4.2.1   Toxic Influence of Silver in Micro-organisms

4.2.1.1   Cell Membrane

Cell membranes in micro-organisms provide a protective function in limiting influx of toxic materials from the medium whilst conserving the intracellular homeostatic balance of electrolytes, metal ion exchange and metabolites. The capacity of Ag$^+$ to bind cell wall components varies according to the organism, its genotype and the expression of anionic moieties on its surface. Ag$^+$ released from silver sulfadiazine, silver nitrate. *etc.* binds readily sulfydryl groups on cysteine as in the enzyme phosphomannose isomerase (PMI). Inhibition of this essential cell wall-related enzyme in the fungal pathogen, *Candida albicans*, led to irreversible cell wall changes and cell death.[53] Recombinant gene technology was employed to demonstrate that mutated fungi expressing alanine rather than cysteine were resistant to silver. *Escherichia coli* PMI which lacked cysteine was similarly resistant to silver antibiotic.

Escherichia coli exposed to nanoparticulate silver (10–$100\,\mu g\,cm^{-3}$) on agar plates developed severe pitting and increased cell wall permeability.[54] Minute particles of silver became embedded within the membrane and penetrated the cell wall leading to 70% bacterial growth inhibition at silver concentrations as low as $10\,\mu g\,cm^{-3}$. Increased permeability in the cell membrane was associated with reduced uptake and exchange of key nutrients including phosphates, mannitol, succinate, glutamine and proline.[55] It is unclear whether impaired phosphate balances are, in part, due to the action of silver as an "uncoupling agent" or inhibitor of bacterial cell respiration.

The use of silver–copper filters in the protection of hospital hot and cold water systems against *Legionella* sp. infection provides further evidence for the action of silver on bacterial cell membranes.[29–31] This aspect is discussed in more detail in Chapter 7, but it is noteworthy here that copper–silver ions act synergistically to evoke lethality.[56] Copper cations form electrostatic bonds with anionic residues on the cell wall, thereby impairing its integrity as a barrier function and enhancing the penetration of cytotoxic concentrations of silver ion. Silver–copper balances of 40 and $400\,\mu g\,L^{-1}$, respectively, have been shown to inhibit planktonic *Legionella* sp.[57]

4.2.1.2 Intracellular Homeostasis

Modak and Fox first demonstrated the cytotoxic properties of silver in their early development of silver sulfadiazine in treating *Pseudomonas aeruginosa* infections.[58] The intracellular binding and influence of silver on bacterial respiratory enzyme synthesis and DNA/RNA metabolism was equivocal at the time, but recent research demonstrates that thiol and sulfydryl groups in membranes and in enzymes readily bind Ag^+.[44,59] Silver readily accumulates in bacterial cells cultivated in the presence of ionisable silver compounds such as silver nitrate and silver sulfadiazine,[60,61] but uptake is strongly dependent on local concentrations of sodium chloride and the metabolic state of the bacterial cell. Silver accumulating within a bacterial cell has been shown to disturb electrolyte balances, notably of sodium and chloride, and influence electron transport.[62] Silver particles deposited in *Escherichia coli, Pseudomonas aeruginosa* and *Cryptococcus albicans* following exposure to silver nitrate were associated with irreversible DNA damage and loss of replicating capacity, degeneration of intracellular proteins.[54]

4.2.1.3 Morphological and Functional Change

Silver has been recognised as a protoplasmic poison in sensitive micro-organisms for many years.[63] Sulfydryl and thiol groups both in bacterial cell walls and enzymes were shown to be vulnerable to denaturation, but the most sensitive enzymic sites for the action of silver lay between cytochrome b and a_3 with less sensitive areas in the region of NADP and succinate dehydrogenase;[64] nucleoproteins and nucleic acids are additional target sites. Silver impairs phosphate metabolism and interacts with base pairs (guanine–cytosine, adenine–thymine) thereby inhibiting DNA replication.[65–67] Degenerative intracellular changes reported in *Escherichia coli* (Gram-negative) or *Staphylococcus aureus* (Gram-positive) exposed to silver nitrate and identified as "electron-light regions" corresponded with condensed DNA molecules and precipitation of electron-dense silver granules.[68] Impaired DNA replication, and anatomical and functional changes in the bacterial membrane bacterial strains were associated with silver precipitation.

Electron microscopy and X-ray analyses indicate that intracellular accumulation of Ag^+ up-regulate specific silver-binding periplasmic proteins (*Sil*-complex) which, according to Simon Silver, represent a first line of defence against silver toxicity.[69] These silver-binding proteins regulated by gene-expression including *SilE* protein exhibit a strong capacity to bind Ag^+ in preference to other metal ions.

4.3   Bacterial Resistance to Silver

The widespread and often indiscriminate use of antibiotics in clinical medicine and the failure to implement basic infection control practices are regarded as major factors in the emergence of drug-resistant bacteria.[70] At one time, microbiologists were of the opinion that bacteria are "unable to develop a resistance to silver ions", but since the late 1970s silver-resistant bacteria have been shown to emerge in a variety of circumstances and environments *"where silver toxicity may be expected to select for resistance"*.[32,71–73] Modak *et al.* were unable to induce silver resistance in *Pseudomonas aeruginosa* exposed to silver sulfadiazine by repeated subculture and considered it to be a rare event.[74] Clinical evidence for bacterial resistance to silver was first appreciated in burns therapy when topical silver nitrate or silver sulfadiazine failed to suppress *Pseudomonas aeruginosa* and other major infections.[34,75] Increased dilutions of silver nitrate in nutrient agar showed reduced inhibition of bacterial growth.[76] Supposed silver-resistant strains of *Pseudomonas aeruginosa* grew on media containing 1.0 mM silver nitrate. However in that work and subsequent studies in Canada, resistance to silver antibiotics in strains of *Proteus, Klebsiella* and *Enterobacteriaceae* declined and became unstable.[77] In neither study were facilities available to examine the molecular or genetic profile of the suspected silver-resistant strains, and it is unlikely that the bacteria contained the characteristic transmissible plasmids marking true silver-resistance[78,79] or the *Sil*-gene complex.[69,72,73,80]

Resistance to silver and other metal ions is complex and almost certainly involves mutational changes and acquisition of specific plasmids (cytoplasmic bodies) and the so-called R-factors (resistance transfer factors), plasmids or transposons (or colloquially the "jumping genes").[63,81,82] Resistance to antibiotics may occur by "intrinsic or "acquired" mechanisms, but the prevalence of silver-resistant pathogenic bacteria and fungi in man and animals is not known. Present knowledge is based upon a relatively small number of resistant bacteria isolated in recent years (Table 4.2).[50,51,66,71]

Table 4.2 Silver-resistant bacteria isolated in human infections.

- *Escherichia coli*
- *Enterobacter cloacae*
- *Klebsiella pneumoniae*
- *Aceniter baumannii*
- *Salmonella typhimurium*
- *Pseudomonas stutzeri*

Silver resistant bacteria or fungi exhibit the following principal characteristics:

- Not inhibited by the presence of Ag^+ in culture medium;
- Mutational change;
- The mutation is stable within a population;
- Resistance is transferable to a susceptible strain by conjugation.[69,72]

The first unequivocal evidence for silver resistance was provided by the burns clinic in Massachusetts General Hospital (Boston) in 1975.[51,52] A strain of *Salmonella typhimurium* isolated from several patients who died with profound septicaemia was resistant to 0.5% silver nitrate and other antibiotics including mercuric chloride, ampicillin, chloramphenicol, tetracycline and sulfonamides. This multiple drug resistance was transferred to sensitive recipient strains of *Escherichia coli* and *Salmonella typhimurium* by *in vitro* mating and conjugation. The silver-resistant strain of *Salmonella typhimurium* (identified as pMGH100) has provided fundamental information on the genetic and molecular basis of resistance and the role of the *Sil*-gene complex.[69,72,73,83] Other studies relating to silver-resistant strains of *Escherichia coli* and *Pseudomonas stutzeri* isolated from burn wound patients in Canada have used molecular techniques to study patterns of resistance conferred by intracellular bodies or plasmids.[78,79,84] These plasmid-mediated patterns of silver resistance were transferable by conjugation.

Dashpande and Choparde identified a silver-resistant strain of *Acinetobacter baumannii* (BL88 strain) as an environmental contaminant.[85] The bacterium was resistant to 13 metallic and 10 non-metal based antibiotics, but the resistance to silver was transmissible to a susceptible strain of *Escherichia coli*. *Acinetobacter baumannii* (BL88) was found to accumulate and retain silver, whereas resistance transmitted to a sensitive strain of *Escherichia coli* was associated with 63% silver ion efflux. The authors employed molecular techniques to qualify the stability of the resistance to silver, cadmium, and other antibiotics.

Morphological changes in silver-resistant *Klebsiella pneumoniae* suggest that more than one molecular mechanism for silver resistance may occur.[86] Kaur and Vadehra claimed that, whereas silver uptake in a silver-resistant strain was three-to-four fold lower than in a sensitive strain, "spheroplasts"[i] of the two strains exhibited similar silver uptake. Silver nitrate led to a complete inhibition of succinate dehydrogenase in the sensitive strain but only 18% inhibition in the resistant strain.

4.3.1 Molecular and Genetic Basis of Bacterial Resistance

Silver resistance in the strain of *Salmonella typhimurium* isolated from burn wound patients in Massachusetts General Hospital (identified as pMGH100)[87] has been attributed to a single plasmid involving seven genes plus two less-recognised "open reading frames" in three discrete transcription units.[69] The plasmid system was stable and transmissible to a sensitive strain of *Escherichia coli* in which it was cloned and sequenced. Simon Silver's group in Chicago deserve credit for identifying the gene products as being homologous to those associated with resistance in bacteria to other metals including arsenic, germanium, cadmium and mercury.[69,83] Current views are that silver resistance in the pMGH100 strain of *Salmonella typhimurium* at least is attributable to a *Sil*-gene complex, with the main gene products as in Table 4.3. In this model, a small periplasmic protein *SilE* provides an initial means of cellular protection by specifically binding silver ion at the bacterial cell surface. Silver that bypasses this initial protective mechanism is selectively pumped out of resistant bacteria through the action of *SilP*. The *Sil*-gene complex is further determined by a two-component system of membrane sensor and transcriptional responder systems—*SilS* and *SilR,* which are co-transcribed. These proteins have been termed "component regulatory membrane kinases" comprising a sensor protein (*SilS*) and a transcriptional responder protein (*SilR*) which are thought to regulate mRNA transcription in the silver resistance pathway. The three

Table 4.3 Gene products of the *Sil*-complex in the silver-resistant strain of *Salmonella typhimurium* (pMGH100).[73]

Gene product	Biological characteristics
Sil-E	A small silver-binding protein located on the bacterial cell surface
Sil-S	A regulatory kinase membrane receptor protein
Sil-R	A transcriptional responder protein that regulates mRNA transcription for silver resistance
Sil-CBA	A complex of three proteins acting as membrane potential-dependant silver and hydrogen ion "anti-porter agents"
Sil-P	A P-type efflux ATP-ase.

[i] Spheroplasts or protoplasts may be prepared from bacteria either by removing the cell wall or by metabolically blocking synthesis of cell wall. Lysozyme, which dissolves the cell wall, and penicillin, which inhibits cell wall synthesis, are most frequently used in preparation of spheroplasts. Spheroplasts in hypertonic or isotonic fluid are often able to grow and multiply on agar media.

protein complex, *SilCBA*, appears to act with *SilP* in the form of parallel efflux pumps serving to protect the intracellular enzymic systems and cell organelles from the denaturing influence of accumulating silver. In this model at least, silver resistance appears to include two discrete and opposing events—silver ion binding and silver efflux. Interestingly, although Ag^+ binds sulfydryl groups on cysteine residues on cell membranes and enzymes, the periplasmic protein *SilE* does not seem to express cysteine. Amit Gupta suggested that the profile of silver resistance seen in the pMGH100 strain of *Salmonella typhimurium* might be employed more widely to provide a basis for determining the incidence and epidemiology of silver-resistant bacteria in burns clinics, wound care facilities and numerous environmental situations where silver antibiotics are still widely employed.[80] In environmental situations, halide ions which avidly bind free Ag^+ are known to influence silver resistance.[88] Thus research with the mutated *Escherichia coli* (J53) strain has demonstrated that resistance to silver nitrate and silver sulfadiazine is diminished in the presence of high chloride, but increased when chloride levels are low.

Investigations into the role of plasmids in silver resistance have centred on a strain of *Escherichia coli* originally isolated from a burn wound patient treated with silver sulfadiazine.[78,79] The silver-resistance plasmids were transmissible to a sensitive strain of the same bacterium, and have been cloned and the genome sequenced. This silver-resistant R1 strain contained two large plasmids identified as 83kb (pJT1) and 77kb (pJT2) and was shown to be resistant to 1 mM silver nitrate. When compared with a silver-sensitive strain (S1) "cured of the 83kb (pJT1) plasmid", there was no significant difference in the silver absorbed by the two strains in culture, but electron microscopy and X-ray analysis showed that whereas the resistant-strain did not retain silver, the silver-sensitive strain exhibited electron-dense silver granules in the cytoplasm.[89] Intracellular immobilisation of silver by phosphate or other precipitants may also reduce the toxic influence of silver ion. Other intracellular characteristics of silver-resistant *Escherichia coli* include a 33% higher emission of hydrogen sulfide and 32% increase in acid-labile sulfydryl than in silver-sensitive bacteria when grown in the absence of silver nitrate. Cell surface hydrophobicity was higher in R1 bacteria and cell-free extracts revealed differences in the protein fractions between R1 and S1 strains.[79] Similar observations have been observed in silver-resistant and silver-sensitive strains of *Pseudomonas stutzeri* but, in that study, greater emphasis was placed on intracellular immobilisation of silver in stable complexes as a mechanism of resistance.[90]

Silver-resistant mutants of *Escherichia coli* lacking specific plasmids have been shown to possess decreased membrane permeability to due to deficiencies in "porins".[91] Five resistant strains exhibiting a lack of the porins, OmpF or OmpH plus OmpC, accumulated four-fold less ^{110m}Ag when cultured in silver nitrate. Treatment of silver-resistant mutant strains with carbonyl cyanide *m*-chlorophenyl hydrazone or glucose increased the permeability of the cell membrane, thereby altering the uptake and efflux of silver ion.

4.4 Biofilm Formation

Biofilm formation represents a second form of resistance to antibiotics. A wide range of nosocomial and opportunistic pathogens have the capacity to biofilm formation and development of a unique form of self-preservation from dehydration, biocides and antibiotics.[92,93] Bacteria and fungi in biofilms become resistant to a host's natural immune responsiveness and form reservoirs from which to infect vulnerable tissues in the body. Biofilms were first observed by Anthonie von Leeuwenhoek (1632–1723) who deserves much credit for developing the early microscope. In recent years, they have become a major focus of clinical interest in view of their ability to adhere to and impair the clinical function of medical devices designed for long-term implantation. Plaque formation on teeth and dental devices are further illustrations of biofilm formation.[94] Considerable efforts are made in the design and antibiotic treatment of catheters, prosthetic devices, bone cements, cardiovascular implants and stents to engineer out risks of biofilm formation and associated impairment of patient health and clinical dysfunction of devices.

Biofilms are defined as assemblages of micro-organisms which transform from a free-living or planktonic state to a sedentary lifestyle.[92,95,96] They undergo mutagenic changes reflecting not only their way of life but resistance to a wide range of antibiotics used in human medicine, including silver. They become irreversibly attached to hard surfaces where they secrete and become enclosed within polysaccharide matrices which may contain minerals, blood, cell debris and environmental contaminants. Biofilms can form on a variety of human tissues including tendon, cartilage, tooth and bone, but of greatest clinical concern is their capacity to infect and adhere to in-dwelling medical devices, prostheses, dental materials, and catheters for intraurethral and central venous cannulation, and where rougher surfaces provide greater opportunity for initial attachment of organisms leading to colonisation.

Recent research has shown that bacteria in biofilms are particularly prone to conjugation and exchange of extra-chromosomal material (especially plasmids), with transmission of resistance.[96] Library screening has revealed that, in *Streptococcus sanguis* at least, 32 recombinant clones encoding for 21 different proteins have been identified, each with potential role in adhesion, colonisation, polysaccharide synthesis and metabolic activities associated with biofilm formation. Considerable attention has been paid in recent years to the circumstances favouring biofilm formation; mechanisms by which bacteria and fungi transform to a sedentary habit and develop a tendency to "creep" along surfaces to obstruct or otherwise impair the clinical function of devices.[97,98]

Biofilm formation is a particular problem in urinary medicine and, as will be seen in Chapter 7, considerable research efforts have been made to devise means of eliminating risks or conditions favouring bacterial infection and catheter-related colonisation.[99,100] Silver, silver oxide, silver alloys, silver sulfadiazine and nanocrystalline silver particles have been variously impregnated into hydrophilic coatings, polymers and other materials used to construct medical catheters.[101,102] Most have been evaluated in laboratory studies with limited evidence obtained to show that silver eluted from the devices inhibits bacterial adhesion; more commonly however, in clinical practice silver has failed to significantly influence either infection or colonisation rates.[103,104]

Mechanisms for resistance to silver or other antibiotics by organisms in biofilm formation are imperfectly understood. Recent research suggests that bacteria, yeast and fungal infections in biofilm are resistant to silver for one or more of the following reasons:

- Mutagenesis leading to silver resistance;
- Inability of silver ion to penetrate the polysaccharide calyx (Ag^+ precipitates or is otherwise bound by materials of the calyx);
- Insufficient silver penetrates bacterial cell membrane;
- Bacteria are not responsive to silver.[92]

Examples of the inadequacy of silver in preventing biofilm formation are discussed in relation to silver-coated heart valve sewing cuffs, intra-urethral catheters and central venous catheters. Results tend to be inconsistent and contradictory. For example, a pilot study to assess the antimicrobial efficacy of a silver coated heart valve cuff showed that *Staphylococcus epidermidis* adhered more to the surfaces of the silver-coated cuff than to untreated surfaces.[105] In contrast, when a silver-coated polyethylene terephthalate fabric was exposed to this organism in culture, significant bacterial inhibition was recorded.[106] Similarly, experimental studies with the St Jude Medical heart valve with a dense layer of metallic silver deposited on individual fibres showed that the silver coating reduced local inflammation and is capable of inhibiting biofilm formation due to *Staphylococcus epidermidis* in a guinea pig model.[107]

General observations on the efficacy of silver impregnation or coating in inhibiting adherence of pathogenic organisms and biofilm formation are difficult to formulate. Whereas in some situations silver is claimed to inhibit bacterial adherence in preliminary *in vitro* cultures, the observations are not reflected in experimental or clinical situations.[108–110] David Stickler was of the opinion that, whereas a few materials including silver show some promise in reducing biofilm formation, the "gold standard" material has yet to be determined.[110]

4.5 Antiviral Action of Silver

The antiviral action of silver and silver-containing compounds is not well-defined and, whilst there is some evidence that the Ag^+ ion binds avidly with viral DNA, mechanisms of action are not well understood.[111] Silver nitrate, silver sulfadiazine and silver sulfathiazole have variously been shown to inhibit DNA viruses of the herpes virus group *in vitro* and *in vivo*[112–116] and silver nanoparticles are known to exhibit promising cytoprotective activities towards HIV-infected T-cells in culture.[111] Limited evidence is available to show that silver-containing agents might be useful in eliminating waterborne infections, but this awaits further development.[117] Other recent developments indicate that metal–thiolate complexes and a silver polysulfonate complex may be effective in inhibiting DNA and RNA viruses.[118]

Experimental studies indicated that 1% silver nitrate inhibits herpes simplex virus type 1 in tissue culture and is effective in protecting against viral conjunctivitis in rabbits, but is significantly less effective against type 2 virus.[115] Herpes simplex type 2 seems to show greater resistance to chemical inactivation and hence presents greater risk in causing neonatal eye infections. The inactivated viruses retained the ability to adsorb to cells, showing adsorption kinetics similar to those of intact viruses but having lost the capacity to elaborate viral antigens and evoke pathogenic changes.[119] Silver nitrate was ineffective against poliovirus, Japanese haemagglutinating virus, vaccinia or adenovirus. Silver sulfathiazole exhibited similar action to silver nitrate in inhibiting herpes virus type 1 and 2, with inhibitory changes related to the concentration of the drug and period of exposure.[116]

Silver sulfadiazine (1%) has also shown commendable antiviral action against the herpes viruses at low concentrations *in vitro* and prevented viral keratoconjunctivitis in rabbits.[113–115] In a clinical evaluation of 42 patients with *Herpes zoster*, silver sulfadiazine eliminated signs of infection and patient discomfort with 72 hours with no obvious toxic changes.[120]

References

1. A. G. Baxter, Louis Pasteur's beer of revenge, *Nat. Rev. Imunol.*, 2001, **1**, 229.
2. R. Koch, Untersuchungen über die Aetiologie der Wundinfectionskrankheiten, W. Vogel, Leipzig, 1878 [Translated by W. Watson Cheyne, *New Sydenham Society, London*, 1880, **13**, 74].
3. R. Munch, Robert Koch, *Microbes Infect.*, 2003, **5**, 69.
4. E. G. D. Murray, A synopsis of the history of medical microbiology, in *Bacterial and Mycotic Infections of Man*, ed. R. J. Dubois, Lippincott, Philadelphia, 1952, pp. 1–13.
5. D. G. Brater and W. J. Daly, Clinical pharmacology in the Middle Ages: principles that presage the 21st century, *Clin. Phar. Therap.*, 2000, **67**, 447.
6. D. Simpson, Paré as a neurosurgeon, *Aust. N. Z. J. Surg.*, 1997, **67**, 540.
7. A. B. G. Lansdown, Pin and needle tract infections: the prophylactic role of silver, *Wounds UK*, 2006, **2**, 51.
8. W. S. Halsted, Treatment for hernia, *Am. J. Med. Sci.*, 1895, **110**, 13.
9. P. Ehrlich, Chemotherapie von Infektionskrankheiten, *Z. Artzl. Fortbild.*, 1909, **6**, 721.
10. P. Ehrlich, Über Chemotherapie: Die Behandlung der Syphilis mit Salvarsan und verwandten Stoffen, *Münch. Med. Wochenschr.*, 1913, 1959.

11. T. Sollemann, Silver, in *A Manual of Pharmacology and its Applications to Therapeutics and Toxicology*, Saunders, Philadelphia, 1942, pp. 1102–1109.

12. B. D. Davis, Principles of sterilization, in *Bacterial and Mycotic Infections of Man*, ed. R. J. Dubois, Lippincott, Philadelphia, 1952, pp. 707–725.

13. K. W. von Nägeli, Leben die oligodynamischen Erscheinungen an lebenden Zellen, *Denkschr. Schweiz. Naturforsch. Ges.*, 1893, **33**, 174.

14. A. J. Clark, General pharmacology, in *Handbuch der Experimentellen Pharmacologie*, ed. A. Heffter, Springer, Berlin, 1937, Vol. **4**, pp. 165–176.

15. R. E. Burrell, A scientific perspective on the use of topical silver preparations, *Ostomy Wound Manage.*, 2003, **49** (Suppl), 19.

16. K. S. F. Credé, *Die Verhütung der Augenentzündung der Neugeborenen, der häufigsten und wichtigsten Ursache de Blindheit*, Hirschwald, Berlin, 1895.

17. Bray Healthcare, *Toughened silver nitrate caustics in medicine: warts, verrucae, cautery, granuloma*, 1997, www.bray.co.uk.

18. H. Crookes, *Use of Colloidal Silver*, London, 1910.

19. C. L. Fox, Silver sulphadiazine—a new topical therapy of *Pseudomonas aeruginosa*, *Arch. Surg*, 1968, **96**, 184.

20. A. B. G. Lansdown, K. Jensen and M. Qvist Jensen, Contreet Foam and Contreet Hydrocolloid: an insight into two new silver-containing dressings, *J. Wound Care*, 2003, **12**, 205.

21. J. B. Wright, D. L. Hansen and R. E. Burrell, The comparative efficiency of two antimicrobial barrier dressings: *in vitro* examination of two controlled release of silver dressings, *Wounds*, 1998, **10**, 179.

22. H. Q. Yin, R. Langford and R. E. Burrell, Comparative evaluation of the antimicrobial activity of Acticoat® antimicrobial barrier dressing, *J. Burn Care Rehabil.*, 1999, **20**, 195.

23. V. Edwards-Jones, Antimicrobial and barrier effects of silver against methicillin-resistant *Staphylococcus aureus*, *J. Wound Care*, 2006, **15**, 285.

24. T. Karlsmark, R. H. Agerslev, S. H. Bendz, J. R. Larsen, J. Roed-Petersen and K. E. Andersen, Clinical performance of a new silver dressing, Contreet Foam, for chronic exuding venous leg ulcers, *J. Wound Care*, 2003, **12**, 9.

25. R. Birringer, Nanocrystalline materials, *Mater. Sci. Eng.*, 1989, **A117** 33.

26. J. P. Guggenbichler, M. Boswald, S. Lugauer and T. Krall, A new technology of micro-dispersed silver in polyurethane in antimicrobial activity in central venous catheters, *Infection*, 1999, **27** (Suppl. 1), S16.

27. U.-C. Hippler, P. Elsner and J. W. Fluhr, A new silver-loaded cellulosic fibre with antifungal and antimicrobial properties, *Curr. Probl. Dermatol.*, 2006, **33**, 165.

28. U. Wollina, M. B. Abdel-Naser and S. Verma, Skin physiology and textiles- consideration of basic interactions, *Curr. Probl. Dermatol.*, 2006, **33**, 1.

29. A. Hambidge, Reviewing efficacy of alternative water treatment techniques, *Health Estate*, 2001, **55**, 23.

30. Z. Liu, J. E. Stout, L. Tedesco, M. Boldin, C. Hwang, W. F. Diven and V. L. Yu, Controlled evaluation of copper-silver ionisation in eradicating *Legionella pneumophila* from hospital water distribution system, *J. Infect. Dis.*, 1994, **169**, 919.

31. U. Rohr, M. Senger and F. Selenka, Effect of silver and copper ions on survival of *Legionella pneumophila* in water, *Zentralbl. Hyg. Umweltmed.*, **198**, 514.

32. E. J. L. Lowbury, Infection associated with burns, *Postgrad. Med. J.*, 1972, **48**, 338.

33. E. J. L. Lowbury, Problems of resistance in open wounds and burns, in *The Rational Choice of Antibacterial Agents*, eds. R. P. Mouton, W. Brumfitt and J. M. T. Hamilton-Miller, Kluwer, London, 1977, pp. 18–31.

34. E. J. L. Lowbury, J. R. Babb, K. Bridges and D. M. Jackson, Topical prophylaxis with silver sulphadiazine and silver nitrate-chlorhexidine creams; emergence of sulphonamide-resistant Gram-negative bacilli, *Br. Med. J.*, 1976, **i**, 493.

35. V. Falanga, Classification for wound bed preparation and stimulation of chronic wounds, *Wound Rep. Regen.*, 2000, **8**, 347.

36. S. Enoch and J. K. Harding, Wound bed preparation: the science behind the removal of barriers to healing, *Wounds*, 2003, **15**, 213.

37. G. S. Schultz, R. G. Sibbald, V. Falanga, E. A. Ayelolo, C. Dowsett, K. G. Harding, M. Romanelli, M. A. Stacey, L. Teot and W. Vanscheidt, Wound bed preparation: a systematic approach to wound management, *Wound Rep. Regen.*, 2003, **11**, 1.

38. A. B. G. Lansdown and A. Williams, Bacterial resistance to silver in wound care and medical devices, *J. Wound Care*, 2007, **16**, 15.

39. A. B. G. Lansdown, A. Williams, S. Chandler and S. Benfield, Absorption and antibacterial efficacy of silver dressings, *J. Wound Care*, 2005, **14**, 161.

40. U. Samuel and J. P. Guggenbichler, Prevention of catheter-related infections: the potential of a new nano-silver impregnated catheter, *Int. J. Antimicr. Agents*, 2004, **23** (I Suppl. S1), S75.

41. S. Saint, R. H. Savel and M. A. Matthay, Enhancing the safety of critically ill patients by reducing urinary and central venous catheter-related infections, *Am. J. Respir. Crit. Care Med.*, 2002, **165**, 1475.

42. D. J. Stickler, G. L. Jones and A. D. Russell, Control of encrustation and blockage of Foley catheters, *Lancet*, 2003, **361**, 1435.

43. P. Thibon, X. Le Coutour, R. Leroyer and J. Fabry, Randomised multicentre trial of the effects of a catheter coated with hydrogel and silver salts on the incidence of hospital-acquired urinary tract infections, *J. Hosp. Infect.*, 2000, **45**, 117.

44. S. Y. Liau, D. C. Read, W. J. Pugh, J. R. Furr and A. D. Russell, Interaction of silver nitrate with readily identifiable groups: relationship to the antibacterial action of silver ions, *Lett. Appl. Microbiol.*, 1997, **25**, 279.

45. R. L. Woodward, Review of the bactericidal effect of silver, *Am. Water Works Assn*, 1963, **55**, 881.
46. W. K. Jung, H. C. Koo, K. W. Kim, S. Shin, S. H. Kim, Y. H. Park Antibacterial activity and mechanism of action of the silver ion in *Staphylococcus aureus* and *Escherichia coli, Appl. Environ. Microbiol.*, 2008, **74**, 2171–2178.
47. M. S. Wysor and R. E. Zollinhofer, On the mode of action of silver sulphadiazine, *J. Microbiol.*, 1972, **38**, 296.
48. T. Klaus, R. Joerger, E. Olsson and C. G. Granqvist, Silver-based crystalline nanoparticles, microbially fabricated, *Proc. Natl. Acad., Sci., U.S.A.*, 1999, **96**, 13611.
49. J. R. Morones, J. L. Elechiguerra and A. Cambacho, The bactericidal effect of silver nanoparticles, *Nanotechnology*, 2005, **16**, 2346.
50. S. Thomas, MRSA and the use of silver dressings: overcoming bacterial resistance, *World Wide Wounds*, 2004, www.worldwidewounds.com/2004/november/Thomas/Introducing-Silver-Dressings.html, accessed 19 November 2009.
51. G. L. McHugh, R. C. Moellering, C. C. Hopkins and M. N. Swartz, *Salmonella typhimurium* resistant to silver nitrate, chloramphenicol, and ampicillin, *Lancet*, 1975, **1** (7901), 235.
52. C. Haefeli, C. Franklin and K. Hardy, Plasmid-determined silver resistance in *Pseudomonas stutzeri* isolated from a silver mine, *J. Bacteriol.*, 1984, **158**, 389.
53. T. N. Wells, P. Scully, G. Paravicini, A. E. Proudfoot and M. A. Payton, Mechanism of irreversible inactivation of phosphomannose isomerise by silver ions and Flamazine, *Biochemistry*, 1995, **34**, 7896.
54. I. Sondi and B. Salopek-Sondi, Silver nanoparticles as antimicrobial agent: a case study on *E. coli* as a model for Gram-negative bacteria, *J. Colloid Interface Sci.*, 2004, **275**, 177.
55. W. J. Scheurs and H. Rosenberg, Effect of silver ions on transport and retention of phosphate by *Escherichia coli, J. Bacteriol.*, 1982, **152**, 7.
56. N. L. Pavey, *Ionisation Water Treatment for Hot and Cold Water Services*, Building Services Research and Information Association, Bracknell, 1996, BSRIA Technical Note TN 6/96.
57. J. E. Stout, Y. S. Lin, A. M. Goetz and R. R. Muder, Controlling *Legionella* in hospital water systems: experience with the super-heat method and copper-silver ionization, *Infect. Control, Hosp. Epidemiol.*, 1999, **19**, 911.
58. S. M. Modak and C. L. Fox, Binding of silver sulphadiazine to the cellular components of *Pseudomonas aeruginosa, Biochem. Pharmacol.*, 1973, **22**, 2391.
59. N. Silvestry-Rodrigues, E. E. Sicairos-Ruelas, C. P. Gerba and K. R. Bright, Silver as a disinfectant, *Rev. Environ. Contam. Toxicol.*, 2007, **191**, 23.
60. R. C. Charley and A. T. Bull, Bioaccumulation of silver by a multispecies community of bacteria, *Arch. Microbiol.*, 1979, **123**, 239.

61. G. M. Gadd, O. S. Laurence, P. A. Briscoe and J. T. Trevors, Silver accumulation in *Pseudomonas stutzeri, Biol. Met.*, 1989, **2**, 168.
62. A. L. Semeykina and V. P. Shulachev, Submicromolar Ag^+ increases passive Na^+ permeability and inhibits respiration-supported formation of Na^+ gradient in Bacillus FTU vesicles, *FEBS Lett.*, 1990, **269**, 69.
63. A. D. Russell and W. B. Hugo, Antimicrobial activity and action of silver, in *Progress in Medicinal Chemistry*, ed. G. P. Ellis and D. K. Luscombe, Elsevier, 1994, Vol. **31**, pp. 351–370.
64. J. Yudkin, Biosensors and biotronics, chemically imaging cells, *Enzymologica*, 1937–1938, **2**, 161.
65. S. K. Zavriev, L. E. Minchenkova, M. Vorliėkova, A. M. Kolchinsky, M. V. Volkenstein and V. I. Ivanov, *Biochim. Biophys. Acta*, 1979, **564**, 212.
66. K. I. Batarseh, Anomaly and correlation of killing in the therapeutic properties of silver (1) chelation with glutamic and tartaric acids, *J. Antimicrob. Chemother.*, 2004, **54**, 546.
67. R. B. Thurman and C. P. Gerba, The molecular mechanisms of copper and silver ion disinfection, *CRC Crit. Rev. Environ. Control*, 1989, **4**, 295.
68. Q. L. Feng, J. Wu, G. Q. Chen, F. Z. Cui, T. N. Kim and J. O. Kim, A mechanistic study of the antibacterial effect of silver ions on *Escherichia coli, J. Biomed. Mater. Res.*, 2000, **52**, 662.
69. S. Silver, Bacterial silver resistance: molecular biology and uses and misuses of silver compounds, *FEMS Microbiol. Rev*, 2003, **27**, 341.
70. P. J. Easterbrook, Superbugs: are we at the threshold of a new Dark Age? *Hosp. Med.*, 1998, **59**, 524.
71. S. L. Percival, P. G. Bowler and D. Russell, Bacterial resistance to silver in wound care, *J. Hosp. Infect.*, 2004, **60**, 1.
72. A. Gupta and S. Silver, Silver as a biocide, *Nature, Biotech.*, 1998, **16**, 888.
73. A. Gupta, K. Matsui, J. F. Lo and S. Silver, Molecular basis for resistance to silver cations in *Salmonella, Nature, Med.*, 1999, **5**, 183.
74. S. M. Modak, L. Sampath and C. L. Fox, Combined topical use of silver sulphadiazine and antibiotics as a possible solution to bacterial resistance in burn wounds, *J. Burn Care Rehabil.*, 1988, **9**, 359.
75. K. Bridges, A. Kidson, E. J. L. Lowbury and M. D. Wilkins, Gentamycin and silver-resistant pseudomonas in a burns unit, *Br. Med. J.*, 1972, **1**, 446.
76. J. S. Cason, D. M. Jackson, E. J. L. Lowbury and C. R. Ricketts, Antiseptic and aseptic prophylaxis for burns; use of silver nitrate and of isolators, *Br. Med. J.*, 1966, **2**, 1288.
77. A. T. Hendry and I. O. Stewart, Silver-resistant *Enterobacteriaceae* from hospital patients, *Can. J. Microbiol.*, 1979, **25**, 915.
78. M. E. Starodub and J. T. Trevors, Silver accumulation and resistance in *Escherichia coli* R1, *J. Med. Microbiol.*, 1989, **29**, 101.
79. M. E. Starodub and J. T. Trevors, Mobilisation of *Escherichia coli* R1 silver-resistance plasmid pJT1 by Tn5-Mob in *Escherichia coli* C600, *Biol. Met.*, 1990, **3**, 24.

80. S. Silver, J.-F. Lo and A. Gupta, Silver cations as antimicrobial agent: clinical uses and bacterial resistance, *APUA Newsletter*, 1999, **17**, 1.
81. G. E. W. Wostenholm and C. M. O'Connor, Drug resistance in microorganisms, *Ciba Symposium*, Little Brown, Boston, 1962.
82. N. Datta, Infectious drug resistance, *Br. Med. Bull.*, 1965, **21**, 255.
83. S. Silver, R. Novick and A. Gupta, Mechanisms of resistance to heavy metals and quaternary amines, in *Gram-positive Pathogens*, ed. V. A. Fischetti, American Society for Microbiology, Washington, 2000, pp. 647–659.
84. R. M. Slawson, H. Lee and J. T. Trevors, Bacterial interactions with silver, *Biol. Met.*, 1990, **3**, 151.
85. L. M. Dashpande and B. A. Chopade, Plasmid-mediated silver resistance in *Acinetobacter baumannii*, *Biometals*, 1994, **7**, 49.
86. P. Kaur and D. V. Vadehra, Mechanism of silver resistance to silver ions in *Klebsiella pneumoniae*, *Antimicrob. Agents Chemother.*, 1986, **29**, 165.
87. S. Silver and L. T. Phung, Bacterial heavy metal resistance: new surprises, *Ann. Rev. Microbiol.*, 1996, **50**, 753.
88. A. Gupta, M. Maynes and S. Silver, Effects of halides on plasmid-mediated silver resistance in *Escherichia coli*, *Appl. Environ. Microbiol.*, 1998, **64**, 5042.
89. R. M. Slawson, E. M. Lohbeier-Vogel, H. Lee and J. T. Trevors, Silver resistance in *Pseudomonas stutzeri*, *Biometals*, 1994, **7**, 30.
90. R. M. Slawson, M. I. Van Dyke, H. Lee and J. T. Trevors, Germanium and silver resistance, accumulation and toxicity in microorganisms, *Plasmid*, 1992, **27**, 72.
91. X. Z. Li, H. Nikaido and K. E. Williams, Silver-resistant mutants of *Escherichia coli* display active efflux of Ag$^+$ and are deficient in porins, *J. Bacteriol.*, 1997, **179**, 6127.
92. R. M. Donlon, Biofilms: microbial life on surfaces, *Emerg. Infect. Dis.*, 2002, **8**, 1.
93. R. M. Donlon and J. W. Costerton, Biofilms: survival mechanisms of clinically relevant microorganisms, *Clin. Microbiol. Rev.*, 2002, **15**, 167.
94. C. Black, I. Allen, S. K. Ford, M. Wilson and R. McNab, Biofilm specific surface properties and protein expression in *Streptococcus sanguis*, *Arch. Oral Biol.*, 2004, **49**, 295.
95. W. G. Characklis and P. Wilderer, *Structure and Function of Biofilms*, John Wiley, New York, 1989.
96. J. M. Ghigo, Natural conjunctive plasmids induce bacterial biofilm development, *Nature*, 2001, **412**, 4442.
97. D. Stickler and G. Hughes, Ability of *Proteus mirabilis* to swarm over urethral catheters, *Eur. J. Clin. Microb. Infect. Dis.*, 1999, **18**, 206.
98. N. Sabbuba, G. Hughes and D. J. Stickler, The migration of *Proteus mirabilis* and other urinary tract pathogens over Foley catheters, *Br. J. Urol. Int.*, 2002, **89**, 55.

99. S. Saint, D. L. Veenstra, S. D. Sullivan, C. Chenoweth and A. M. Fendrick, The potential clinical and economic benefits of silver alloy urinary catheters in preventing urinary tract infection, *Arch. Intern. Med.*, 2000, **160**, 2670.
100. D. G. Maki and P. A. Tabbyah, Engineering out the risk of infection with urinary catheters, *Emerg. Infect. Dis.*, 2001, **7**, 1.
101. K. Davenport and F. X. Keeley, Evidence for the use of silver-alloy-coated urethral catheters, *J. Hosp. Infect.*, 2005, **60**, 298.
102. D. G. Maki, S. M. Stolz, S. Wheeler and L. A. Mermel, Prevention of central venous catheter-related bloodstream infections by use of an anti-septic-impregnated catheter, *Ann. Intern. Med.*, 1997, **127**, 257.
103. J. J. Bong, P. Kite, M. H. Wilco and M. J. McMahon, Prevention of catheter related bloodstream infection by silver iontophoretic central venous catheters: a randomised controlled trial, *J. Clin. Pathol.*, 2003, **56**, 731.
104. A. Bach, H. Eberhardt, A. Frick, H. Schmidt, B. W. Bottiger and E. Martin, Efficacy of silver-coating central venous catheters in reducing bacterial colonization, *Crit. Care Med.*, 1999, **27**, 456.
105. G. Cook, J. W. Costerton and R. O. Darouische, Direct confocal microscopy studies on the bacterial colonisation *in vitro* of a silver-coated heart valve sewing cuff, *Int. J. Antimicrob. Agents*, 2000, **13**, 169.
106. U. Klueh, V. Wagner, S. Kelly, A. Johnson and J. D. Bryers, Efficacy of silver-coated fabric to prevent bacterial colonisation and subsequent device-based biofilm formation, *J. Biomed. Mater. Res.*, 2000, **53**, 621.
107. B. L. Illingworth, K. Tweden, R. F. Schroeder and J. D. Cameron, *In vivo* efficacy of silver-coated (Silzone) infection-resistant PVE fabric against a biofilm-producing bacteria, *Staphylococcus aureus*, *J. Heart Valve Dis.*, 1998, **7**, 524.
108. J. W. Leung, G. T. Lau, J. J. Sung and J. W. Costerton, Decreased bacterial adherence to silver-coated stent material: an *in vitro* study, *Gastrointest. Endosc.*, 1992, **38**, 338.
109. A. M. Mulligan, M. Wilson and J. C. Knowles, Effect of increasing silver content in phosphate-based glasses on biofilms of *Streptococcus sanguis*, *J. Biomed. Mater. Res.*, 2003, **67A**, 401.
110. D. J. Stickler, Biomaterials to prevent nosocomial infections: is silver the gold standard? *Curr. Opin. Infect. Dis.*, 2000, **13**, 389.
111. L. Lu, R. W. Sun, R. Chen, C. K. Hul, C. M. Ho, J. M. Luk, G. K. Gau and C. M. Che, Silver nanoparticles inhibit hepatitis virus replication, *Antivir. Therap.*, 2008, **13**, 253.
112. T. W. Chang, Anti-herpes viral activity of silver sulphadiazine, *J. Cutan. Pathol.*, 1975, **2**, 320.
113. T. W. Chang and L. Weinstein, Prevention of herpes keratoconjunctivitis in rabbits by silver sulphadiazine, *Antimicrob. Agents Chemother.*, 1975, **8**, 677.
114. T. W. Chang and L. Weinstein, *In vitro* activity of silver sulphadiazine against Herpes virus hominis, *J. Infect. Dis.*, 1975, **132**, 79.

115. V. R. Coleman, J. Wilkie, W. E. Levinson, T. Stevens and E. Jawetz, Inactivation of Herpes virus hominis types 1 and 2 by silver nitrate *in vitro* and *in vivo*, *Antimicrob. Agents Chemother.*, 1973, **4**, 259.

116. W. Stozkowska and M. Wroczyñska-Palka, Studies on the antiviral activity of silver sulphathiazole, *Med. Dosw. Mikrobiol.*, 1999, **51**, 167.

117. H. Mahnel and M. Schmidt, Effect of silver compounds on viruses in water, *Zentralbl. Bakteriol. Mikrobiol. Hyg. B*, 1986, **182**, 381.

118. D. E. Bergstrom, X. Lin, T. D. Wood, M. Witvrouw, S. Ikeda, G. Andrei, R. Snoek, D. Schols and E. De Clercq, Polysulphonates derived from metal thiolate complexes as inhibitors of HIV-1 and various other enveloped viruses *in vitro*, *Antivir. Chem. Chemother.*, 2002, **13**, 185.

119. F. Shimizu, Y. Shimizu and K. Kumagai, Specific inactivation of herpes simplex virus by silver nitrate at low concentrations and biological activities of the inactivated virus, *Antimicrob. Agents Chemother.*, 1976, **10**, 57.

120. L. F. Montes, G. Muchinik and C. L. Fox, Response to varicella zoster virus and herpes zoster to silver sulphadiazine, *Cutis*, 1986, **38**, 363.

CHAPTER 5

Silver in Medical Devices: Technology and Antimicrobial Efficacy

5.1 Introduction

Medical devices have become increasingly important in medical research and human healthcare in recent years and their impact on nursing, patient well-being and improved quality of life is now recognised. Major advances in the technology of biomaterials and their compatibility with human tissues have led to the introduction of more than 8000 new medical devices in the past ten years with accent on such areas as wound care, catheterisation, dentistry, ortho-paedics and cardiovascular medicine. They range from simple wound dressings and bandages to life-maintaining surgically implanted devices, diagnostic equipment and materials for minimal invasive surgery.

It is appreciated that infection presents a major recurrent problem in wound care and in the implantation of surgical devices and that opportunist pathogens from a patient's own body flora or from the nosocomial environment can be life-threatening. Problems associated with infections vary greatly according to:

- the type and virulence of the infection;
- duration of infection;
- increasing resistance to common antibiotics;
- the age, state of health and immuno-competence of the individual;
- the site of infection and the inherent protective mechanisms;
- the nature and intention of the medical device and its site of implantation (*e.g.* catheter related infections);
- the device and its susceptibility to device-related infections (notably bio-films) and their capacity to impair clinical function;
- opportunities for pelagic organisms (bacteria, fungi) to form biofilms on hard tissues or on exposed surfaces of medical devices.

Issues in Toxicology No. 6
Silver in Healthcare: Its Antimicrobial Efficacy and Safety in Use
By Alan B. G. Lansdown
© Alan B. G. Lansdown 2010
Published by the Royal Society of Chemistry, www.rsc.org

Silver has acquired a worldwide acceptance as a broad spectrum antibiotic. Laboratory and clinical studies have shown it to be safe in long-term use and efficacious in controlling most pathogenic bacteria, fungi and parasitic infections including the methicillin-resistant strains of *Staphylococcus aureus* (MRSA) which have proved fatal in many aged and immuno-compromised people. Technology is now available for uniformly coating the surfaces of or impregnating polyurethanes, silastic and other polymers for use in medical devices. The silver content is stable and ionisation in the presence of body fluids and secretions is controlled.

Classification of medical devices and definitions has been made according to such factors as:

- how long the device is intended to be in continuous contact with the body;
- whether the intervention is invasive or surgically invasive;
- whether the device is "implantable" or "active";
- whether the device contains a substance which in its own right is considered to be a medicinal substance or exhibits an ancillary action.

Occasionally, however, medical devices containing silver lie on the borderline of classification as medicines. As a generalisation, medicines exhibit "pharmacological, immunological or physiological action in the human body and are actively metabolised".[i] Medical devices containing ionisable silver and releasing Ag^+ as a means of achieving antibiotic action should be justifiably identified as "bioactive" materials.

Contrary to many claims made in the literature, silver is a highly bioactive metal and, irrespective of its route of exposure, some is absorbed into the body and as discussed earlier is metabolised in the form of protein complexes to many tissues and eliminated *via* the liver and kidney.[1,2] Silver absorbed intracellularly by pinocytic or other active process promotes and binds metallothioneins.[3-5] Although some silver is retained for long periods in the skin and eye in the form of argyria/argyrosis and complexes with hydroxyapatite in bone, silver is not commonly regarded as a cumulative poison in the same way as lead or mercury.[6] Silver does interact immunologically, and patients with known contact sensitisation/allergy to the metal can be expected to exhibit immuno-reactive responses on contact through wound dressings, silver-treated catheters or prosthetic implants.[7,8] On rare occasions, very high levels of silver accumulate in body fluids, but as far as is known it does not exhibit any pharmacological action in the human body.

5.2 Wound Care and Management

Silver has a history of use in wound care extending back over more than 400 years, many years before the classical studies of Robert Koch and Louis

---

[i] J. H. Powers, *Regulatory considerations in drug and device development.* US Food and Drug Administration, 1976, Amendments to the Federal Food, Drugs and Cosmetics Act, 1951.

Pasteur who identified and classified bacteria as transmissible agents of disease
in the 1880s. William Stewart Halsted (1852–1922) may have been the first to
recognise the merits of silver foil in protecting post-operative wounds from
pathogenic infections, but much of the fundamental knowledge on the action of
silver on bacteria and on preparation of the wound bed came later. As a generalisation, the development of silver as an antibiotic in wound care (Table 5.1)
and its central role in wound management reflects the state of knowledge of
microbiology at the time, technological advances in biomaterials and the
chemistry of silver, and understanding of the biological events comprising the
four principal phases of wound healing. In modern day wound dressings, silver-
release patterns and dressing design are tailored to treat wounds according to

Table 5.1 Sequential phases in the development of silver as an antibiotic in
skin wound dressings.

Period	Knowledge of bacteriology	Silver therapies in wound care
Up to 1800	Minimal knowledge of causes or spread of disease.	Metallic silver foil, silver surgical instruments, silver wire sutures.
1800–1940	Development of the microscope, identification and classification of bacteria as a cause of human diseases, Jenner's studies on vaccination, value to silver in water purification.	Metallic silver, silver nitrate (also as lunar caustic, sticks *etc* as escharotic agents in treatment of warts, granulations *etc.*), Silver arsphenamine for syphilis, Colloidal silver as strong or mild silver proteins.
1940–1970	Rapid development of knowledge on bacterial and fungal infection, mechanisms of resistance to drugs, development of sulfonamides and penicillins, research into mechanisms of human immunity, Development of electron microscopy and knowledge of bacterial structure.	Development of silver sulfadiazine as an effective therapy for burns wound infections, Regulation of colloidal silver anti-infective on safety grounds and lack of efficacy, Research into the biology and mechanisms of wound repair (including the importance of moisture control).
1970–present day	Knowledge of genetical and molecular mechanisms of bacterial resistance, Research into the biology of and resistance patterns in biofilms, Profiles of infection in wounds and their implication in diabetic and chronic wounds, Infection as a central feature in "wound bed preparation", Role of growth factors and cytokines in wound healing.	Development of sustained silver-release wound dressings, Wound assessment as a criterion for selection of silver containing dressings in care and management, Inclusion of an ionisable silver salt in established dressings to provide a means of controlling infection.

the level of infection, but increasingly comply with recommendations of expert committees in achieving appropriate "wound bed preparation".[9-13]

Modern sustained silver (Ag^+) release wound dressings can be expected to comprise:

- an ionisable silver source to provide antimicrobial action over a defined period;
- absorbent material to control exudates and provide a medium for moisture control;
- materials with minimal bioactivity which provide a matrix for the dressing or means of applying it to a wound to provide comfort for a patient;
- fibres, fabrics and materials as structural components, adhesives, backings, *etc.*

5.2.1 Silver Metal and Silver Nitrate

Metallic silver and silver nitrate were probably introduced into medicine and surgery during the 15th century although accurate records from this period are few. The first documented evidence of the value of silver in wound care appears in documents relating to the French surgeon Ambrose Paré (1517–1590) who as barber-surgeon to the French courts of Francis I, Henry II, Charles IX and Henry III studied gunshot wounds which at the time were thought to be "poisoned". In 1545, Paré provided the first rational basis for wound management, bandaging and cauterisation; he is believed to have used silver sutures in ligation of arteries following amputations and silver implants in dentistry. Paré is regarded by many as the "father of surgery" and king's surgeon (Figure 5.1).[14,15]

William Halsted as chief surgeon of the Johns Hopkins Hospital in Baltimore, USA, established the practice of using silver foil as a protection against infections in closed wounds—a practice that was still popular amongst plastic surgeons in Europe until the 1980s.[16] Halsted was a general surgeon with many interests including surgery of the mammary and thyroid glands, and cancer research but, unlike many surgeons of his time, he recognised the importance of experimental surgery. Halstead did much to establish the principles of aseptic surgery, wound management and the value of silver wire in surgery for hernia and intestinal surgery. Surgeon's hands and a patient's skin were disinfected with permanganate and then bichloride of mercury; Halsted introduced rubber gloves for surgery.[17,18] Many of Halsted's ideas are fundamental in present concepts of wound bed preparation, though this is not generally recognised in the literature.[9-13]

At the beginning of the 20th century, soluble salts of such metals as mercury, copper, antimony and bismuth were variously beneficial in treating syphilis, protozoan parasites and other life-threatening infections.[19] Even as recently as the 1950s, mercury as $HgCl_2$ and silver as $AgNO_3$ formed the most effective antibacterial compounds preventing the growth of many organisms at

Figure 5.1   Ambrose Paré, 1517–1590: 'The Father of Surgery'.

concentrations of less than 1 part per million (ppm).[20] Karl von Nägeli (1893) considered that certain metals exhibited an oligodynamic action and were lethal to susceptible bacteria at extremely high effective dilutions and that comparatively few ions exerted a remarkable effect on bacterial cells.[21] More recently, experiments by Clark in 1937 suggested that the concentration of metal ions exhibiting a lethal effect in bacteria, trypanosomes and yeasts might be equivalent to the estimated number of enzyme protein molecules in each cell.[22] Silver was regarded as a relatively ineffective antiseptic in the form of silver proteinates (colloidal silver proteins) in view of their relatively low ionisation patterns. Arsenic, bismuth and antimony salts were preferable in treating syphilis.[20] In contrast, silver nitrate has a high rate of ionisation in aqueous media and is an efficacious antibiotic for control of infections in a wide range of open skin wounds and burns in 0.5% solution or compresses.[23]

Paul Ehrlich evaluated the efficacy of many metal ions on protozoal and bacterial infections prevalent at the time and introduced arsenic-containing therapies of arsphenamine (Salvarsan) and silver arsphenamine (Silver-Salvarsan) for spirochaetes and trypanosome infections.[19,24] Although silver arsphenamine undoubtedly showed effective antimicrobial action, its value in clinical use was overshadowed by its profound toxicity and ability to evoke symptoms of argyria following intravenous injection.[25,26]

Silver nitrate is possibly the longest known and most efficacious antibiotic introduced for skin wound care. Its discovery possibly dates from the time of early alchemists like Gerber in 730AD or the alchemist-monk Basilius Valentinus in the 15th century, but a British barber-surgeon, John Woodall, who became first surgeon general to the East India Company (1617), was reputedly first to discover the medicinal properties of silver nitrate (also known as lunar caustic, *lapis infernalis* or *pierre infernale*).[27] He is accredited with using silver nitrate in treating surgical wounds in seamen; as such, silver nitrate became known as "surgeon's mate". Its caustic and astringent properties were effective in controlling infections in limb amputations and major surgery, although details of Woodall's procedures are unclear. Surgeons of the 17th and 18th century are recorded as using the astringent and caustic action of silver nitrate in cautery, as an effective therapy for venereal diseases, and in the ablation of warts, granulations and verrucae, possibly in much the same way as Strong Silver Nitrate (40–95% in sticks, pencils) (Avoca®, Bray Healthcare) is used today.[19,28]

The astringency and irritancy of silver nitrate to the skin and other tissues is proportional to its concentration and duration of its application, with aqueous concentrations of $<5\%$ being preferred for wound cleansing and antisepsis. The skin surface turns white initially but later darkens gradually as the bioactive Ag^+ precipitates as silver sulfide or silver chloride on the surface of the skin or in the outer layers of the stratum corneum.[19] Earlier pharmacologists believed that precipitation of black deposits in the outer layers of the stratum corneum permitted an easy control of its antiseptic action.[29] Others perceived its action as extending "quite deeply", and that soluble double salts of silver albuminates and silver chloride formed in the tissues".[19] We now know that Ag^+ does complex with albumins and macroglobulins in exudates and tissue fluids, and that the inert precipitates provide a safety mechanism against the toxicity of xenobiotic Ag^+ cations.[3]

Greater understanding of the cellular action of silver nitrate in wounded skin is provided by experimental studies.[3] At low concentrations (0.01, 0.1 or 1.0%), silver nitrate was not an irritant in rat skin and black deposits were located superficially and in wound debris without noticeable penetration (Figures 5.2 and 5.3). The wounds treated with silver nitrate healed more rapidly than controls and discolourations of the skin and hair shafts declined without adverse effects. Improved healing was attributed to the induction of metallothioneins I and II by silver absorbed into metabolically active cells of the wound margin and activation of zinc and copper ions, both of which have mitogenic properties in epidermal and dermal cell populations.[30] Percutaneous absorption of silver was low and not pathogenic to intact epidermal tissues (Table 5.2). Wounds healed normally.

The use of silver nitrate in antimicrobial chemotherapy declined up to about 1960 as newer and safer products became available and undesirable discolourations of the skin were avoided (see Chapter 8). In part, silver nitrate was replaced through development of a range of mild and strong silver proteins (colloidal silver products) (Table 5.3), which were claimed to be less irritant.

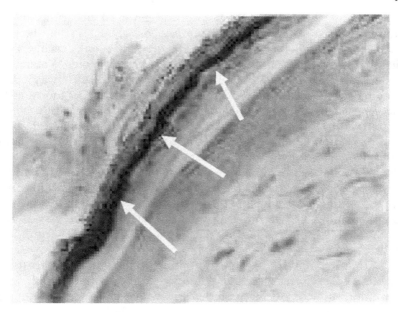

Figure 5.2   Deposition of silver sulfide precipitate in outer layers of the stratum corneum (arrowed) in rat skin treated for seven days with 1.0% silver nitrate (× 40 objective).

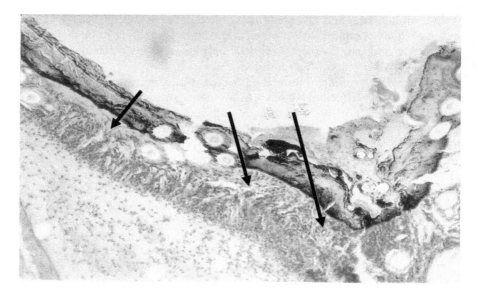

Figure 5.3   Deposition of silver sulfide precipitates (arrowed) in wound debris and inflammatory cells in outer aspects of a surgical rat wound treated with 1.0% silver nitrate for three days (× 16 objective).

Table 5.2 Silver ion uptake (mg g^{-1})* by wounded or intact rat skin treated daily with 1.0% silver nitrate daily for ten days.[3]

Silver nitrate	Intact skin	Incisional wound site
0.01% w/v	0.010 ± 0.001	0.036 ± 0.020
0.10%	0.092 ± 0.005	0.213 ± 0.050
1.0%	0.593 ± 0.072	1.395 ± 0.385
Control	0.012 ± 0.001	0.055 ± 0.003

*Wet tissue weight

Table 5.3 Colloidal silver proteins available in 1942 for wound care.

Silver Protein	Silver content	Product
Strong silver protein	7–8.5% (highly ionised)	Protargol type
		Protargol
		Protargentum
Mild silver protein	19–30% (mildly ionised)	Argyrol type
		Argyrol
		Collargol
		Silvol
		Solargentum
Colloidal silver halides	Variable	Colloidal silver chloride
		Colloidal silver iodide

However, in these colloidal silver proteins: "... the mass of silver does not exist as free ions, and therefore does not precipitate chloride or proteins and is therefore non-corrosive, relatively non-astringent and non-irritant".[19] Their antiseptic action was attributed to release of silver ions in free form or adsorbed to colloids.[31–34]

Surprisingly, the major part of the silver bound in colloidal form was inactive and did not contribute to the antimicrobial properties of the products, even after prolonged contact. These products were phased out of many international pharmacopoeias in the 1960s, but a new range of colloidal silver proteins have been introduced in recent years, although their value in wound care is not substantiated scientifically or medically. Like silver nitrate, they are accompanied by profound risk of argyria.[35,36] Their exact composition is rarely known. The US Food and Drug Administration (FDA) proposed in 1996 to establish that all over-the-counter (OTC) drug products containing colloidal silver ingredients for internal or external use are "not generally as safe and effective, and are misbranded".[37] At that time at least, the FDA was not aware of any substantial scientific evidence that OTC colloidal silver products are effective in treating the disease states claimed.

Carl Moyer and his colleagues revitalised interest in 0.5–1.0% silver nitrate as a broad spectrum antibiotic in the treatment of large burns, which are particularly prone to life-threatening *Staphylococcus aureus*, β-haemolytic streptococci, *Escherichia coli* and *Pseudomonas aeruginosa* infections.[38] Whilst effective in controlling these infections, silver nitrate is intensely irritating and its prolonged use is precluded by its tendency to promote metabolic acidosis (through its action in inhibiting carbonic anhydrase) and trace metal ion (and electrolyte) imbalances through prolonged irrigation. On the other hand, Lowbury considered that the most effective control measures for infection in open wounds and burns could be obtained by application of creams, solutions and *tulle gras* dressings containing chlorhexidine or silver nitrate, against which these pathogenic bacteria did not acquire resistance.[23] In his experience, the incidence of *Pseudomonas aeruginosa* infection declined from about 70% to 3% in patients treated with 0.5% silver nitrate compresses. Topical chemoprophylaxis with 0.5% silver nitrate in clinical trials was preferred to 1% silver sulfadiazine and 0.2% chlorhexidine digluconate in protecting against Gramnegative bacilli, and *Pseudomonas aeruginosa* and *Proteus* sp. were less often found in patients treated with silver nitrate compresses.[39]

5.2.2 Silver Sulfadiazine (Silvadene, Flamazine, Flammazine)

Silver sulfadiazine was developed by Charles Fox in 1968 as a prophylactic therapy for *Pseudomonas aeruginosa* infections in burn wounds.[40] His original intention was to combine the efficacy of silver nitrate with the antibiotic properties of soluble sodium sulfonamides, which had proved especially effective in the treatment of wounds and burns during World War II. In addition, the new drug would be safer and more widely acceptable than other antibiotics available at the time including mafenide acetate and silver nitrate–mafenide ointment, chlorhexidine digluconate and Sulfamylon.

Sulfadiazine is a weak acid that reacts with silver nitrate to form a white fluffy silver complex salt that is sparingly soluble in water but which ionises in body fluids to achieve the oligodynamic properties of silver nitrate, but combining the antibiotic properties of silver ion and sulfonamide moieties. Application of silver sulfadiazine to human skin leads to the build up of a reservoir of silver ions to provide a sustained antimicrobial effect with efficacy claims of action against at least 95% of known pathogenic strains, notably *Pseudomonas aeruginosa, Escherichia coli, Enterobacter cloacae, Proteus morganii, Staphylococcus aureus* and *Staphylococcus epidermidis*.[41–43]

Clinical trials in burn wound clinics have shown that 1% silver sulfadiazine cream applied topically with or without chlorhexidine digluconate for at least seven days is effective in reducing infections and reducing patient mortality without the adverse effects associated with silver nitrate.[44–47] Whilst some clinicians are of the opinion that silver sulfadiazine impairs re-epithelialisation of wounds, these effects are possibly marginal or of minimal clinical consequence in relation to its antibiotic efficacy.[48–50] Clinical experience has shown that bacterial resistance to silver can be an occasional problem in using silver sulfadiazine cream in wound care and burn clinics, but greater problems have been experienced with sulfonamide-resistant organisms.[23] This led to a temporary lapse in the use of the drug in the 1980s, but nowadays silver sulfadiazine is an essential medicament in burn wound and chronic ulcer clinics where it may be used with sustained silver release dressings or other therapies designed to achieve maximal wound bed preparation.[51]

In attempts to maximise the antimicrobial efficacy and delivery of silver sulfadiazine to skin wounds, several new presentations have been devised in recent years. These presentations range from artificial dermis, textiles and dressings of the sustained silver-release variety, some of which contain bioactive materials like hydrocolloids, growth factors and materials known to promote wound repair mechanisms.[52–56] Some of these devices are still at an experimental stage, whereas others have entered clinical trials or become accepted as highly efficacious and beneficial wound dressings as in the case of Urgotul® (Urgo, France) (Figure 5.4). The sustained silver-release dressings are discussed below, but new products increasingly reflect the need to observe principles of wound bed preparation in addition to providing a suitable carrier system for silver. As expected the inclusion of silver sulfadiazine as a silver vehicle increases the antimicrobial range of a product beyond that of the silver (Ag^+) congener.

A drug delivery system devised for delivery of silver sulfadiazine to wounds used liposomal encapsulation.[57–59] Human keratinocytes in culture were shown by electron microscopy to engulf and break down liposomes to release their antibiotic content.[59] Experimental studies in rat wounds have demonstrated delivery of silver sulfadiazine by injecting liposomally encapsulated drug beneath a non-adherent wound dressing. Decreased bacterial cell counts were reported and preliminary evidence is provided for a clinically safe and efficacious route of therapy.

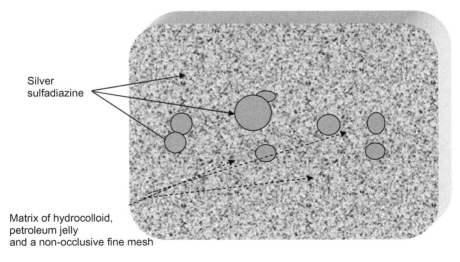

Silver
sulfadiazine

Matrix of hydrocolloid,
petroleum jelly
and a non-occlusive fine mesh

Figure 5.4 Composition of a non-adherent lipocolloids wound dressing releasing silver sulfadiazine in the presence of wound fluids (Urgotul® SSD, Urgo, Chenôve, France).

Experimental studies provide details of the cellular management of silver and its toxicology.[3] Unlike silver nitrate, topical application of 1% silver sulfadiazine cream to rodent wounds did not evoke dark discolouration of the tissues but precipitation of silver as colourless protein complexes was appreciated by tissue analysis.[3] In his early development studies, Charles Fox applied $30\,\text{mM}\,\text{kg}^{-1}$ silver sulfadiazine ointment to scald wounds in mice contaminated with *Pseudomonas aeruginosa* and noted that their mortality was significantly reduced and post-burn damage was greatly reduced.[40] Comparative studies demonstrated that $30\,\text{mM}\,\text{kg}^{-1}$ silver sulfadiazine ointment was superior to 0.5% silver nitrate or 10% Mafenide in terms of improved wound healing, absence of discolouration and toxic changes. Fox later demonstrated that zinc sulfadiazine therapy was comparable to silver sulfadiazine in alleviating infection and would provide a suitable alternative for use in burn wound patients with silver allergy.[60] In experiments with incisional wounds induced in the very thin skin of the mouse ear, neovascularisation and re-epithelialisation patterns were more rapid with 1.0% silver sulfadiazine than with 0.5% silver nitrate and other topical antiseptic agents.[61] As in our own studies, improved healing in wounds is attributable to minimal local irritancy and metallothionein-associated excitation of zinc metabolism in the regenerative epithelia.[3]

Porcine skin might afford a more suitable experimental animal model for investigating the clinical and behavioural action of silver sulfadiazine in view of its similar anatomical structure and wound healing patterns to human.[62–64] However, application of 1% silver sulfadiazine cream to skin wounds on the New Hampshire pig was associated with a delayed re-epithelialisation patterns, though the therapy had the potential to preserve viable tissue; wounds showing an abundance of myofibroblasts and formation of granulation tissue.[65] Re-epithelialisation revealed an unusual pseudocarcinomatous pattern with inflammatory cell infiltration preceding normalisation. As in later clinical studies, levels of argyraemia increased from wounding for approximately 15 days but then declined to near zero as wounds healed.[2] Lower levels of Ag^+ penetrated systemically from liposome formulations (M. J. Hoekstra, personal communication).

Silver sulfadiazine was first prepared in 1968 with the intention of treating burn wounds at risk of life-threatening *Pseudomonas aeruginosa* infection, but has since proved to be an efficacious topical antibiotic for controlling wound infections in situations other than burns including chronic ulcers, various dermatological lesions, as well as infections associated with catheters, textiles and endotrachial tubes.[40,66–68] At a time of renaissance in antimicrobial chemotherapy, Fox re-examined the antimicrobial action of several metal sulfonamides including those of zinc, cerium and silver and which exhibited similar levels of action.[60,69] Whilst they exhibited similar patterns of antimicrobial efficacy, their mechanisms of action differed leading Fox to conclude that their unique effects on bacterial cell biology pointed to their specific roles in topical chemotherapy. Effective therapy required the active drug to remain sufficiently long in the tissues to eliminate infections without impairing wound regeneration and repair.[70]

5.2.3 Cerium Nitrate and Cerium Sulfadiazine

Charles Fox evaluated the microbiological properties of the rare earth metal cerium. He found that cerium sulfadiazine exhibited interesting antimicrobial effect, and when used in combination with silver sulfadiazine, produced greater effect than silver sulfadiazine alone. In preliminary clinical studies, Gram-positive organisms predominated in patients treated topically with cerium sulfadiazine, whereas those exposed to silver sulfadiazine alone harboured predominantly Gram-negative flora. Clinical and laboratory studies indicated that by modifying 1% silver sulfadiazine by incorporating a cerous salt, the antiseptic effect was greatly increased without greater toxic risk. (Cerium sulfadiazine was shown to be very safe and with intraperitoneal LD_{50} in mice of $>1000\,mg\,kg^{-1}$ compared to silver sulfadiazine of $140\,mg\,kg^{-1}$.) Cerium ion (Ce^{3+}) is not readily absorbed percutaneously.[70]

However, of far greater significance is the action of cerium salts in enhancing a patient's immuno-competence in the face of thermal injury.[71] At the time silver sulfadiazine was introduced, clinicians were not fully aware that thermal damage in the skin was associated with release of pharmacologically active factors acting on vascular permeability and inhibiting neutrophil chemotaxis and T-cell function.[72,73] Patients with severe burns can be expected to experience extensive local trauma, but in addition factors released from the wound may exert pharmacological and immunosuppressive effects in tissues remote from the wound site, possibly leading to fatality.[74] Experimental studies have demonstrated that the immunosuppressive factor released in tissues subject to thermal injury is a heat-induced cross-linking complex of six skin cell membrane-associated lipoproteins. The low molecular weight lipoprotein complex (LPC) (3×10 kDa) has been shown to impair normal homeostatic processes leading to cellular damage, in much the same way as thermal injury itself.[75] LPC has further been shown to inhibit the growth of interleukin-2-dependant lymphocytes in tissue culture and is considered to be at least 1000-fold more immunosuppressive than many common bacterial toxins on a molecular basis.[76,77] *In vivo* studies in mice have shown that LPC administration greatly increased the sensitivity of the animals to the lethal effects of *Pseudomonas* sp. infection and immune responsiveness to sheep lymphocytes. Cerium nitrate (with or without silver sulfadiazine) can suppress delayed hypersensitivity reactions to such reagents as 2,4-dinitroaminobenzene and restore lymphocytic homeostasis.[77–79] In summary, whereas silver nitrate and silver sulfadiazine therapies are effective against many infections, they have minimal effect on the immune status of patients. By combining cerium nitrate–silver sulfadiazine, clinicians have the opportunity of reducing infections and increasing patient survival through the rare property of the cerium ion (Ce^{3+}) in binding to and neutralising the LPC (Figure 5.5).[78]

The commercially available product, Flammacerium® (Solvay Pharmaceuticals, Paris), has been used extensively in the treatment of mixed depth paediatric and adult scald/burn injuries. It has been found particularly useful in treating wounds that stand an improved chance of healing conservatively on

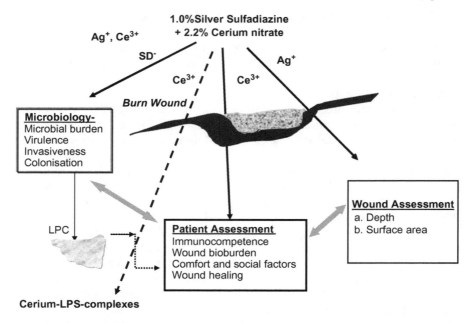

Figure 5.5 The burn wound, assessment, and chemotherapy with 1.0% silver sulfadiazine + 2.2% cerium nitrate (Flammacerium®, Solvay Pharmaceuticals, Paris).

the basis of depth or site (poor graft take on back) in life-threatening situations.[73,78–83] An added advantage in using cerium is its propensity to interact with tissue calcium. Wounds treated with cerium nitrate develop a tough mineralised eschar, which not only protects wound repair and acts as a barrier against infection, but is conveniently excised when healing is complete to reveal newly re-epithelialised tissue.[84]

5.2.4 Sustained Silver-release Dressings

Sustained silver-release wound dressings are now well-established in the care and management of acute and chronic skin wounds and burns.[85] They are safe, readily applied and efficacious in controlling a wide range of infections.[8,51,86] Silver is presented as an antibiotic in many wound dressings nowadays with focus on the amount of silver released and patterns of elution of Ag^+ into the wound bed.[87]

The dressings differ greatly in their total silver content, the nature of the ionisable silver source, patterns of Ag^+ release in the presence of wound exudates and moisture, and in their presumed mode of pharmacological and antibacterial action. In each case, exposure to wound fluid or moisture triggers the release of the so-called "bioactive silver ion" for bactericidal and fungicidal action, neutralisation of any toxins produced, and elimination of antisocial odours.[88] Many of the more complex dressings have an excellent record of improving patient comfort and quality of life and elimination of antisocial features. Modern wound dressings are increasingly tailored to engineer out risks of infection and to improve the clinical condition of the wound bed—now thought to be critical in the management of indolent and difficult-to-heal wounds. Great importance these days is attached to improving the "quality of life" of patients.

Three principal forms of sustained-silver release dressings have been produced in the past 25 years.

1. Simple mono-laminate dressings;
2. Complex dressings designed to control exudates and the moisture balance in the wound but releasing silver ion into the wound bed to control pathogenic infections;
3. Absorptive dressings which absorb excess moisture (exudates) and infections into the fabric where silver ion exerts its antimicrobial effect.

Although absorbent dressings act by retaining the larger part of their ionisable silver content within the fabric of the dressing, inevitably some Ag^+ will elute into the wound bed to exert bactericidal action by secondary intention. In each case, the wound dressings release more silver than is required for antibiotic action which in tissue culture at least is expected to be 10–40 ppm (60 ppm for particularly virulent organisms).[89,90] The nature of the silver source in dressings is frequently not revealed, or is expressed in terms of "ionic silver" or an unspecified silver source. The widely differing solubilities and ionisation properties of different silver compounds used in sustained silver-release dressings are critical in delivery of Ag^+ to the wound bed and its rate of microbial action (Table 5.4). Burrell[89] emphasised that a large proportion of the silver released from these antimicrobial dressings is bound to cell debris, proteins and anions in the wound bed and is thus unavailable for antimicrobial action.

Silver ion release profiles for wound dressings evaluated *in vitro* using a Franz cell or similar model have shown that products currently available in Europe and the USA to be of three main types:

1. High silver content—rapid ionisation and Ag^+ release;
2. Modest to high silver content but with more sustained ionisation and Ag^+ release;
3. Low silver content—designed for wounds with low risk of infection or as barrier dressings for post-operative wounds and acute lesions.

Ideally, nurses and clinicians planning to use silver-release wound dressings will consider the duration and overt condition of the wound (colour, surface condition, haemorrhage, *etc.*), its location on the body, the behaviour of the damaged tissue (exudates, odours, pain) and wound history. Laboratory investigations provide supportive information for clinicians and tissue viability nurses,[91–96] and pathologists have a central role in aiding practitioners in

Table 5.4 Solubilities of silver and silver salts used in antibiotic wound
            dressings.

Silver source	Solubility at 24 °C	Rate of ionisation
Metallic silver	<1 µg/mL	Very slow
Nanocrystalline silver	70–100 µg/mL	Fast
Oxides	*ca.* 28 µg/mL	Moderate
Nitrate	2570 g/L	Rapid
Chloride	0.002 g/L	Very slow
Sulfate	8.3 g/L	Very slow
Phosphate	0.00644 g/L	Very slow
Acetate	11.11 g/L	Moderate
Ag-Zirconium salts	n/a	n/a

n/a = not available

Table 5.5 The pathologist in wound care.

Microbiologist	Clinical chemist	Histopathologist
Wound micro-biological burden	Composition of the exudate	Assessment of the depth and composition of wounded tissue
Microbiological sensitivities to antibiotics including silver	Presence of xenobiotic materials derived as perfusates from the plasma or components of the wound dressing	Signs of toxic reactions within the wound bed, including adverse responses to materials eluting from wound dressings or therapies
Presence of biofilms	Identification of bacterial/fungal cell toxins	Mechanisms of action of silver on wound repair, including macrophage responses
Responses to therapy including antibiotics in wound dressings	Confirmation of a patient's nutritional and disease status	Deposition of silver as insoluble precipitates in wound bed

wound assessment with particular reference to microbiological burdens, identification of silver resistant organisms, assessment of the wound exudates and biopsy to clarify the wound depth (Table 5.5).[97] Silver allergies should be substantiated.

5.2.4.1 Monolaminate Dressings

Thin film silver-release dressings were introduced in the 1970s as flexible and minimally adhesive products providing a moisture–air permeable barrier for post-operative and acute wounds with a low risk of infection. Transparent dressings like Arglaes™ (Maersk Medical, Medline Industries Inc.) and

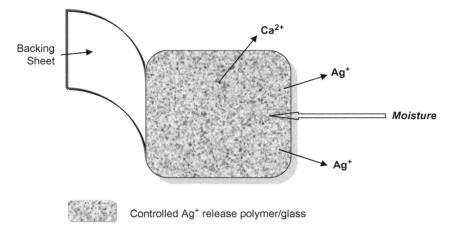

Figure 5.6 Monolaminate absorbant dressing incorporating silver-release technology and alginate.

Avance™ (SSL International, Mölnlycke Healthcare) allow practitioners to observe patterns of healing and to detect adverse responses to silver ion released into the wound bed. The dressings were designed for minimally infected post-surgical wounds and experience has shown that both are accompanied by very low toxic risk.

Arglaes™ and Arglaes™ Island dressings incorporate an ionisable silver compound in the form of a water-soluble glass comprising a furnace-fused complex of calcium and sodium phosphates (Figure 5.6). The products release silver over up to seven days to provide an efficacious antimicrobial barrier over initial phases in the wound healing cascade, notably during haemostasis, inflammation and proliferative stages.[98] Calcium ion (Ca^{2+}) released from Arglaes™ barrier film dressings with a calcium alginate pad interacts with platelet phospholipids providing essential factors in the coagulation of blood in the haemostatic phase of healing. Formation of the blood clot with interaction of Factor XIII provides a stabilising effect on the wound.[99] Avance™ is described as a hydropolymer (polyurethane foam) dressing with an adhesive border with a silver compound "bonded" into it as in "Bacti-Shield" technology (SSL International).[100] These dressings were designed to maximise Ag^+ release for treating lacerations, abrasions, post-operative lesions, minor burns and possibly chronic wounds and leg ulcers but differ in their silver content, rates of silver ionisation and gas/moisture permeability. Clinical experience indicates that much of the silver content of film dressings ionises rapidly in the first few hours of application, with a modest or low Ag^+ release pattern over the following few days.[101] In static dissolution tests, silver released from Arglaes™ (2.5 cm) containing 0.127 mg maximised within 2.5 d and exhibited a plateau of Ag^+ release of approximately 30 ppm for at least 24 h, and with an aqueous concentration of $20\,\mu g\,Ag^{+}\,mL^{-1}$ after 24 h ($2.5\,cm^{2}$ dressing sample immersed in 5 mL simulated exudate).[98,102] Thin film dressings have limited capacity for fluid management in chronic wounds, but the hydrophilic gel adherent to the wound surface should provide a modest barrier function for acute wounds.

Medline Industries Inc. has introduced an antibiotic powder based on Arglaes™ technology with the intention of reducing the bio-burden in deeper and tunnelling wounds with heavy exudates. The formulation may be beneficial in managing wounds of widely varying depth, dimension and location. A further innovation is PolyMem® Wound Care Dressing (QuadraFoam) (Ferris Mfg. Corp., Burr Ride, IL), with nanosilver silver particles impregnated into a polyurethane matrix containing a cleanser, moisturiser and superabsorbent copolymer to aid wound exudates management. Ferris Mfg. Corp. has also introduced a wound "filler", PolyMem Silver Rope, for tunnelling wounds to add to the current PolyMem® Silver cavity/undermining and cover dressings.

Case studies have demonstrated the advantages of monolaminate film dressings in treating indolent and chronic wounds and their efficacy as a barrier against acute infections. Authors have also claimed that they improved healing patterns and reduced wound pain, with no obvious adverse effects or difficulty at dressing changes at intervals of 1–3 d.[100,103] Arglaes™ Island dressings are not recommended for third degree burn wounds and surgical implants. Experimental studies have demonstrated that silver deposited in the wound site from Arglaes™ Powder has a tendency to form darkish precipitates, some of which are absorbed by wound macrophages and a small proportion reach the systemic circulation to be eliminated *via* the liver and kidney (Figure 5.7).[104] (The dressings darken on exposure to light.)

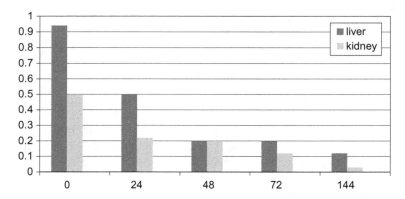

Figure 5.7   Silver concentrations ($\mu g\,g^{-1}$) in liver and kidney following application of Arglaes™ Powder (5 mg) to surgical wounds for four days (15 mm linear wounds).[104]

5.2.4.2   Multilaminate Dressings

Complex multilaminate dressings comprise:

- an ionisable source of silver;
- hydrophilic matrix containing bioactive materials;
- non-bioactive materials as adhesives, backing materials, *etc.*

Like the film dressings, they are designed for direct application to the wound surface where wound exudates "activate" silver to release Ag^+ into the wound bed, and possibly absorption into the systemic circulation (Figure 5.8). A balance is observed between the level of ion released, its sufficiency for bactericidal/fungicidal action and its toxicity threshold. In addition to the inherent risks of silver allergies, rare instance of contact hypersensitivity may be encountered to non-silver components of dressings including:

- Fabrics – rayon, polyesters, polyurethane, alginates, viscose, *etc.*;
- Activated charcoal;
- Silicones;
- Glycerine;
- Petrolatum;
- Nutrients and growth factors.

Contraindications for using silver release dressings also include MRI scans, contact with electrodes or conduction gels used in electronic monitoring procedures, X-radiography, ultrasonic therapy and diathermy. Clinical evaluations

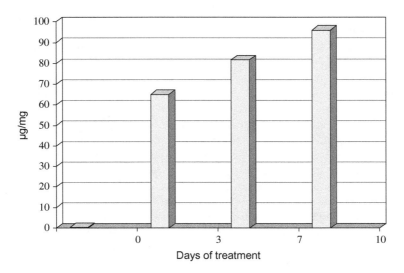

Figure 5.8   Silver in wound scale in a patients treated with a sustained silver-release wound dressing. Silver was lost progressively as the wound healed.

of dressings concentrate mostly on their ability to control pathogenic infections and the bacterial balance in wounds, and their capacity to advance wound healing.[105,106] Dressings have been routinely assessed clinically on their ability to control offensive odours, excessive exudation, maceration, patient discomfort and to improve the quality of life.

The antimicrobial capacity of wound dressings is not determined by the amount of silver compound contained but on:

1. The solubility of the compound;
2. Its rate of ionisation in wound fluids;
3. The proportion of the available silver ionising within the lifetime of the dressing.

As discussed by Burrell,[89] Ag^+ has a strong capacity to bind and precipitate with albumins and macroglobulins in the wound bed, anions (*e.g.* chloride and phosphate), wound debris and cell membrane receptors at the wound margin, thereby reducing the amount of free ion available for antibacterial or antifungal action. Thus, whilst silver nitrate is highly soluble and ionises rapidly in water, silver chloride and silver sulfate are only sparingly soluble and a local saturation point is rapidly attained. In dressings containing minimally soluble silver chloride ions, saturation occurs at about 1.3×10^{-5} M (the ionisation constant K_{sp} is exceedingly low at 1.8×10^{-10}). Silver sulfate is also sparingly soluble, but manufacturers claim that the level of ionisation in a wound is sufficient to supply the oligodynamic concentrations necessary for antibiotic action.

Great advantage is taken of advances in the nanotechnology of silver in the production of antibiotic dressings. The solubility and ionisation is proportional to the surface area of particles.[107] Whereas metallic silver has a solubility of less than $1 \mu g \, mL^{-1}$, nanocrystalline silver with particle size of <20 nm has a solubility of $70–100 \mu g \, mL^{-1}$ and appreciably higher than most other salts of silver, other than nitrate.[89] Silver sulfadiazine co-ordinating three silver atoms has low solubility in water (<1 part in 10 000) but ionises in wound fluids to provide a reservoir of silver ions.[42]

Acticoat® (Westaim Biomedical Inc., Toronto) was possibly the first wound dressing to take advantage of silver nanotechnology and extensive research in *in vitro* experimental and clinical situations has established its highly efficacious barrier function in chronic wound therapy with wound healing advancement.[108,109] Acticoat® developed through Silcryst™ technology is a trilaminate dressing consisting of an absorbent rayon and polyester core with upper and lower layers of high density polythene mesh coated with nanocrystalline silver using a physical vapour deposition (magnetron-sputtering process).[110] The silver coating is $<1 \mu m$ thick (0.2–0.3 mg Ag per mg) and is a binary alloy of 97% silver and 3% oxygen. Antimicrobial efficacy has been established against a broad spectrum of infections in experimental and clinical situations, and the dressing is commensurate with good healing practice in burn and chronic wound centres.[110–113] *In vitro* silver release patterns showed that, after an exponential rise in ionisation in 4–6 h, silver in solution plateaued at 70 ppm for at least 24 h.[98] Acticoat® is compatible with enzymic debriding agents, skin equivalents and autograft situations. In more recent years, Silcryst™ technology has been employed in production of Acticoat® Moisture Control (comprising a foam layer, waterproof layer and silver coated layer), Acticoat® Absorbent (with an alginate layer) and Acticoat® 7 (two absorbent rayon/polyester layers and three layers of nanocrystalline silver-coated material) with varying levels of exudate absorbency and efficacious for up to seven days).[85] The nanocrystalline silver barrier dressing has been shown to promote healing in indolent wounds and to reduce inflammatory changes,[114] and is safe in burn wound therapy.[115] Maximal systemic argyraemia was seen in 30 burn patients with postoperative wounds of 12% body surface area was 56.8 μg L^{-1} over nine days.

Different technology is employed in the Actisorb® Silver 220 and related dressings (Johnson & Johnson, Somerville, NJ) which incorporate an activated charcoal cloth impregnated with metallic silver sealed in a nylon fabric sleeve with recommendations for treating malodorous chronic wounds. Through the action of the charcoal, they absorb wound exudates, micro-organisms, bacterial and fungal toxins and proteins from the wound bed into a hydrophilic matrix where they are exposed to ionised silver.[116,117] Actisorb® dressings containing appreciably less ionisable silver than the Acticoat® products (0.15% by weight chemically and physically bound within the matrix of carbon fibres) are clinically effective in the therapy of chronic wounds and ulcers infected with MRSA and other common pathogens creating a conducive milieu for healing and pain and odour relief.[118,119] *In vitro* studies show that the activated charcoal is capable of binding lipopolysaccharide and endotoxins released from *Escherichia coli* and inhibiting endotoxin release from *Pseudomonas aeruginosa*, thereby inhibiting wound sepsis and impaired healing.[120] Clinical experience shows that Actisorb® Silver 220 can be folded as a "filler" dressing for deep cavitaceous wounds and wounds in difficult areas.[51,121]

Alginates have a long history in wound care and are increasingly used in silver delivery systems in wound care dressings. Alginates are produced by brown seaweeds (mainly *Laminaria* sp.) and chemically are composed of linear unbranched polymers containing β-(1 → 4)-linked D-mannuronic acid (M) and α-(1 → 4)-linked L-guluronic acid (G) residues (Figure 5.9). The flexible polysaccharide chains are cross-linked by calcium ions, which in aqueous media exchange for sodium ions; the free Ca^{2+} participate in haemostasis in acute wounds and post-surgery.[122–126] The residual sodium alginate chains provide a matrix for platelet aggregation but then biodegrade within the wound with no harmful effects.[127] Alginates used in wound care differ greatly in their composition (mannose and glucuronic acid residues), but are mostly hydrophilic and beneficial in exudate management and promotion of epidermal regeneration.

Aquacel® Ag (ConvaTec, Skillman, NJ) is an example of a hydrophilic dressing with sodium carboxymethylcellulose and 1.2% silver in ionised form

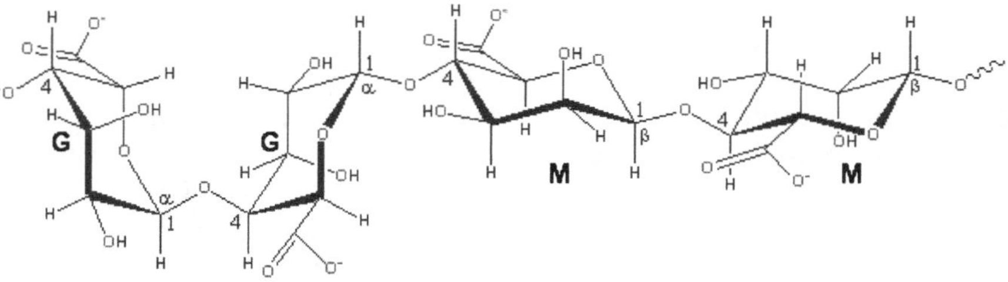

Figure 5.9   Structural unit of alginate polysaccharide fibre.

permitting sustained silver release but with principal recommendations for exudate management, prolonged antimicrobial action and wound bed preparation. This hydro-fibre dressing containing metallic silver with high absorbency maximises conditions the wound bed for maximal repair.[51] Ca^{2+} released aids haemostasis and alginate fibres provide a matrix for blood clotting and release of platelet-derived growth factor (PDGF).[128,129] Being principally an absorbent dressing, Aquacel® Ag has a low silver content (8.3 mg 100 cm^{-2}) and releases low concentrations of Ag^+ into the wound bed.[5,51] Aquacel® Ag has been used successfully in management of non-ischaemic diabetic foot ulcers, improving healing and bioburden.[130]

Silvercel® (Johnson & Johnson Wound Management) is a hydroalginate wound dressing which combines the properties of alginate and carboxymethylcellulose to improve fluid/exudate management in chronic wounds and provide a barrier against opportunist and nosocomial pathogens.[131,132] Silvercel® has been shown to be beneficial in the therapy of moderate to severe exudating wounds and, under simulated conditions, releases silver until a "saturation point" of 15–25 ppm is reached. As Ag^+ becomes bound or is used to provide antibiotic action, so further ionisation is activated. The silver-coated fibres of the dressing provide a reservoir of ions to provide sustained antibiotic action for at least 14 d.

Silver sulfadiazine delivery systems in wound care are numerous and compatible with damaged human tissues. Delivery systems involving polyurethane foam and lipocolloid formulations offer interesting possibilities of introducing the drug to chronic and difficult-to-heal wounds with few complications other than contact allergies to silver, sulfadiazine or components of the dressings. Urgotul® SSD (Urgo, Dijon, France), with a total silver content of 3.75% silver sulfadiazine in a polyester fabric with petroleum jelly and cohesion polymer, has proved an efficacious bacterial barrier and wound healing aid in critically colonised leg ulcers, epidermolysis bullosa lesions and paediatric wounds and second degree burns with minimal complication.[133–135] As for 1% silver sulfadiazine cream, Urgotul® SSD exhibits commendable antibacterial action *in vitro* against named pathogens,[86] and its toxicity for dermal fibroblasts in culture was low.[136] Normal cell morphology was substantiated using confocal laser microscopy to examine the influence of extracts of the dressing on cytoskeletal proteins including β-actin and β-tubulin.

Silver-containing dressings are well-established in the care and management of a wide range of acute and chronic wounds, varicose ulcers and burns with few reported toxic sequalae other than occasional cases of delayed sensitivity to silver[137] and rare incidences of bacterial resistance.[138] Despite claims made in most clinical trials citing commendable levels of antimicrobial action and improved wound healing, the Cochrane Collaboration is concerned that appropriate "criteria for patient and wound evaluation" are lacking and that more clinical and laboratory data are needed to establish the relative values of the numerous dressings available. In Professor Leaper's opinion, the nano-crystalline silver dressings appear to give the highest sustained release of silver (Ag^+) to wounds without clear evidence of toxic risk but more evidence is required to support clinical claims.[137,139]

5.2.5 Sutures, Staples and Clips

Silver wire has possibly been used in surgery for closing skin wounds since the days of the ancient Greek and Roman empires and there are records showing that Galen (*ca.* 130–200AD) used silver and gold as ligatures and clips in wound surgery and in the treatment of aneurisms.

Notable surgeons such as John Hunter (1728–1793) used silver clips in vascular and general surgery though the American gynaecologist, J. Marion Sims, possibly did most to establish the value of silver sutures in general surgery.[140] In 1846, in Montgomery, Alabama, Sims attempted repair of a vesicovaginal fistula in a country woman who suffered a retroversion of the uterus in a fall from a horse. He employed fine silver sutures and silver catheter to correct her fistula successfully. Sims travelled to Europe and his intervention rapidly gained international acceptance. In 1858 Sims published *Silver Sutures in Surgery*, which recounts his early experiences in the use of silver sutures to avoid sepsis and takes full account of their introduction and acceptance by the surgical profession.[141]

Reuben Ottenberg (1882–1959) recognised the value of silver in vascular surgery and described arterial surgery using tiny silver rings held in place by fine silver wire in arterial anastomosis, but he seemed to be more interested in blood transfusion and compatibility.[142,143] Silver sutures and staples are still used in surgery today[144] but have been largely superseded by stainless steel, silk and synthetic fibres. Recent views suggest that, whereas the use of silver wire retention sutures are contraindicated on account of increased operating room time and expense, they also incur potential risks of caught viscera, adhesions, leakage of intraperitoneal fluid through exit wounds, mild infection on removal of the sutures, pain despite sedation, cutting of viscera, and a residual cross-hatched scar. The technique is justified, however, by the prevention of abdominal wound disruption, ease in visualisation of the wound, facility in providing equal distribution and periodic adjustment of wound tension, re-approximation of superficial tissues in the event of wound breakdown, and elimination of need for any external supportive device.[145]

Recent research in biomaterials science has led to the production of anti-microbial sutures for use in wound closures. Although still largely at an experimental stage, natural (silk) and synthetic fibres (nylon) coated with or impregnated with metallic silver, silver oxide or other silver compound have shown commendable antimicrobial action in *in vitro* bacterial cultures and biocompatibility following implantation in animal models.[146–149] BioGlass® (Novabone, Florida) developed as a system comprising $-SiO_2-CaO-P_2O_5$, and found to be entirely biocompatible with human tissues in the formation of mechanically strong bonds to bone and soft tissues, has been modified by addition of silver oxide (Ag_2O) to provide an antibacterial glass for use in surgical sutures. Research at Imperial College in London has demonstrated that "composite sutures" can be prepared using a slurry dipping technique with commercially available silk sutures (Mersilk®, Ethicon Inc., Somerville, NJ) being exposed to BioGlass® silver by immersion. The silver-coated sutures have been characterised by scanning electron microscopy to determine their thickness, morphology, homogeneity and coating structure (Figures 5.10 and 5.11).[150–152] They were biocompatible in experimental surgery and rodent wounds healed with minimal scar tissue or inflammatory change.[153] Macro-phages were implicated in metabolism of released silver ion. Elsewhere, braided nylon sutures coated with a silver compound were shown to be biocompatible in rat gluteal muscle implantation and exhibited antimicrobial action *in vitro* against named pathogens; the effect was enhanced by application of a weak

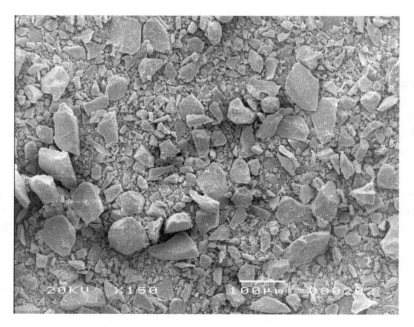

Figure 5.10   Bioglass® powder doped with silver oxide for use in sutures (low power magnification).[150]

Figure 5.11 Scanning electron micrograph of a silk suture (Mersilk®, Ethicon, Edinburgh) treated with Bioglass® silver.[150]

direct current.[146,154] In this *in vitro* model, Ag$^+$ release reached a peak of 85.3 µg mL^{-1} after the first hour and 305.7 µg mL^{-1} after 6 h, Ag$^+$ release increasing with the current applied and the duration of application.[154]

5.3 Medical Catheters

5.3.1 Catheter-related Infections

Implantable devices such as catheters are indispensible in the management of critically ill and chronic patients for the administration of electrolytes, drugs, intraparenteral nutrients and blood transfusion as well as for the drainage of excess body fluids and secretions.[155] They are essential for vascular access in patients undergoing renal haemodialysis.[156]

Considerable efforts have been made in recent years to develop antibiotic catheters for mid- or long-term implantation in patients with urinogenital diseases.[157] Maki reported that more than five million urinary catheters were implanted annually in patients in acute care hospitals and extended care facilities worldwide, and that catheter-associated urinary tract infections are the most commonly encountered problem in hospital infections.[158] Catheter-associated urinary tract infections (CAUTIs) possibly account for more than 40% of all institutionally acquired infections and are the second most common

Figure 5.12 A 19th century silver catheter. *By courtesy of the Royal College of Surgeons of England.*

cause of nosocomial bloodstream infections. They can be a cause of morbidity and mortality if unchecked.

The technology for catheter production and introduction of antimicrobial measures has changed dramatically over the past 200 years in the light of new skills in vascular surgery and introduction of materials with improved qualities of biocompatibility and safety in use. New devices have been developed with the intention of engineering out risks of infection, improving catheter care and management, and maximising cost : clinical efficiency. Silver has a long record of use in medical catheters and records exist of metallic silver catheters being used in naval surgery in Nelson's time when urinogenital disease and gonorrhoea were common (Figure 5.12).[159] It is unlikely that the antibiotic value of silver was appreciated at that time when naval surgeons had an unreserved reputation as being "... rough *sawbones* types, only able to hack off limbs and pull teeth".[157] Nowadays, with major advances in plastics technology and development of polymer chemistry, catheters are manufactured of high quality tubing with ultra-smooth surfaces and internationally recognised standards of safety.

Modern catheter design focuses on low-cost technology, wide spectrum and long-lasting antimicrobial action, and safety in use.[160] Polymers have been developed to provide chemically stable and flexible tubing, low in plasticisers, and composed of mechanically suitable materials for administration of drugs or fluids, or to provide drainage as clinically required. Antibiotics such as chlorhexidine, gentamycin, minocycline, vancomycin, flucloxicillin, rifampicin, silver sulfadiazine, silver oxide and metallic silver (including nanocrystalline forms) have been used singly or in combination in attempts to provide efficacious antibiotic properties. Other antibiotic measures used include modification of catheter surfaces using hydrophilic coatings to improve contact with human epithelial surfaces and to enhance delivery of ionic silver or other antibiotic.[161]

Major differences exist in the design of catheters according to their intended use. The main criteria reviewed by international organisations regulating medical devices, *e.g.* FDA, Medicines and Healthcare products Regulatory Agency (MHRA) *etc.* include:

- Biocompatibility with human tissues;
- Chemically defined and stable in the presence of human secretions, excretions and body fluids (*i.e.* they do not degrade to release pharmacologically or physiologically active agents);
- Retain their physicochemical properties throughout the expected lifespan of the devices.

Where technological innovations in catheter design introduce antimicrobial materials, design features direct that maximal release of the active principle (*i.e.* Ag^+) is timed to coincide with the period of highest risk of infection and takes account of expected pathogenesis of infections and any toxins that they produce.[158,162,163] In view of the expectation that catheters implanted into the body will be exposed to infection on both outer and inner surfaces, new devices will be treated ideally either by full impregnation of the polymers with metallic silver or other ionisable silver compound, or coating both surfaces to provide long-lasting protection.[164,165]

Central venous catheters are commonly exposed to infection during implantation or through daily use and maintenance therapy. Catheter-related bloodstream infections develop where infections migrate from contaminations at catheter hubs, bacterial or fungal cell colonies at insertion sites, and infections borne by infusates or haematogenous seeding.[166] In a similar way, infections affecting dialysis catheters migrate from portals of entry and migrate along catheter tracts to cause life-threatening peritonitis.[167] The severity of catheter-related infections is invariably related to the duration of placement, the number of procedures performed and the number of lumina in catheter design.

Infections associated with intra-urethral catheters are reported to be the most common source of nosocomial infection in acute care hospitals and extended care facilities.[168] They develop through insertion of the catheters through unhygienic practice and inappropriate management, but more commonly are a feature of prolonged exposure to contaminated urine and hospital-related infections. Over a long period, catheter-related infections of the urinary tract may progress to severe inflammatory changes, pyelonephritis, renal and cystic lithiasis, urinary tract infections, bacteraemia and eventually fatality.[169,170] Catheter infection and adhesions commonly involve bacterial adhesion to endoluminal or external surfaces *via* endoluminal or periluminal routes. Infections (notably *Proteus* sp.) creep between the urethral epithelium and the external surface of the catheter.[171–173] Infections such as *Proteus, Providencia* and *Morganella* colonise catheter tips and, in the presence of bacterial urease, evoke calcium phosphate deposition leading catheter blockage and patient distress.[172] Biofilms are notoriously resistant to most antibiotics including silver!

5.3.2 Central Venous Catheters

Catheters impregnated with metallic silver or coated with silver, silver oxide or silver sulfadiazine have been manufactured for intravenous access. The technologies include impregnation of polyurethane or other suitable polymer using ion beam technology (ion beam assisted deposition), sputter coating or surface coating with silver antibiotics in a bonding process, microdispersion of microparticles of silver, and deposition of silver alloy on catheter surfaces. Silver-impregnated catheter cuffs have been used to reduce the migration of bacteria such as *Proteus* sp. down the tunnel tract.[156] Experimental studies and clinical trials have been conducted using levels of bacteraemia, local inflammatory reactions and patient compliance as a guide to the suitability and antimicrobial efficacy of devices; whatever the technology, ionisable silver or another silver compound should be distributed homogeneously throughout the catheter material. Tobin considered that, for catheters prepared by techniques such as ion beam or sputter coating, a coating of 1 μm thickness should be sufficient to provide antibiotic protection.[156,174]

Ion beam assisted deposition (IBAD) is a method of coating or impregnation of devices to achieve a gradual release of Ag^+ and long-lasting antimicrobial action. The technique, which can be applied to catheters and other medical devices, has been developed by Spire Biomedical (Spi-Argent®; Bedford, MA) and has been defined as a thin film deposition process that combines evaporation with concurrent ion beam bombardment in a high vacuum environment (10^{-4} to 10^{-6} torr) (Figure 5.13). Catheter materials are placed in silver vapour created by electron beam evaporation and individual coating atoms condense and adhere to the surface to form a coating in the presence of high energy ions (100–2000 eV). The primary advantage of the technology is that surface coatings achieve excellent adhesion and high film quality.[156]

Varying antimicrobial efficacies have been demonstrated with IBAD technology in *in vitro* and clinical studies. Thus, SPI-ARGENT-II silver coated catheters developed by Spire Services Technology Systems (Milan, Italy) failed to significantly influence colonisation by three strains of pathogenic bacteria or *Candida albicans* in an *in vitro* system.[175] On the other hand, Tobin and Bambauer reviewed several *in vivo* studies in animal models and in clinical implants, and in each case substantiated that silver coatings *did* reduce bacterial colonisation and occurrence of catheter-related infections without evidence of local inflammation or toxic changes.[156] Silver-coated catheters did not exhibit deposits

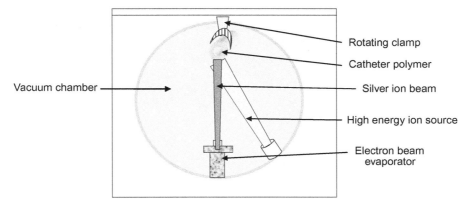

Figure 5.13 Diagram of the apparatus used in ion beam assisted deposition (IBAD) of silver coating on medical catheters (Spire Biomedical Technology) using concurrent ion bombardment of catheter material.

of plasma proteins or blood cells on outer treated surfaces after two months implantation. Blood silver levels attained in human patients increased from a mean of 1.3 to $6.9 \mu g\, dL^{-1}$ (acute catheters) and from 3.4 to $19.6 \mu g\, dL^{-1}$ (long term catheters), but declined to baseline levels after catheters were removed. The antimicrobial properties of the IBAD technology were confirmed in multi-centre studies with catheters implanted for up to 880 d.[176,177] Experimental studies have demonstrated that both IBAD and silver impregnation methods of catheter treatment are biocompatible with mammalian tissues and that implantation is not associated with local irritancy or foreign body reactions.[177–179]

Goldschmidt *et al.* conducted a randomised study of silver alloy coated central venous catheters and, whilst the infection rate was reduced by 50%, there was no significant reduction in catheter colonisation or associated sepsis.[180] Disappointing observations were made also in a study of 268 patients in which a total of 304 single-lumen iontophoretic catheters were inserted and observed over 5449 d.[166] Catheter-related bloodstream infections were not significantly reduced and catheters showed a 71% incidence of endoluminal colonisation. *In vitro* studies have shown that IBAD technology did not significantly influence survival of common pathogens and is of dubious value.[166]

Catheters available for central vascular use these days are commonly coated in silver sulfadiazine and chlorhexidine digluconate.[159] Some authors are of the opinion that chlorhexidine digluconate is a more efficacious antibiotic than silver and is largely responsible for reducing catheter-related systemic infections and colonisation. However, Maki *et al.* conducted a clinical trial in 158 patients, implanting a total of 403 catheters and reported a five-fold reduction in catheter-related infections and reduced catheter colonisation.[181] The chlorhexidine–silver sulfadiazine combination was well tolerated by all patients, although some erythema was noted at catheter insertion points.

Although Tobin and Bambauer measured increases in blood silver following implantation of IBAD catheters, patterns of Ag^+ release from central venous catheters *in situ* is not well documented.[156] Guggenbichler *et al.* calculated that an Erlanger polyurethane catheter containing 0.8% mass of Ag per g would release 1.5×10^{15} silver particles of approximate 50 nm diameter (equivalent to $0.5\,\mu g\,Ag^+$ per 24 h) *in vitro*.[165] By these calculations, this would result in a surface equivalent of 0.45 m̊ metallic silver per gram of polythene. They argued that:

> *"It has to be technically feasible to distribute billions of submicron silver particles throughout a polymer matrix to provide a sufficiently large active surface area to give the antimicrobial action desired. A catheter with a single lumen of 1.8 mm diameter is expected to release 1 μg Ag^+ per metre of catheter in 24 hours".*

Antimicrobial concentrations of silver are expected to be of the order of $10^{-9}\,mol\,L^{-1}$ but this does not allow for the unavailability of Ag^+ that binds irreversibly to anionic moieties in the microenvironment.[165]

5.3.3   Intra-Urethral Catheters

Catheters used for urinary tract and bladder drainage are of two main types:[169]

- those inserted intra-urethrally which, under normal circumstances, should not involve tissue damage;
- those implanted by the supra-pubic route.

Theoretically, there are fewer microbes on the abdominal wall than in the region of the perineum and risks of catheter-related infections are lower. Supra-pubic catheters allow a patient greater freedom of movement but are liable to cause cellulitis, leakage and haematoma at points of percutaneous insertion. Surgical implantation of both types of catheter carries the risk of infection by nosocomial and patient-related organisms, and reduction of risks of CAUTIs and bacteruria requires strict codes of hygiene and management.[168,169,182] Nevertheless, urologists accept that virtually all urinary catheters will become colonised to some extent both on endoluminal and extraluminal surfaces, and that infections such as *Pseudomonas, Proteus, Providencia, Morganella, Enterobacteriaceae, etc.* will lead to biofilm formation with the calcareous concretions leading to an eventual blockage of drainage and patient distress.[171,172,188]

As noted above, catheter technology has advanced from the 1800s and use of silver metal catheters to the current practice of using latex, polyurethane and silicone rubber coated or impregnated devices with ionisable silver—possibly with a second antibiotic such as chlorhexidine to inhibit bacterial adhesions and biofilm formation. Akiyama and Okamoto are credited with introducing the first sustained silver ion release urinary catheters in 1979 which effectively reduced levels of bacteruria without adverse effects in 102 short- and long-term patients.[184] They recognised the oligodynamic properties of silver as an antibiotic and coated catheters with silver powder and silver-plated connectors to protect patients from CAUTIs. More recent work *in vitro* and *in vivo* has seen experimentation with iontophoretic methods of silver treatment of catheters, production of silver alloy coating and techniques for impregnation of catheter materials with nanocrystalline silver.[185,186] Liedberg and Lundeberg conducted early experiments with a Foley catheter coated with silver on endoluminal and extraluminal surfaces (Bard Ltd) in 60 patients and reported a significant reduction in CAUTIs. They demonstrated that bacteria failed to adhere and colonise silver-coated catheters and that, in animal experiments, the silver was not locally toxic.[187,188] Queries as to the possibility that Ag^+ release might trigger symptoms of local argyria in the urethral epithelium seem to be unfounded.[189]

The Bardex® I.C. technology for silver coating Foley type catheters introduced in 1996 comprises a bilaminate coating of inner and outer surfaces of catheters with hydrogel to absorb moisture (C.R. Bard Inc., Covington, GA). The hydrated hydrogel forms an interface between the catheter and the urethral wall, conforming closely with its contours to prevent migration of pathogens; moisture absorbed triggers release of Ag^+ for bactericidal action from a silver, gold and palladium alloy (Figure 5.14). This Bacti-Guard® technology is claimed, in multicentre trials, to be a cost-effective and efficacious means of reducing bacteruria and biofilm formation without toxic complications.[168,187,188,190–194] However, on occasions, evaluation of the efficacy of the technology was complicated by systemic infections and age-related conditions.

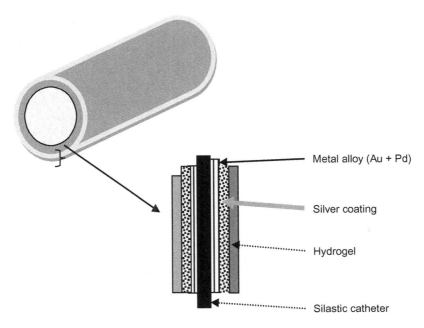

Figure 5.14 Hydrogel technology in silver-release medical catheters to overcome catheter-related infections.

Adhesion of Gram-positive and Gram-negative bacteria to surfaces of silicone and latex catheters coated with Bard hydrogel–silver technology was significantly reduced.[195]

A second type of silicone Foley catheter coated with silver hydrogel technology was introduced by Tyco Healthcare (Mansfield, MA). The manufacturers claimed reduced risks of hypersensitivity to their product which was coated with a lubricious hydrophilic gel containing an inorganic silver-releasing polymer. Clinical trials in 161 patients showed a reduction of 65% CAUTIs with the Tyco/Kendall bacteriostatic tubing/collection bag technology over a six-month timeframe.[196] Other attempts at introducing multi-metal systems such as a silver–copper coating seem not to have been followed by clinical experiments, despite some promising *in vitro* observations.[197]

Despite the numerous claims for the efficacy of silver hydrogel and other silver technologies in reducing CAUTIs, questions remain concerning the safety of the materials used in the catheters and analyses used in estimating the benefits claimed.[173,198,199] Although latex is inexpensive as a catheter material, latex allergy is prohibitive for its widespread use in urinary drainage devices and in the surgical gloves of operators.[200,201] On the other hand, silastic catheters are minimally allergenic and are at low risk of causing urethritis and possibly urethral stricture. They are the choice catheter for use in cases of latex allergy and long-term implantation.[169,199]

On balance, silver technology is efficacious in reducing bacteruria and urinary tract infections in short-term patients[182]—silver alloy coating methods being more effective than silver oxide catheters.[202] It is unclear on the basis of meta-analyses conducted whether silver will reduce the clinically more important outcomes of catheterisation including CAUTIs, bacteraemia or mortality in patients with long-term catheterisation.[182,189] Evaluation of the effects of silver impregnation or coating with silver in the longer term has received mixed response. Small numbers of patients and wide variations in the severity of urinary problems have complicated statistical evaluations.[199] Whereas catheters coated with silver sulfadiazine with chlorhexidine and triclosan antibiotics offer long-term resistance against antibiotics,[203] it is still unclear whether silver technology alone offers sufficient protection.

5.3.4   Intracerebral Catheters

Intracerebral catheters of the Silverline range (Spiegelberg, Hamburg) commonly used these days for external ventricular drainage and measurement of intracranial pressure in cases of hydrocephaly are composed of polyurethane or silicone polymer tubing impregnated with 2% nanoparticulate silver and an insoluble silver salt. *In vitro* antibacterial tests in roll cultures have shown that the technology is efficacious against named human pathogens but suggest that the mechanism of action is biphasisic.[204] A combination of a fast-acting silver salt with the slow ionisation of metallic silver is held to optimise the action of antimicrobial effect over several hours or days.

Catheter-related infections in the cerebrospinal fluid are potentially life-threatening as well as a cause of catheter dysfunction. Rickham and Holter in their initial experiments with intracerebral catheters impregnated silastic tubing with 15% precipitated silver and investigated their use in chest radiograms.[205] They observed no toxic symptoms but did not report microbiological findings. Preliminary clinical trials of the Spiegelberg catheter have demonstrated statistically significant protective action afforded by the silver technology against CSF infection, but coagulase-negative staphylococci were isolated from the tips of five out of 19 silver-treated catheters.[206] No adverse effects attributable to elution of Ag^+ in to brain tissue along the catheter tract were observed. Bayston and colleagues have also gained experience in the use of Spiegelberg catheters, but questions are still asked concerning the safety of the device in treating juvenile hydrocephaly and possible risks of bacterial resistance to silver emerging.[207,208]

5.4 Orthopaedic Devices

5.4.1 Introduction

Silver has been used in orthopaedic surgery from at least the Middles Ages when it was commonplace for surgeons to use silver plates to repair bone fractures and in facial reconstruction. In orthopaedic surgery, metallic silver has been used in surgical instruments and devices where insertion into the body is accompanied by risk of infections and subsequent biofilm formation.[209] Implantable devices using metallic silver or silver-coating technology include simple wires, pins and screws, fracture fixation plates and total joint prostheses for hips, knee, shoulders and ankles. These days, however, stainless steel, titanium and titanium alloys and cobalt-base alloys have largely replaced silver in orthopaedics; they are biocompatible, do not deteriorate with prolonged usage and provide rigidity of function. There is also increasing doubt whether silver or silver coating is beneficial in overcoming bacterial adhesion of biofilm formation.[210,211] Silver external fixation pins are still used in fracture clinics and new antibiotic bone cements are being developed.[212,214]

Pin tract infection is a frequent complication of external bone fixation and can present difficulties in 2–30% of operations. Clayton Parkhill (1897) was possibly the first surgeon to use silver-coated metals to fix fractures and is supposed to have used steel clamps heavily plated with silver to "secure the antiseptic action of the device".[215,216] Parkhill believed that the silver coating created an unfavourable environment on the surface of his fixation devices for bacterial adhesion and colonisation, which influenced the stability of the fixation. This concept has been more widely developed by others who have used silver plated screws, pins and clamps—all of which are prone to infection and biofilm formation with common pathogens such as *Staphylococcus aureus, Staphylococcus epidermidis, etc.*[217,218] In a prospective study, silver-coated screws were implanted in the lower limb of patients to fix femoral or tibial diaphyseal fractures.[212] Although there was a tendency for the silver coating to inhibit bacterial adhesion, the effect was not significantly different to that seen with commercially available stainless steel screws. Some silver eluted into the systemic circulation but toxic changes were not reported. Similar observations were made in experimental studies with stainless steel pins coated by the SPI-ARGENT silver ion beam assisted deposition (Spire Corp., Bedford, MA).[213] When implanted into sheep iliac crest and exposed to *Staphylococcus aureus*, 62% of the silver-coated pins were infected compared with 84% of uncoated stainless steel pins; however, silver coating technology was beneficial in aiding anchorage of the pins and decreasing the incidence of biofilm formation. The results of these *in vivo* studies have been confirmed in part by *in vitro* experiments where silver-coated pins were exposed to infections in tissue culture containing human serum proteins; however, the extent to which released Ag^+ bound to these serum proteins in this model is not known.[219] Where external fixator pins were coated with colloidal-silver or a polyurethane–argentum sleeve (PAS®; Orthomed, Lauteral) and inserted into sheep tibiae and exposed to *Staphylococcus aureus*, the silver treatment showed no significant effect in reducing infection within four weeks and colloidal silver was ineffective in reducing pin tract infections.[220]

Other experimental studies suggest that silver coating of metals as a general rule does not significantly inhibit biofilm formation,[221,222] but this may vary according to the type of coating used, the experimental model (*in vitro vs. in vivo*) and the extent to which the silver coating ionises. Stainless steel coated with a silver zeolite complex containing 2.5% Ag and 14% zinc (mildly antibacterial as Zn^{2+}) did reduce bacterial adhesion,[223,224] but a 2 mm thick titanium/silver coating by a physical vapour deposition process releasing 0.5–2 parts per billion (ppb) Ag^+ was efficacious against *Staphylococcus aureus* and *Klebsiella pneumoniae* without evidence of cytotoxicity in osteoblasts or epithelial cells.[225]

Bioglass® technology (Novabone, Florida) originally developed as a non-toxic biodegradable composite for bone repair and dental graft applications has been developed with inclusion of silver oxide (Ag_2O) with wide opportunities as an antibiotic tissue engineering scaffold.[147–152] With the exception of risks of silver contact sensitivity, these products have been found to be entirely biocompatible with mammalian tissues and show broad spectrum bacteriostatic and bactericidal action *in vitro* and *in vivo*.

5.4.2   Bone Cements

Bone cements based on polymethyl methacrylate (PMM) are regarded as the gold standard for anchoring artificial joints and bone prostheses but, like most biomaterials used in medical devices, they are prone to infection and biofilm formation.[226] Buchholz and Engelbrecht were probably the first to introduce antibiotics into bone cements and evaluated the suitability of gentamycin, vancomycin and fosfomycin in their preliminary *in vitro* experiments.[227] Whilst each exhibited commendable efficacy against common pathogens, they failed to gain acceptance on account of their inability to provide adequate protection in total joint arthroplasties, their toxicity and lack of biocompatibility with human tissue. The prevalence of gentamycin-resistant bacteria is well known.[228] Silver might have been discounted as a suitable antibiotic for bone cements on account of its toxicity to osteoblasts in culture and claims that it exhibited a time-dependant low biocompatibility with human tissues,[229] but silver-resistant bacteria are seemingly not common and exposure of common pathogens to silver does not favour emergence of silver-resistant strains.[138] Whereas Cortizo *et al.* claimed that $0.5\,mg\,mL^{-1}$ silver in tissue culture inhibited the growth of osteoblasts in culture and reduced mitotic indices,[229] these observations are not substantiated by *in vivo* studies where Ag-PMM intramuscular implants in rabbits were not associated with evidence of abnormal reactions and there were minimal foreign body responses.[230,231]

Spadaro and Becker have published much to establish the role of silver as an antimicrobial in medical devices and pioneered the use of electrical currents to generate Ag^{+} to promote antimicrobial action.[231–233] In their early experiments, they evaluated four salts and a chloride silver powder as an antibiotic in Simplex-P Radiopaque Bone Cement (Howmedica Inc.) and examined their antimicrobial action against common pathogens in bacterial inhibition plates.[231] The Ag-PMM complexes (1% w/w) were shown to be biocompatible with mammalian tissues in intramuscular implants in rabbits for up to 12 weeks. Antimicrobial action based on the duration of action against *Staphylococcus aureus* was greatest for silver sulfate, other silver compounds being ineffective after 1–8 d. Zones of bacterial inhibition increased with silver concentration (0.05–1.0%). More recent studies have confirmed the antimicrobial action of a high porosity and nanoparticulate 1.0% Ag-PMM in microplate cultures of *Pseudomonas aeruginosa, Staphylococcus aureus, Enterococcus faecium* and three strains of MRSA.[234,235]

Bone cement impregnated with metallic silver or other ionisable silver compounds was widely employed as a prophylactic against deep infections in total hip arthroplasty. A case is reported of a lady implanted with a Christiansen total hip arthroplasty who developed serious neurological problems after five years. Her prosthesis was loose and she showed high levels of silver in her blood and tissues.[236,237] The toxic implications of silver in this case are reported in Chapter 9 of this book, but it is noteworthy here that the silver released from the PMM cement containing 3.0% Ag was sufficient to protect against aerobic and anaerobic infections, and macrophages were extensively involved in the sequestration and metabolism of silver accumulating in soft tissues. The proportion of the silver content of the bone cement ionising in the five years since implantation of the prosthesis is not known, but we can speculate that with 1000-fold increases of serum silver much of the 3% contained in the cement had ionised.

5.5 Cardiovascular: Heart Valves and Stents

Advances in vascular surgery in the 20th century have seen the introduction of synthetic prosthetic material, heart valves, vascular grafts and sewing cuffs.[238]

Infections associated with medical prostheses can account for major morbidity and increased mortality. The benefits of including silver as a means of protecting against infections has been debated over many years, but despite the introduction of many new technologies, there is still ambiguity through lack of objective clinical appraisal and debate over cost–benefit. As in the catheters for central venous implantation, silver-containing devices should protect against device-related infections, be biocompatible with tissues at the implant site, and should not release plasticisers or other materials in the lifetime of the device.

A novel bioprosthetic heart valve incorporating a silver-coated polyester fabric as a sewing cuff was introduced as a means of protecting against prosthetic valve endocarditis and dystrophic mineralisation.[239] Thus stented porcine mitral valves coated with a silver technology (Silzone™, St Jude Epic™) with an ethanol anti-calcification treatment were evaluated clinically by implantation in a sheep model. Blood silver measured post-operatively was <40 ppm and of minimal toxicological consequence. Silver concentrations in liver and kidney were raised without toxic changes. Whilst the experimental implant studies predicted the clinical suitability of the device for human cardiac valve replacement, clinical studies in human patients failed to substantiate this view. One 72-year-old patient suffering from mitral valve disease suffered from acute mitral failure four months after surgery with partial detachment of the prosthetic valve, severe inflammatory changes and foreign body reactions,[240] whilst others have shown severe thromboembolic complications leading to withdrawal of the device.[241] It is unclear whether or to what extent silver was contributory to the lesions reported, but clearly much research has to be undertaken before silver-coated or other cardiac devices are approved as safe for human use.[239]

Other attempts at presenting mechanical heart valve implants using silver technology have shown promising biocompatibility with human cells in culture and efficacy against prosthetic valve infective endocarditis, but have failed so far to provide evidence for cost : efficacy benefits *in vivo*.[242–244] Thus terephthalate textiles have been silver-coated and shown to protect against infective endocarditis *in vitro* and in sheep implant models, and Dacron coated with silver awaits validation.[245]

5.6   Acupuncture Needles

Silver-coated and silver/gold needles have a long history in acupuncture and the ancient oriental tradition (Hari therapy) is still widely practiced today.[246,247] Needles used vary in size according to practitioners and are implanted in the skin for several days, but very little is published on the release of Ag^+ at sites of insertion except when local symptoms of argyria appear.[248,249] Infection does not seem to be a problem with implanted acupuncture needles, but electron microscopy has shown migration of Ag^+ and its deposition as fine brown granules about eccrine glands, blood vessels and nerve fibres in the papillary region of the dermis of argyric patients. Chrysiasis due to accumulation of dark deposits of gold complexes are additional complications of acupuncture, but the discolourations are of cosmetic significance rather than signs of toxicity.

5.7 Discussion

Silver is incorporated into a wide range of medical devices with the prime objective of overcoming infection. Technologies developed over the past 30 years provide a controlled and sustained release of Ag^+ over the expected lifetime of devices, with minimisation of local responses or adverse reactions in patients. Illustrations are provided to demonstrate that silver as a coating or impregnate is entirely compatible with human tissues and that, on the rare occasions where complications have arisen, they have been attributable to materials other than silver present in the devices. Fundamental defects in design of devices or in their placement in the human body do occur as in the case of the silver-treated sewing cuff of the mechanical heart valve that was withdrawn on safety grounds.

The wide and varied range of sustained silver-release wound dressings represents a success story. The tissue viability nurse and wound care manager have a choice of well-established dressings tailored as a chemoprophylaxis for acute and chronic wounds and burns where nosocomial and opportunist pathogens can be life-threatening. Silver-release patterns and non-silver components of dressings have been specifically tailored according to the chronicity of the lesions and severity of infection as determined by laboratory tests, but in each case design features are such that they comply closely with the current state of knowledge and opinion on wound bed preparation.

Development of silver-containing medical devices has mostly followed the logical sequence of biomaterials research and the biotechnology of silver and silver alloys, followed by preclinical research in animal models, and clinical trials in representative panels of patients. Hydrogel technology as applied to polyurethane, silastic and other polymers used in in-dwelling catheters for central venous and intraurethral implantation, ion beam assisted silver deposition, enhancement of silver ionisation using electric currents; silver Bioglass® technology in antibiotic sutures and bone cements are examples of recent innovations in laboratory research with clear clinical implications. In each case, manufacturers have appreciated the importance of:

- the chemical source of Ag^+, its rate of ionisation and environmental and clinical factors affecting it (constitution of body fluids, temperature, pH, *etc.*);
- silver binding substances in the medium;
- presentation of the silver-bearing products to the human body (wound, implant, *etc.*) and the clinical condition of epithelia and endothelia, and the expected state of health of patients;
- compatibility of silver and Ag^+ with other chemical constituents of the human body, nutrients and key trace metals such as zinc, copper and calcium.

Experience has shown that, whilst there is substantive evidence that silver can control infections and reduce bacterial adhesions and colonisation in skin wounds, catheters, external fixation pins, bone cements, *etc.* and retard biofilm formation, Ag^+ release rarely leads to a "germfree state". In part, this may be due to the existence of silver-resistant micro-organisms, but more importantly it is expected to be attributable to insufficiency of "free" Ag^+ for antimicrobial action. The ion is highly active in the presence of inorganic and organic moieties, cell surface receptors and tissue debris, and precipitation or irreversible binding inevitable reduces the amount available for biocidal action. As Burrell pointed out, 40–60 ppm Ag^+ might be an ideal objective for controlling pathogenic infections but prediction of the amount of free ion available in any clinical situation is extremely difficult.[89] Rarely, the situation of "too much silver" does arise, but whereas patients may develop symptoms of argyria on rare occasions, silver toxicity *per se* is not seen.

Studies in animal models can provide important safety data for regulatory purposes but there are widespread and tenable arguments regarding their suitability as a surrogate models for the human body in the evaluation of silver in medical devices. Ultimate success or failure of any silver-containing device rests entirely on the outcome of multicentre, random controlled and statistically valid meta-analyses in human patients.

5.8   Further Reading

W. S. Halstead, Ligature and suture material; the employment of fine silk in preference to catgut and the advantages of transfixion of tissues and vessels in control of haemorrhage—also an account of the introduction of gloves, guttapercha tissue and silver foil, *J. Am. Med. Assoc.*, 1913, **60**, 1119.

References

1. A. T. Wan, A. J. Conyers, C. J. Coombs and J. P. Masterton, Determination of silver in blood, urine and tissues of volunteers and burn patients, *Clin. Chem.*, 1991, **37**, 1683.
2. C. J. Coombs, A. T. Wan, J. P. Masterton, R. A. J. Conyers, J. Pedersen and Y. T. Chia, Do burn patients have a silver lining?, *Burns*, 1992, **18**, 179.
3. A. B. G. Lansdown, B. Sampson, P. Laupattarakasem and A. Vuttivirijana, Silver aids healing in the sterile wound: experimental studies in the laboratory rat, *Br. J. Dermatol.*, 1997, **137**, 728.
4. A. B. G. Lansdown, B. Sampson and A. Rowe, Sequential changes in trace metal, metallothionein and calmodulin concentrations in healing skin wounds, *J. Anat.*, 1999, **195**, 375.
5. A. B. G. Lansdown, Silver 2: toxicity in mammals and how its products aid wound repair, *J. Wound Care*, 2002, **11**, 173.
6. W. Zheng, Toxicology of choroid plexus: species reference to metal-induced neurotoxicities, *Microsc. Res. Tech.*, 2001, **52**, 89.
7. A. B. G. Lansdown, Silver in healthcare: antimicrobial effects and safety in use, *Curr. Probl. Dermatol.*, 2006, **33**, 17.

8. A. B. G. Lansdown and A. Williams, Bacterial resistance to silver in wound care, *J. Wound Care*, 2007, **16**, 15.
9. G. S. Schultz, R. G. Sibbald, V. Falanga, E. A. Ayello, C. Dowsett, K. Harding, M. Romanelli, M. C. Stacey, L. Teot and W. Vanscheidt, Wound bed preparation a systematic approach to wound management, *Wound Rep. Reg.*, 2003, **11**, 1.
10. G. S. Scultz, D. J. Barillo, D. W. Mozingo and G. A. Chin, Wound bed preparation and a brief history of TIME, *Int. Wound J.*, 2004, **1**, 19.
11. V. Falanga, The chronic wound, failure to heal, in *Cutaneous Wound Healing*, eds. V. Falanga, Martin Dunitz, London, 2001, pp. 155–164.
12. V. Falanga, Introducing the concept of wound bed preparation, in, *An International Forum on Wound Care*, 2001, **16**, pp. 1–4.
13. S. Enoch and K. G. Harding, Wound bed preparation: the science behind the removal of barriers to healing, *Wounds*, 2003, **15**, 213.
14. J. Forrai, History, Ambrose Paré: the father of surgery, *Clin. Pesq. Odontol. Curitiba*, 2006, **2**, 447.
15. M. Shah, Premier Chirurgien du Roi: the life of Ambrose Paré (1510–1590), *J. R. Soc. Med.*, 1992, **85**, 292.
16. W. S. Halsted, Ligature and suture material; also an account of introduction of rubber gloves, gutta percha tissue and silver foil, *J. Am. Med. Assoc.*, 1913, **55**, 1119.
17. W. G. MacCallum, *William Stewart Halstead, Surgeon*, Johns Hopkins Press, Baltimore, MD, 1930.
18. J. Cameron, William Stewart Halsted: our surgical heritage, *Ann. Surg.*, 1997, **225**, 445.
19. T. Sollemann, Silver, in *A Manual of Pharmacology and its Applications to Therapeutics and Toxicology*, Saunders, Philadelphia, 1942, pp. 1102–1109.
20. B. C. Davis, Principles of sterilization, in *Bacterial and Mycotic Infections of Man*, ed. R. J. Dubois, Lippincott, Philadelphia, 1952, pp. 707–735.
21. K. W. von Nägeli, Leben die oligodynamischen Erscheinungen an lebenden Zellen, *Denkschr. Schweiz. Naturforsch. Ges.*, 1893, **33**, 174.
22. A. J. Clark, General pharmacology, in *Handbuch der experimentellen Pharmakologie*, ed. A. Heffter, Springer, Berlin, 1937, Vol. 4, pp. 165–176.
23. E. J. L. Lowbury, Problems of resistance in open wounds and burns, in *The Rational Choice of Antibacterial Agents*, eds. R. P. Mouton, W. Brumfitt and J. M. T. Hamilton-Miller, Kluwer, London, 1977, pp. 18–31.
24. P. Ehrlich, *Paul Ehrlich, The Nobel Prize in Physiology or Medicine 1908*, Copyright The Nobel Foundation, 1908.
25. L. E. Gaul and A. H. Staud, Clinical spectroscopy. Seventy cases of generalized argyrosis following organic and colloidal silver medication, *J. Am. Med. Assoc.*, 1935, **104**, 1387.
26. Anon, Silver arsphenamine, *Cal. State J. Med.*, 1921, **19**, 304.
27. G. Hazlewood, 'The Surgeon's Mate', 1617. John Woodall: from barber surgeon to surgeon-general, in *The Proceedings of the 12th Annual History*

of Medicine Days, ed. W. A. Whitelaw, Faculty of Medicine, University of Calgary, Alberta, 2003, pp. 117–137.

28. British Medical Association and Royal Pharmaceutical Society of Great Britain, *British National Formulary*, BMJ Group and RPS Publishing, London, Vol. **58**, 2009.

29. W. Lubinsku, Silbernitrat oder Silbereiweiss, *Berl. Klin. Wochenschr.*, 1914, **51**, 1643.

30. A. B. G. Lansdown, Metallothioneins: potential therapeutic aids for wound healing in the skin, *Wound Rep. Regen.*, 2002, **10**, 130.

31. J. D. Pilcher and T. Sollmann, Organic, protein and colloidal silver compounds: their antiseptic efficiency and silver-ion content as a basis for their classification, *J. Lab. Clin. Med.*, 1923, 301.

32. H. Schlee, Excess of sodium chloride in complex silver preparations with special reference to their medicinal use, *Biochem. Z.*, 1924, **148**, 383.

33. M. Kusonski, On the bactericidal action of the hydrosol of silver, *J. Biochem*, 1922, **1**, 381.

34. K. von Neergaard, Determination of silver in molecular solution in presence of colloidal silver, *Arch. Exp. Path. Pharm.*1923, **100**, 162.

35. N. S. Tomi, B. Kränke and W. Aberer, A silver man, *Lancet*, 2004, **363**, 532.

36. S. H. Gulbranson, J. A. Hud and R. C. Hansen, Argyria following use of dietary supplements containing colloidal silver protein, *Cutis*, 2000, **66**, 3737.

37. Department of Health and Human Services, US Food and Drug Administration, Over-the-counter drug products containing colloidal silver ingredients or silver salts, *Fed. Regist.*, 1996, **61**, No. 200 (October 15, 1996), 53685.

38. C. A. Moyer, L. Brentano, D. L. Gravens, H. W. Margraf and W. W. Monafo, Treatment of large human burns with 0.5% silver nitrate solution, *Arch. Surg.*, 1965, **90**, 799.

39. E. J. L. Lowbury, K. Bridges and D. M. Jackson, Topical prophylaxis with silver sulphadiazine and silver nitrate–chlorhexidine creams: emergence of sulphonamide-resistant Gram-negative bacilli, *Br. Med. J.*, 1976, **I**, 493.

40. C. L. Fox, Silver sulphadiazine—a new topical therapy for *Pseudomonas aeruginosa* in burns, *Arch. Surg.*, 1968, **96**, 184.

41. H. S. Carr, T. J. Wlodkowski and H. S. Rosenkranz, Silver sulphadiazine: *in vitro* antibacterial activity, *Antimicrob. Agents Chemother.*, 1973, **4**, 585.

42. C. T. Dollery, Silver sulphadiazine, in *Therapeutic Drugs*, Churchill Livingstone, Edinburgh, 1991, Vol. **2**.

43. D. S. Cook and M. F. Turner, Crystal and molecular structure of silver sulphadiazine (N^1-pyrimidin–ylsulphanilamide), *J. Chem. Soc., Perkin Trans. 2*, 1975, 1021.

44. C. P. Sawhney, R. K. Sharma, K. R. Rao and R. Kaushish, Long-term experience with 1 per cent silver sulphadiazine cream in the management of burn wounds, *Burns*, 1989, **15**, 403.

45. W. W. Monafo and M. West, Current treatments recommendations for topical burn therapy, *Drugs*, 1990, **40**, 364.
46. R. J. Inman, C. F. T. Snelling, F. J. Roberts, K. Shaw and J. C. Boyle, Prospective comparison of silver sulphadiazine 1% plus chlorhexidine digluconate 0.2% (Silvazine) and silver sulphadiazine 1% (Flamazine) as prophylaxis against burn wound infection, *Burns*, 1984, **11**, 35.
47. C. F. T. Snelling, A. R. Ronald, W. R. Waters, D. S. Yaworski, K. Drulak and M. Sunderland, Comparison of silver sulphadiazine and gentamycin for topical prophylaxis against burn wound infections, *Can. Med. Assoc. J.*, 1978, **119**, 466.
48. H. S. Stern, Silver sulphadiazine and healing of partial thickness burns: a prospective trial, *Br. J. Plast. Surg.*, 1989, **42**, 581.
49. J. F. Hansbrough, B. Achauer, J. Dawson, H. Himel, A. Luterman, H. Slater, S. Levenson, A. Salzberg, W. B. Hansbrough and C. Doré, Wound healing in partial-thickness burn wounds treated with collagenase ointment *versus* silver sulphadiazine cream, *J. Burn Care Rehabil.*, 1995, **16**, 241.
50. J. B. Bishop, L. G. Phillips, T. A. Mustoe, A. J. VanderZee, L. Weissema, D. E. Roach, J. P. Heggers, D. P. Hill, E. L. Taylor and M. C. Robson, A prospective randomised evaluator-blinded trial of two potential wound healing agents for the treatment of venous stasis ulcers, *J. Vasc. Surg.*, 1992, **16**, 251.
51. A. B. G. Lansdown, A. Williams, S. Chandler and S. Benfield, Silver absorption and antibacterial efficacy of silver dressings, *J. Wound Care*, 2005, **14**, 131.
52. K. Kawai, S. Suzuki, Y. Tabata, T. Taira, Y. Ikada and Y. Nishimura, Development of an artificial dermis preparation capable of silver sulphadiazine release, *J. Biomed. Mater. Res.*, 2001, **57**, 346.
53. Y. Kuroyangi, E. Kim and N. Shioya, Evaluation of a synthetic wound dressing capable of releasing silver sulphadiazine, *J. Burn Care Rehabil.*, 1991, **12**, 106.
54. L. Miller, J. Hansbrough, H. Goldfarb, P. Kealey, J. Saffle, M. Kravitz and P. Silverstein, Sildmac: a new delivery system for silver sulphadiazine in the treatment of full-thickness burn injuries, *J. Burn Care Rehabil.*, 1990, **11**, 35.
55. C. H. Fang, P. Nathan, E. C. Robb, J. W. Alexander and B. G. MacMillan, Prospective clinical trial of Hydron, a synthetic dressing, in delivery of antimicrobial drug to second-degree burns, *J. Burn Care Rehabil.*, 1987, **8**, 206.
56. C. W. Cruse and S. Daniels, Minor burns: treatment using a new drug delivery system with silver C.I. sulphadiazine, *South. Med. J.*, 1989, **82**, 1135.
57. Price, J. W. Horton and C. R. Baxter, Topical liposome delivery of antibiotics in soft tissue infection, *J. Surg. Res.*, 1990, **49**, 174.
58. R. Naeff, Feasibility of topical liposome drugs produced on an industrial scale, *Adv. Drug Deliv. Rev.*, 1996, **18**, 343.

59. M. Schaller, H. C. Korting and M. H. Schmidt, Interaction of cultured human keratinocytes with liposomes encapsulating silver sulphadiazine: proof or the uptake of intact vesicles, *Br. J. Dermatol.*, 1996, **134**, 445.
60. C. L. Fox, T. N. Rao, R. Azmeth, S. S. Gandhi and S. D. Modak, Comparative evaluation of zinc sulphadiazine and silver sulphadiazine in treating burn wound infection, *J. Burn Care Rehabil.*, 1990, **11**, 112.
61. D. Kjolseth, J. M. Frank, J. H. Barker, G. L. Anderson, A. I. Rosenthal, R. D. Ackland, D. Schluschke, F. R. Campbell, G. R. Tobin and L. J. Weiner, Comparison of the effects of commonly used wound agents on epithelialization and neovascularisation, *J. Am. Coll. Surg.*, 1994, **179**, 305.
62. G. D. Winter, Formation of the scab and the rate of epithelization of superficial wounds in the skin of the young domestic pig, *Nature (London)*, 1962, **193**, 293.
63. G. D. Winter and J. T Scales, Effect of air drying and dressings on the surface of a wound, *Nature (London)*, 1963, **197**, 91.
64. G. D. Winter, Movement of epidermal cells over the wound surface, in *Advances in Biology of the Skin*, eds. W. Montagna and R. E. Billingham, Pergammon Press, Oxford, 1964, Vol. V, pp. 113–117.
65. M. J. Hoekstra, P. Hupkens, R. P. Dutrieux, M. M. C. Bosch, T. A. Brans and R. W. Kreis, Comparative burn wound model in the New Hampshire pig for the histopathological evaluation of local therapeutic regimens: silver sulphadiazine cream as a standard, *Br. J. Plast. Surg.*, 1993, **46**, 585.
66. L. Berra, T. Kolobow, P. Laquerriere, B. Pitts, S. Bratami, J. Pohlmann, C. Marelli, P. Brambilasca, F. Villa, A. Baccarelli, S. Bouthors, H. Stelfox, L. Bigatello, J. Moss and A. Presenti, Internally-coated endotrachial tubes with silver sulphadiazine in polyurethane to prevent bacterial colonisation: a clinical trial, *Intensive Care Med.*, 2008, **34**, 1030.
67. E. M. Mintz, D. E. George and S. Hsu, Silver sulphadiazine therapy in widespread bullous disorders: potential toxicity, *Dermatol. Online J.*, 2008, **14**, 19.
68. J. I. Greenfield, L. Sampath, S. J. Popilskis, S. R. Brunnert, S. Stylianos and S. Modak, Decreased bacterial adherence and biofilm formation on chlorhexidine and silver sulphadiazine impregnated catheters implanted in swine, *Crit. Care Med.*, 1995, **23**, 894.
69. C. L. Fox, W. W. Monafo, V. H. Ayvazian, A. M. Skinner, S. Modak and J. C. Stanford, Topical chemotherapy for burns using cerium salts and silver sulphadiazine, *Surg. Gynecol. Obstet.*, 1977, **144**, 668.
70. C. L. Fox, S. Modak, J. W. Stanford and P. L. Fox, Metal sulphonamides as antibacterial agents in topical therapy, *Scand. J. Plast. Reconstruct. Surg.*, 1979, **13**, 89.
71. W. W. Monafo and V. H. Ayvazian, Topical therapy, *Surg. Clin. N. Am.*, 1978, **58**, 1157.
72. J. L. Ninnemann, A. N. Ozkan and J. J. Sullivan, Hemolysis and suppression of neutrophil chemotaxis by a low molecular weight component of human burn patient sera, *Immunol. Lett.*, 1994, **42**, 211.

73. J. A. Clarke, Burns: a review, *Br. Med. Bull.*, 1999, **55**, 885.
74. B. G. Sparkes, J. W. Gyorkos and R. M. Gorczynski, Comparison of the endotoxins and cutaneous burn toxins as immuno-suppressants, *Burns*, 1990, **16**, 123.
75. G. A. Schoenberger, Burn toxins isolated from mouse and human skin, *Monogr. Allergy*, 1975, **9**, 72.
76. B. Kremer, M. Allgower and M. Graf, The present status of research on burn toxins, *Intensive Care Med.*, 1981, **7**, 77.
77. B. G. Sparkes, Mechanisms of immune failure in burn injury, *Vaccine*, 1993, **11**, 504.
78. V. M. Petersen, J. F. Hansbrough, X. W. Wang, R. Zapata-Sirvant and J. A. Boswick, Topical cerium nitrate prevents post-burn immuno-suppression, *J. Trauma*, 1985, **25**, 1039.
79. R. L. Zapata-Sirvant and J. F. Hansbrough, Post-burn immuno-suppression in an animal model: III Maintenance of splenic helper and suppressor lymphocyte populations by immuno-modulating drugs, *Surgery*, 1985, **97**, 721.
80. W. Boeckx, N. Blondeel, K. Vandersteen, C. Wolf-Peeters and A. Schmitz, Effect of cerium nitrate sulphadiazine on deep dermal burns: a histological hypothesis, *Burns*, 1992, **18**, 456.
81. J. P. Garner and P. S. Heppell, The use of Flammacerium in British burns units, *Burns*, 2005, 379.
82. Scheidegger, B. G. Sparkes, N. Luscher, G. A. Schoenberger and M. Allgower, Survival in major burn injuries treated with bathing in cerium nitrate, *Burns*, 1992, **18**, 296.
83. A. B. G. Lansdown, S. R. Myers, J. A. Clarke and P. O'Sullivan, A reappraisal of the role of cerium in burn wound management, *J. Wound Care*, 2003, **12**, 113.
84. C. G. de Gracia, An open study comparing topical silver sulphadiazine and topical silver sulphadiazine-cerium nitrate in the treatment of moderate and severe burns, *Burns*, 2001, **27**, 67.
85. *Wound Care Handbook 2009–2010: The Comprehensive Guide to Product Selection*, MA Healthcare, London, 2009, www.woundcarehandbook.com.
86. J. J. Castellano, S. M. Shafii, F. Ko, G. Donate, T. E. Wright, R. J. Mannari, W. G. Payne and D. J. Smith, Comparative evaluation of silver-containing antimicrobial dressings and drugs, *Int. Wound J.*, 2007, **4**, 114.
87. R. J. White and K. F. Cutting, Exploring the effects of silver in wound management: what is optimal?, *Wounds UK*, 2006, **18**, 307.
88. S. Thomas, B. Fisher, P. J. Fram and M. J. Waring, Odour absorbing dressings, *J. Wound Care*, 1998, **7**, 246.
89. R. E. Burrell, The scientific perspective on the use of topical silver preparations, *Ostomy Wound Manage.*, 2003, **49**, 19.
90. J. M. Hamilton-Miller, S. Shah and C. Smith, Silver sulphadiazine: a comprehensive *in vitro* reassessment, *Chemotherapy*, 1993, **39**, 405.
91. K. F. Cutting and K. G. Harding, Criteria for identifying wound infection, *J. Wound Care*, 1994, **3**, 198.

92. S. E. Gardner, R. A. Franz and B. N. Doebbeling, The validity of the clinical signs and symptoms used to identify localised chronic wound infection, *Wound Rep. Regen.*, 2001, **9**, 178.

93. A. Kingsley, The wound infection continuum and its application to clinical practice, *Ostomy Wound Manage.*, 2003, **49** (7A), Suppl. 1–7.

94. J. R. Mekkes, M. A. Loots, A. C. van der Wal and J. H. D. Bos, Causes, investigation and treatment of leg ulceration, *Br. J. Dermatol.*, 2003, **148**, 388.

95. R. J. White, The wound infection continuum, *Br. J. Nurs.*, 2002, **11** (2 Suppl), 7.

96. D. Gray, R. J. White, A. Kingsley and P. Cooper, Using the wound infection continuum to assess wound bioburden, *Wounds UK*, 2005, **1**(2 Suppl), S15.

97. A. B. G. Lansdown, The role of the pathologist in wound management, *Br. J. Nurs.*, 2007, **16**, S24.

98. J. B. Wright, D. L. Hansen and R. E. Burrell, The comparative efficacy of two antimicrobial barrier dressings: *in vitro* examination of two controlled release of silver dressings, *Wounds*, 1998, **10**, 179.

99. G. C. Troy, An overview of hemostasis, *Vet. Clin. North Am. Small Anim. Pract.*, 1988, **18**, 5.

100. K. Ballard and F. McGregor, Avance: silver hydropolymer dressing for critically colonised wounds, *Br. J. Nurs.*, 2002, **11**, 206.

101. R. G. Sibbald, P. Chapman and J. Contreras-Ruiz, The role of bacteria in pressure ulcers, in *Science and Practice of Pressure Ulcer Management*, ed. M. Romanelli, Springer, London, 2006, pp. 139–163.

102. L. Ovington, Nanocrystalline silver: where the old and familiar meets a new frontier, *Wounds*, 2001, **13** (Suppl. B), 5.

103. T. Morgan, C. Evans and K. G. Harding, A study to measure patient comfort and acceptance of Avance™, a new polyurethane foam dressing containing silver as an antimicrobial agent when used to treat chronic ulcers, presented at the 11th European Wound Management Association conference, Dublin, 2001.

104. A. B. G. Lansdown, Experimental evaluation of a new silver-containing antimicrobial dressing ARGLAES™ healing of skin wounds, presented at the European Pressure Ulcer Association Meeting, Amsterdam, 2000.

105. M. C. Robson, Wound infection, a failure of wound healing through an imbalance of bacteria, *Surg. Clin. North Am.*, 1997, **77**, 537.

106. A. Bellingeri and D. Hoffman, Debridement of pressure ulcers, in *The Science and Practice of Pressure Ulcer Management*, ed. M. Romanelli, Springer, London, 2006, pp. 129–137.

107. L. Ovington, Bacterial toxins and wound healing, *Ostomy Wound Manage.*, 2003, **49** (7A Suppl), 8.

108. R. H. Demling and M. D. L. De Santi, The rate of re-epithelialization across meshed skin grafts is increased with exposure to silver, *Burns*, 2002, **28**, 264.

109. C. Dowsett, An overview of Acticoat dressing in wound management, *Br. J. Nurs.*, 2003, **12**, S44.
110. E. E. Tredgett, H. A. Shankowski, A. Groenveld and R. E. Burrell, A matched-pair, randomised study evaluating the efficacy and safety of Acticoat silver-coated dressing for the treatment of burn wounds, *J. Burn Care Rehabil.*, 1998, **19**, 531.
111. D. W. Voigt and C. N. Paul, The role of silver in wound healing: Pt. 2: Why is nanocrystalline silver superior? The use of Acticoat as silver-impregnated telfa dressings in regional burn and wound care center, *Wounds*, 2001, **13**, 11.
112. J. B. Wright, K. Lam, A. G. Buret, M. E. Olsen and R. E. Burrell, Early healing events in a porcine model of contaminated wounds: effects of nanocrystalline silver on matrix metalloproteinases, cell apoptosis and healing, *Wound Rep. Regen.*, 2002, **10**, 141.
113. R. G. Sibbald, P. Coutts, A. Browne and S. Coehlo, Acticoat®, a new ionized silver-coated dressing: its effect on bacterial load and healing rates, presented at Wound Healing Society Educational Committee conference, Toronto, 2000.
114. R. G. Sibbald, J. Contreras-Ruiz, P. Coutts, M. Fierheller, A. Rothman and K. Woo, Bacteriology, inflammation, and healing: a study of nano-crystalline silver dressings in chronic venous leg ulcers, *Adv. Skin Wound Care*, 2007, **20**, 549.
115. E. Vlachau, E. Chipp, E. Shale, Y. T. Wilson, R. Papini and B. N. S. Moiemen, The safety of nanocrystalline silver dressings on burns: a study of systemic silver absorption, *Burns*, 2007, **33**, 979.
116. R. J. White, A charcoal dressing with silver infection: clinical evidence, *Br. J. Nurs.*, 2001, The Silver Suppl., Part 2, 408.
117. E. Scanlon and C. Dowsett, Clinical governance in control of wound infection and odour, *Br. J. Nurs.*, 2001, The Silver Suppl. Part 2, 9.
118. G. Müller, Y. Winkler and A. Kramer, Antibacterial activity and endotoxin-binding capacity of Actisorb® Silver 220, *J. Hosp. Infect.*, 2003, **53**, 211.
119. J. Furr, A. D. Russell, T. D. Turner and A. Andrews, Antibacterial activity of Actisorb Plus, Actisorb and silver nitrate, *J. Hosp. Infect.*, 1994, **27**, 201.
120. Z. Metzger, D. Nitzan, S. Pitaru, T. Brosh and S. Teicher, The effect of bacterial toxins on the early tensile strength of healing surgical wounds, *J. Endod.*, 2002, **28**, 30.
121. A. B. G. Lansdown, A guide to the properties and uses of silver dressings in wound care, *Prof. Nurse*, 2005, **20**, 41.
122. S. D. Blair, C. M. Backhouse, R. Harper, J. Matthews and C. N. McCollum, Comparison of absorbable materials for surgical haemostasis, *Br. J. Surg.*, 1988, **75**, 969.
123. R. H. Joy and J. R. Murray, A new calcium alginate haemostatic dressing, *Dent. Pract.*, 1989, **27**, 25.
124. K. S. Sirimanna, Calcium alginate (Kaltostat 2g) for nasal packing after trimming turbinates—a pilot study, *J. Laryngol. Otol.*, 1989, **103**, 1067.

125. S. A. Kelley, M. G. Dickson and D. T. Sharpe, Calcium alginates as a temporary recipient bed dressing prior to the delayed application of split skin grafts, *Br. J. Plast. Surg.*, 1988, **41**, 445.
126. H. Hinchley and J. R. Murray, Calcium alginate dressings in community nursing, *Pract. Nurse*, 1989, **2**, 264.
127. A. B. G. Lansdown and M. J. Payne, An evaluation of the local reaction and biodegradation of calcium sodium alginate (Kaltostat) following subcutaneous implantation in the rat, *J. R. Coll. Surg. Edinb.*, 1994, **39**, 284.
128. P. G. Bowler, S. A. Jones, M. Walker and D. Parsons, Microbicidal properties of a silver-containing hydrofiber dressing against a variety of burn wound pathogens, *J. Burn Care Rehabil.*, 2004, **2**, 192.
129. K. G. Harding, P. Price, B. Robinson, S. Thomas and D. Hoffman, Cost and dressing evaluation of hydrofiber and alginate dressings in managing community-based patients with chronic leg ulceration, *Wounds*, 2001, **13**, 229.
130. E. B. Jude, J. Appelqvist, M. Spraul and J. Martini, Prospective randomised controlled study of hydrofiber dressing containing ionic silver or calcium alginate dressings on non-ischaemic diabetic foot ulcers, *Diabet. Med.*, 200, **24**, 280 (Silver Dressing Study Group).
131. L. Teot, G. Maggio and S. Barrett, The management of wounds using Silvercel hydroalginate, *Wounds UK*, 2005, **1**, 70.
132. S. Maume and D. Vallet, Evaluation of a silver-release hydroalginate dressing in chronic wounds with signs of local infection, *J. Wound Care*, 2005, **14**, 411.
133. I. Lazareth, Z. Ourabah, P. Senet, H. Carier, A. Sauvadet and S. Ohbot, Evaluation of a new silver foam dressing in patients with critically colonised leg ulcers, *J. Wound Care*, 2007, **16**, 129.
134. H. Carsin, D. Wasermann, M. Pannier, R. Dumas and S. Bohbot, A silver sulphadiazine-impregnated lipocolloid wound dressing to treat second-degree burns, *J. Wound Care*, 2004, **13**, 145.
135. A. Letouze, V. Voinchet, B. Hoecht, K. C. Muenter, F. Vives and S. Bohbot, Using a new lipocolloids dressing in paediatric wounds; results of French and German clinical studies, *J. Wound Care*, 2004, **13**, 221.
136. F. X. Bernard, C. Barrault, F. Juchaux, C. Laurensou and L. Alpert, Stimulation of the proliferation of human dermal fibroblasts *in vitro* by a lipocolloid dressing, *J. Wound Care*, 2005, **14**, 215.
137. D. J. Leaper, Silver dressings: their role in wound management, *Int. Wound J.*, 2006, **18**, 307.
138. A. B. G. Lansdown and A. Williams, Bacterial resistance to silver in wound care and medical devices, *J. Wound Care*, 2007, **16**, 15.
139. M. Collier, Silver dressings: more evidence is needed to support their widespread clinical use, *J. Wound Care*, 2009, **18**, 77.
140. J. M. Sims, On the treatment of vesico-vaginal fistula, *Am. J. Med. Sci.*, 1852, **2**, 677.
141. J. M. Sims, *Silver Sutures in Surgery*, S. S. and S. W. Wood, New York, 1858.

142. R. Ottenberg, The Treatment of hemolytic Streptococcus infections and the newer applications of sulphanilamide, *Bull. N. Y. Acad. Med.*, 1938, **14**, 453.
143. M. H. Bass, In memoriam, Reuben Ottenberg, 1882–1959, *J. Mt. Sinai Hosp. N. Y.*, 1959, **26**, 421.
144. A. H. Bashir, Wound closure by skin traction: an application of tissue expansion, *Br. J. Plast. Surg.*, 1987, **40**, 582.
145. W. L. Old and T. L. Stokes, Preventing disruption of abdominal wounds, *South. Med. J.*, 1979, **72**, 545.
146. W. C. Tsai, C. C. Chu, S. S. Chiu and J. Y. Yao, *In vitro* quantitative study of newly made antibacterial braided nylon sutures, *Surg. Gynecol. Obstet.*, 1987, **165**, 207.
147. M. Bellantone, H. D. Williams and L. L. Hench, Broad spectrum bactericidal activity of Ag_2O-doped bioactive glass, *Antimicrob. Agents Chemother.*, 2002, **46**, 1940.
148. J. Pratten, S. N. Nazhat, J. J. Blaker and A. R. Boccaccini, *In vitro* attachment of Staphylococcus epidermidis to surgical sutures with and without Ag-containing bioactive glass coating, *J. Biomater. Appl.*, 2004, **19**, 47.
149. J. J. Blaker, A. R. Boccaccini and S. N. Nazhat, Thermal characterisation of silver-containing bioactive glass-coated sutures, *J. Biomater. Appl.*, 2005, **20**, 81.
150. J. J. Blaker, S. N. Nazhat and A. R. Boccaccini, Development and characterisation of silver-doped bioactive glass-coated sutures for tissue engineering and wound healing applications, *Biomaterials*, 2004, **25**, 1319.
151. M. Wang, L. L. Hench and W. Bonfield, Bioglass/high density polyethylene composite for soft tissue applications: preparation and evaluation, *J. Biomed. Mater. Res.*, 1998, **42**, 577.
152. J. J. Blaker, S. N. Nazhat and A. R. Boccaccini, Development and characterisation of silver-doped bioactive glass-coated sutures for tissue engineering and wound applications, *Biomaterials*, 2004, **25**, 1319.
153. A. B. G. Lansdown, J. J. Blaker, I. Thompson, P. Pavan, A. R. Boccaccini and L. L. Hench, Bioglass® and silver-Bioglass®—a novel approach using bioactive scaffolds and silver doped sutures in acute skin wounds, presented at 13th Annual Meeting of the European Tissue Repair Society, Amsterdam, 2003, Abstract P.063.
154. C. C. Chu, W. C. Tsai, J. Y. Yao and S. S. Chu, Newly made antibacterial braided nylon sutures, I. *In vitro* quantitative and *in vivo* preliminary biocompatibility study, *J. Biomed. Mater. Res.*, 1987, **21**, 1281.
155. T. Bechert, P. Steinrücke and J.-P. Guggenbicher, A new method for screening anti-infective biomaterials, *Nature, Med.*, 2000, **6**, 1053.
156. E. J. Tobin and R. Bambauer, Silver coating of dialysis catheters to reduce bacterial colonisation and infection, *Ther. Apher. Dial.*, 2003, **7**, 504.
157. S. Saint, R. H. Savel and M. A. Matthay, Enhancing the safety of critically ill patients by reducing urinary and central venous catheter-related infections, *Am. J. Respir. Crit. Care Med.*, 2002, **165**, 1475.

158. D. G. Maki, Nosocomial bacteraemia, an epidemiological review, *Am. J. Med.*, 1981, **70**, 719.
159. J. C. Goddard, Genitourinary medicine and surgery in Nelson's navy, *Postgrad. Med. J.*, 2005, **81**, 413.
160. D. Roe, B. Karandikar, N. Bonn-Savage, B. Gibbins and J.-B. Roullet, Antimicrobial surface functionalization of plastic catheters by silver nanoparticles, *J. Antimicrob. Chemother.*, 2008, **61**, 869.
161. T. Elliott, Role of antimicrobial central venous catheters for the prevention of associated infections, *J. Antimicrob. Chemother.*, 1999, **43**, 441.
162. D. G. Maki, Risk factors for nosocomial infection in intensive care, "Devices vs. nature" and goals for the next decade, *Arch. Intern. Med.*, 1989, **149**, 30.
163. J. M. Schierholz, C. Fleck, J. Beuth and G. Pulverer, The antimicrobial efficacy of a new central venous catheter with long term broad spectrum activity, *J. Antimicrob. Chemother.*, 2000, **46**, 45.
164. C. Von Eiff, N. Lindner, R. A. Proctor, W. Winkelmann and G. Peters, Development of gentamycin-resistant small colony variants of *S. aureus* after implantation of gentamycin chains in osteomyelitis as a possible cause of recurrence, *Z. Orthop. Ihre Grenzgeb.*, 1998, **136**, 268.
165. J.-P. Guggenbichler, M. Böswald, S. Lugauer and T. Krall, A new technology of micro-dispersed silver in polyurethane induces antimicrobial activity in central venous catheters, *Infection*, 1999, **27** (Suppl.1), S16.
166. J. J. Bong, P. Kite, M. H. Wilco and M. J. McMahon, Prevention of catheter-related bloodstream infection by silver iontophoretic central venous catheters: a randomised controlled trial, *Clin. Pathol.*, 2003, **56**, 731.
167. J. H. Crabtree, R. J. Burchette, R. A. Siddiqi, I. T. Huen, L. L. Hadnott and A. Fishman, The efficacy of silver-ion implanted catheters in reducing peritoneal dialysis-related infections, *Perit. Dial. Int.*, 2003, **23**, 368.
168. D. G. Maki and P. A Tambyah, Engineering out the risk of infection with urinary catheters, *Emerg. Infect. Dis.*, 2001, **7**, 342.
169. D. D. Cravens and S. Zweig, Urinary catheter management, *Am. Fam. Physician*, 2000, **61**, 369.
170. G. Finer and D. Landau, Pathogenesis of urinary tract infections with normal female anatomy, *Lancet Infect. Dis.*, 2004, **4**, 631.
171. D. J. Stickler, N. S. Morris and T. J. Williams, An assessment of the ability of a silver-releasing device to prevent bacterial contamination of urethral catheter drainage system, *Br. J. Urol.*, 1997, **80**, 539.
172. D. J. Stickler, G. L. Jones and A. D. Russell, Control of encrustation and blockage of Foley catheters, *Lancet*, 2003, **361**, 1435.
173. P. Thibon, X. Le Coutour, R. Leroyer and J. Fabry, Randomised multicentre trial of the effects of a catheter coated with hydrogel and silver salts on the incidence of hospital-acquired infections, *J. Hosp. Infect.*, 2000, **45**, 117.
174. L. Rimondi, M. Fini and R. Giardino, The microbial infection of biomaterials: a challenge for clinicians and researchers, a review, *J. Appl. Biomater. Biomech.*, 2005, **3**, 1.

175. G. Kampf, B. Dietze, C. Gross-Siestrup, C. Wendt and H. Martiny, Microbicidal activity of a new silver-containing polymer, SPI-ARGENT II, *Antimicrob. Agents Chemother.*, 1998, **42**, 2440.

176. R. Bambauer, P. Mestres, R. Schiel, R. Klinkmann and P. Sioshansi, Surface treated catheters with ion-beam based process evaluation in rats, *Artif. Organs*, 1997, **21**, 1039.

177. R. Bambauer, P. Mestres, R. Schiel, J. M. Schneidewind-Muller, J. M. Bambauer and P. Sioshansi, Large bore catheters with surface treatments versus untreated catheters for blood access, *J. Vasc. Access*, 2001, **2**, 97.

178. P. Kathuria, H. L. Moore, R. Mehrotra, P. F. Prowant, R. Khanna and Z. J. Twardowski, Evaluation of healing and external tunnel histology of silver-coated peritoneal catheters in rats, *Adv. Perit. Dial.*, 1996, **12**, 203.

179. L. C. Fung, A. E. Khoury, S. I. Vas, C. Smith, D. G. Oreopoulos and M. W. Mittelman, Biocompatibility of silver-coated peritoneal dialysis catheter in porcine model, *Perit. Dial. Int.*, 1996, **16**, 398.

180. H. Goldschmidt, U. Hahn, H.-J. Salwender, R. Haas, B. Jansen, P. Wolbring, R. Rinck and W. Hunstein, Prevention of catheter-related infections by silver coated central venous catheters in oncological patients, *Zentralbl. Bakteriol.*, 1995, **283**, 140.

181. D. G. Maki, S. M. Stolz, S. Wheeler and A. Mermel, Prevention of central venous catheter-related bloodstream infection by use of an antiseptic-impregnated catheter, a randomised clinical trial, *Ann. Intern. Med.*, 1997, **127**, 257.

182. S. Saint, D. L. Veenstra, S. D. Sullivan, C. Chenoweth and A. M. Fendrick, The potential clinical and economic benefits of silver alloy urinary catheters in preventing urinary tract infections, *Arch. Intern. Med.*, 2000, **160**, 2670.

183. N. A. Sabbuba, D. J. Stickler, E. Mahenthralingam, D. J. Painter, J. Parkin and R. C. L. Feneley, Genotyping demonstrates that the strains of *Proteus mirabilis* from bladder stones and catheter encrustation of patients undergoing long-term catheterisation are identical, *J. Urol.*, 2004, **11**, 1925.

184. H. Akiyama and S. Okamoto, Prophylaxis of in-dwelling catheter infection: clinical experience with a modified Foley catheter and drainage system, *J. Urol.*, 1979, **121**, 40.

185. W. Rosch and S. Lugauer, Catheter-associated infections in urology: possible use of silver-impregnated catheters and Erlanger silver catheter, *Infection*, 1999, **27** (Suppl. 1), S74.

186. U. Samuel and J.-P. Guggenbichler, Prevention of catheter-related infections: the potential of a new nano-silver impregnated catheter, *Int. J. Antimicrob. Agents*, 2004, **23** (Suppl. 1), S75.

187. H. Liedberg, Catheter induced urethra inflammatory reaction and urinary tract infection. An experimental and clinical study, *Scand. J. Urol. Nephrol.*, 1989, **124** (Suppl. 1).

188. H. Liedberg and T. Lundeberg, Silver alloy coated catheters reduce catheter-associated bacteruria, *Br. J. Urol.*, 1990, **65**, 379.

189. T. Cymet, Do silver alloy catheters increase the risk of systemic argyria? *Arch. Intern. Med.*, 2001, **161**, 1014; S. Saint, in reply, *ibid*, 1015.

190. R. A. Bologna, L. M. Tu, M. Polansky, H. D. Fraimow, D. A. Gordon and K. E. Whitmore, Hydrogel/silver ion-coated urinary tract catheter reduces nosocomial urinary tract infection rates in intensive care unit patients a multicentre study, *Urology*, 1999, **54**, 982.

191. T. B. Karchmer, E. T. Giannetta, C. A. Muto, B. A. Strain and B. M. Farr, A randomised crossover study of silver-coated urinary catheters in hospitalised patients, *Arch. Intern. Med.*, 2000, **160**, 3294.

192. R. Plowman, N. Graves, J. Esquivel and J. A. Roberts, An economic model to assess the cost and benefits of the routine use of silver alloy coated urinary catheters to reduce the risk of urinary tract infections in catheterised patients, *J. Hosp. Infect.*, 2001, **48**, 33.

193. K. K. Lai and S. A. Fontecchio, Use of silver-hydrogel urinary catheters on the incidence of catheter-associated urinary tract infections in hospitalised patients, *Am. J. Infect. Control*, 2002, **30**, 221.

194. D. G. Aherne, T. Grace, M. J. Jennings, R. N. Borazjani, K. J. Boles, L. J. Rose, R. B. Simmons and E. N. Ahanotu, Effects of hydrogel/silver coatings on *in vitro* adhesion to catheters of bacteria associated with urinary tract infections, *Curr. Microbiol.*, 2000, **41**, 120.

195. T. Newton, J. M. Still and E. Law, A comparison of the effect of early insertion of standard latex and silver-impregnated latex Foley catheters on urinary tract infections in burn patients, *Infect. Control Hosp. Epidemiol.*, 2002, **23**, 217.

196. K. Foster and G. Smith, Clinical effectiveness of the silver-impregnated Foley with bacteriostatic tubing/bag system compared with the standard latex catheters and standard drain systems, *Am. J. Infect. Control*, 2005, **33**, E133.

197. R. J. McLean, A. A. Hussain, M. Sayer, P. J. Vincent, D. J. Hughes and T. Smith, Antibacterial activity of a multilayer silver-copper surface film catheter material, *Can. J. Microbiol.*, 1993, **39**, 895.

198. E. L. Lawrence and I. G. Turner, Kink, flow and retention properties of urinary catheters Pt. 1: conventional Foley catheters, *J. Mater. Sci.*, 2005, **17**, 147.

199. J. Brosnahan, A. Jull and C. Tracy, Types of urethral catheters for management of short-term voiding problems in hospitalised patients, *Cochrane Database System Review*, 2004, **1**, CD004013.

200. Health and Safety Executive (HSE), Latex allergies, www.hse.gov.uk/latex/, accessed 17 November 2009.

201. S. Woodward, Complications of allergies to latex urinary catheters, *Br. J. Nurs.*, 1997, **6**, 786.

202. S. Saint, J. G. Elmore, S. D. Sullivan, S. S. Emerson and T. D. Koepsell, The efficiency of silver alloy-coated catheters in preventing urinary tract infection: a meta-analysis, *Am. J. Med.*, 1998, **105**, 236.

203. T. A. Gaonkar, L. A. Sampath and S. M. Modal, Evaluation of the antimicrobial efficacy of urinary catheters impregnated with antiseptics in

an *in vitro* urinary tract model, *Infect. Control Hosp. Epidemiol.*, 2003, **24**, 506.

204. R. Zschaler, Testing of the antimicrobial effect of catheter tubing with the roll culture method, www.spiegelberg.de/home/documents/Zschaler.pdf, accessed 17 November 2009.

205. P. P. Rickham, A new silver-impregnated silastic type C catheter for use with the Holter valve in the treatment of hydrocephalus, *Dev. Med. Child Neurol.*, 1970, **12** (Suppl), 14–1.

206. P. Lachner, R. Beer, G. Broessner, R. Helbok, K. Galiano, C. Pleifer, B. Pfausler, C. Brenneis, C. Huck, E. Engelhardt, A. A. Obwegeser and E. Schmutzhard, Efficiency of silver nanoparticles–impregnated external ventricular drain catheters in patients with acute occlusive hydrocephalus, *Neurocrit. Care*, 2008, **8**, 360.

207. R. Bayson, A. Mills, S. M. Howdle and W. Ashraf, Comment on: the increasing use of silver-based products as antimicrobial agents: a useful development or cause for concern?, *J. Antimicrob. Chemother.*, 2007, **59**, 587.

208. R. Bayston, W. Ashraf and L. Fisher, Prevention of infection in neuro-surgery: role of "antimicrobial" catheters, *J. Hosp. Infect.*, 2007, **65**, 39.

209. A. B. G. Lansdown, Pin and needle tract infection: the prophylactic role of silver, *Wounds UK*, 2006, **2**, 51.

210. ASM International, *Handbook of Materials for Medical Devices*, ed. J. R. Davies, American Technical Publishers Ltd, Hitchin, UK, 2007.

211. R. Kingston and M. G. Walsh, The evolution of hip replacement surgery, *Irish Med. J.*, 2001, **94**, 1.

212. A. Massé, A. Bruno, M. Bosetti, A. Biasibetti, M. Cannas and P. Gallinaro, Prevention of pin tract infection in external fixation with silver coated pins: clinical and microbiological results, *J. Biomed. Mater. Res.*, 2000, **5**, 600.

213. C. A. Collinge, G. Goll, D. Seligson and K. J. Easley, in tract infections: silver vs. uncoated pins, *Orthopaedics*, 1994, **17**, 445.

214. M. Bellantone, N. J. Coleman and L. L. Hench, Bacteriostatic action of a novel four-component bioactive glass, *J. Biomed. Mater. Res.*, 2000, **51**, 484.

215. B. Browne, A. Levine, J. Jupiter and P. Trafton, *Skeletal Trauma*, W. B. Saunders, Philadelphia, 1998.

216. C. Parkhill, A new apparatus for the fixation of bones after resection and in fractures with a tendency to displacement, *Trans. Am. Surg. Ass.*, 1897, **15**, 251.

217. H. Shintani, Modification of medical device surface to attain anti-infec-tion, *Trends Biomater. Artif. Organs*, 2004, **18**, 1.

218. M. Bosetti, A. Massé, E. Tobin and M. Cannas, Silver coated materials for external fixation devices: *in vitro* biocompatibility and genotoxicity, *Biomaterials*, 2002, **23**, 887.

219. M. A. Wassall, M. Santin, C. Isalberti, M. Cannas and S. P. Denyer, Adhesion of bacteria to stainless steel and silver-coated orthopaedic external fixation pins, *J. Biomed. Mater. Res.*, 1997, **36**, 325.

220. C. Meyer, J. Kessler, V. Alt, S. Wenisch, B. Hartmann, H. G. Schiefer and R. Schnettler, Antimicrobial effect of silver-coated external fixator pins. *Osteo. Trauma Care*, 2004, **12**, 81.
221. E. Sheehan, J. McKenna, K. J. Mulvall, P. Marks and D. McCormack, Adhesion of *Staphylococcus* to orthopaedic metals, an *in vivo* study, *J. Orthop. Res.*, 2004, **22**, 39.
222. L. M. Coester, J. V. Nepola, J. Allen and J. L. Marsh, The effects of silver coated external fixation pins, *Iowa Orthop. J.*, 2000, **26**, 48.
223. M. M. Cowan, K. Z. Abshire, S. L. Houk and S. M. Evans, Antimicrobial efficacy of a silver-zeolite matrix coating on stainless steel, *J. Ind. Microbiol. Biotechnol.*, 2003, **30**, 102.
224. K. R. Bright, C. P. Gerba and P. A. Rusin, Rapid reduction of *Staphylococcus aureus* populations on stainless steel surfaces by Zeolite ceramic coatings containing silver and zinc ions, *J. Hosp. Infect.*, 2002, **52**, 307.
225. A. Ewald, S. K. Gluckermann, R. Thull and U. G. Bureck, Antimicrobial titanium/silver PD coatings on titanium, *Biomed. Eng. Online*, 2006, **5**, 22.
226. G. D. Christensen, L. Bakldassarri and W. A. Simpson, Colonisation of medical devices by coagulase-negative staphylococci, in *Infection Associated with Indwelling Medical Devices*, ed. F. A. Waldvogel and A. L. Bisno, ASM Press, Washington, 1994, pp. 45–78.
227. H. W. Buchholz and H. Engelbrecht, Depot effects of various antibiotics mixed with Palacos resins, *Chirurg.*, 1970, **41**, 511.
228. B. Thornes, P. Murray and D. Bouchier-Hayes, Development of resistant strains of *Staphylococcus epidermidis* on gentamycin-loaded bone cement *in vivo*, *J. Bone Joint Surg.*, 2002, **84**, 758.
229. M. C. Cortizo, M. Fernández, L. de Mele and A. M. Cortizo, Metallic dental material: biocompatability in osteoblast-like cells, *Biol. Trace Elem. Res.*, 2004, **100**, 151.
230. R. Dueland, J. A. Spadaro and B. A. Rahn, Silver antibacterial cement, *Clin. Orthop. Rel. Res.*, 1982, **169**, 264.
231. J. A. Spadaro, D. A. Webster and R. O. Becker, Silver polymethyl methacrylate antibacterial bone cement, *Clin. Orthop. Rel. Res.*, 1979, **143**, 266.
232. R. O. Becker and J. A. Spadaro, Treatment of orthopaedic infections with electrically generated silver ions, *J. Bone Joint Surg.*, 1978, **60A**, 871.
233. J. A. Spadaro and R. O. Becker, Some specific cellular effects of electrically injected silver and gold ions, *Bioenergetics*, 1976, **3**, 49.
234. V. Alt, T. Becherrt, P. Steinrücke, M. Wagener, P. Seidel, E. Dingeldein, E. Domann and R. Schnettler, *In vitro* testing of antimicrobial activity of bone cement, *Antimicrob. Agents Chemother.*, 2004, **48**, 4084.
235. V. Alt, P. Steinrücke, M. Wagener, P. Seidel, E. Dingeldein, D. Scheddin, E. Domann and R. Schnettler, Nanoparticulate silver: a new antimicrobial substance for bone cement, *Orthopade*, 2004, **33**, 885.
236. H. Vik, K. J. Andersen, K. Julshamn and K. Todnem, Neuropathy caused by silver absorption from arthroplasty cement, *Lancet*, 1985, **1**(8433), 872.

237. E. Sudemann, H. Vik, M. Rait, K. Todnem and K.-J. Andersen, Systemic and local silver accumulation after total hip replacement using silver-impregnated bone cement, *Med. Prog. Technol.*, 1994, **20**, 179.

238. R. O. Darouiche, Anti-infective efficacy of silver-coated medical prostheses, *Clin. Infect. Dis.*, 1999, **29**, 1371.

239. D. Langanki, M. F. Ogle, J. D. Cameron, R. A. Lirtzman, R. F. Schroeder and M. W. Mirsch, Evaluation of a novel bioprosthetic heart valve incorporating anti-calcification and antimicrobial technology in a sheep model, *J. Heart Valve Dis.*, 1998, **7**, 633.

240. P. Tozzi, A. Al-Darweeesh, P. Vogt and F. Stumpe, Silver-coated prosthetic valve: a double bladed weapon, *Eur. J. Cardiothorac. Surg.*, 2001, **19**, 729.

241. Medicines Healthcare Products Regulatory *Agency, AN 1999(06) Thromboembolic Complications Involving Silzone Mechanical Heart Valves*, MHRA, London, 1999, Ref. 04/01/98122131, www.mhra.gov.uk/Publications/Safetywarnings/MedicalDeviceAlerts/Advicenotices/CON008871, accessed 19 November 2009.

242. T. Ueberrueck, L. Meyer, R. Zippel, G. Nestler, T. Wahlers and I. Gastinger, Healing characteristics of a new silver-coated, gelatine impregnated vascular prosthesis in the porcine model, *Zentrabl. Chir.*, 2005, **130**, 71.

243. K. S. Tweeden, J. D. Cameron, A. J. Razzouk and W. R. Holmberg, Biocompatibility of silver-modified polyester for antimicrobial protection of prosthetic valves, *J. Heart Valve Dis.*, 1997, **6**, 553.

244. S. A. O'Connor, Re: efficacy of silver in preventing bacterial infection of vascular Dacron graft material, *Eur. J. Vasc. Endovasc. Surg.*, 2004, **27**, 565.

245. M. Strathman and J. Wingender, Use of an oxonol dye in combination with confocal laser scanning microscopy to monitor damage to *Staphylococcus aureus* during colonisation of silver-coated vascular grafts, *Int. J. Antimicrob. Agents*, 2004, **24**, 234.

246. H. McPherson and A. Asgar, Acupuncture needle sensations associated with D Qi: a classification based on experts' ratings, *J. Altern. Complement. Med.*, 2006, **12**, 633.

247. L. Yi-Kai, A. Xueyan and W. Fu-Gen, Silver needle therapy for intractable low back pain at tender point after removal of nucleus pulposus, *J. Manipulative Physiol. Ther.*, 2000, **23**, 320.

248. Y. Tanita, T. Kato, K. Hanada and H. Tagami, Blue macules of localised argyria caused by implanted acupuncture needles, *Arch. Dermatol.*, 1985, **121**, 1550.

249. H. Suzuki, S. Baba, S. Uchigasaki and M. Murase, Localised argyria with chrysiasis caused by implanted acupuncture needles, *J. Am. Acad. Dermatol.*, 1993, **29**, 833.

CHAPTER 6

Silver as an Antibiotic in Water Systems

6.1 Introduction

Silver has been used for purification of water for more than 2000 years and probably represents the first recorded use of the metal as an antibiotic. Numerous references are seen in the literature referring to the use of silver vessels, pitchers and cups to purify and transport drinking water for the nobility of ancient dynasties of Sumeria, Rome, Greece, Egypt and Macedonia. The Phoenicians are recorded as having transported drinking water in silver pitchers in their long voyages throughout the Mediterranean Sea and beyond.

In different ways, silver has been used widely in water preservation and sterilisation up to the present time, notably in the wake of the problems arising from outbreaks of Legionnaire's disease in the 1980s, cleansing of domestic water systems, hospital hot and cold water supplies, and in swimming baths. Technological advances following the introduction of the Katadyn water treatments in 1928 have included development of silver–copper ionisation filters (Figure 6.1), electrolytically modulated silver ionisation, silver–carbon filters, zeolite ceramic silver–zinc coatings and numerous commercial water treatments. Silver-based technologies have proved superior to other purifications, including heating and chloride methods, and are controllable and without human health risk.

6.2 Silver in Natural Water Sources, its Distribution and Environmental Influence

Silver is a naturally occurring metal and a normal constituent of inland water sources, estuaries, seas and oceans. The silver content of inland waterways, streams and rivers varies greatly according to geographical area, the distribution of silver-bearing ores such as argentite and stephanite, the proximity to silver-related industries and silver residues in factory effluents and sludges.[1] The

Issues in Toxicology No. 6
Silver in Healthcare: Its Antimicrobial Efficacy and Safety in Use
By Alan B. G. Lansdown
© Alan B. G. Lansdown 2010
Published by the Royal Society of Chemistry, www.rsc.org

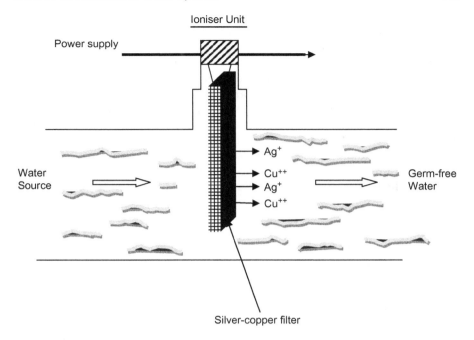

Figure 6.1   Silver–copper ion purification system for water.

Public Health Service (1990) reviewed silver contamination in various water sources in the United States and indicated that surface water, rivers and lakes contain between 0.2 and 2.0 parts per billion (ppb) silver, but recorded that silver levels of hundred fold higher can be expected in waters leaching silver-bearing deposits and in mining areas.[i] Drinking water supplies from ground-water or surface water resources were estimated to contain silver levels of up to 30–80 ppb. Estimates of waterborne silver vary according to the acidity of the conditions and seasonal conditions. Low pH favours increased leaching of silver from deposits.[2,3] A survey of studies conducted from the early 1960s through to 1988 showed that silver was detectable in only 22% of water samples and that mean silver levels of $2 \, \mu g \, L^{-1}$ were recorded.[4]

Considerable variations are recorded in the levels of silver seen in various water sources in the United States. In each case, silver released from natural and industrial sources, factory waste or water purification plants occurs in the form of inorganic salts or as organic complexes. Some adsorbs to organic debris (including humic complexes), or remains as suspended particulate matter or bound in micro-organisms.[4]

These days it is usual that drinking water will be filtered and particulate matter removed. Silver discharged into rivers and streams tends to settle down into the sediment and remain there for many years according to the acidity,

---

[i]This information was published in a report for the Agency for Toxic Substances and Disease Registry (ATSDR), 1990, Atlanta, GA.

salinity and local levels of iron, manganese and organic material.[5] Where water is retained in silver vessels (as was the custom from many years ago), it should be expected that silver ion will be progressively eluted into the drinking water (metallic silver has an estimated solubility in water of $<0.1\,\mu g\,mL^{-1}$). The amount of silver released will be proportional to the surface area of the silver vessel exposed, the acidity of the water and the temperature. None of the estimates of silver content in drinking water in the United States or elsewhere identify the proportion of silver contamination attributable to the use of silver in water purification. Recent estimates suggest that, in the United States at least, silver levels in unpolluted stream and river areas are as low as $0.01\,\mu g\,L^{-1}$ although, in urban and industrialised areas, the silver content may be tenfold higher (Federal Water Pollution Control Act).

6.3   Legionnaires' Disease

The expressions "Legionnaires' disease" or "Legionellosis" are used to define infectious diseases attributable to *Legionella pneumophilia* and related bacteria.[6] The bacteria are waterborne organisms found naturally in low numbers in rivers, lakes, streams and reservoirs. Legionellosis is a potentially fatal form of pneumonia; older people and those subject to illness, immunosuppressive therapy and smoking are particularly vulnerable. Offending bacteria are a public health concern when they infect hospital water systems, cooling towers, swimming baths and evaporative condensers. Inhalation of the bacteria in aerosol or water droplets is a cause of infection. *Legionella* sp. are commonly associated with sludge and moulds in water supplies and are tolerant of water temperatures of 20–50 °C; above 50 °C they become dormant, and at 60 °C they do not survive. In the UK, the Health and Safety Executive (HSE) has documented guidelines and a Code of Practice detailing identification of *Legionella* infections, their management and prevention. The principal objectives of this programme include:

- control of hot and cold water systems;
- plant or systems containing water likely to exceed 20 °C, and which might release aerosols during routine operation or maintenance;
- cooling towers;
- evaporative condensers.[6]

Although silver is not specifically identified in these guidelines and regulatory documents, metallic silver, silver–copper filters and silver zeolite complexes are widely used and efficacious in the prevention and control of *Legionella* sp. in hospital water systems, domestic drinking water supplies and swimming baths.[7–9]

6.4   Katadyn and Silver–Copper Technology in Water Purification

Bacteria, mycobacteria, fungal spores, yeasts, protozoa and many viruses are waterborne infections and some presenting as opportunist pathogens may lead to major health problems. Edit Raulin (1869) may have been first to describe the ability of silver to cleanse water when he observed that *Aspergillus niger* would not grow in silver vessels.[10] The first and probably best known commercial method for water purification was the Katadyn process introduced in 1928 by Professor Alexander Krause in Germany. Krause later moved to Switzerland where his invention, Katadyn®, is now patented for "making water drinking water".

Katadyn silver technology has been described as a spongy preparation of metallic silver or silver coating on sand or impregnated into filter material.[11] The preparation contained small quantities of a second metal such as palladium or gold which was held to potentiate the bioactivity of the silver ions. Limited evidence showed that Katadyn with silver ion concentrations of 0.006–0.5 parts per million (ppm) was efficacious in eliminating *Escherichia coli* infection within 2 to 24 h depending in the level of infection and temperature. Sykes was of the opinion that pathogenic bacteria were more sensitive to Katadyn, but that bacterial spores, moulds and protozoa were resistant.[11] *Bacillus mesentericus* spores were shown to survive in Katadyn (300 mL water + 20 g silver) for more than five weeks. The Electro-Katadyn process introduced later included electrodes to stimulate the ionisation of the silver in the purification of swimming baths. The method used either two silver electrodes or a silver electrode and an electrode of another metal. Reports on the efficacy of this modification are conflicting. The Katadyn trade mark has been adopted commercially in the production of water purification devices which are claimed to combine water filtering technology with a dual mode filtering system and three proven Katadyn technologies:

- A silver impregnated ceramic disk pre-filter;
- A pleated glass fibre primary filter;
- A replaceable carbon granulate core (Katadyn Products Inc., Switzerland).

Other water purification methods used in the early 1900s employed the oligodynamic properties of silver and used soluble silver salts, colloidal silver preparations and suspensions to eliminate waterborne infections.[12] Although silver has been used as an antibiotic for many years, it is a relatively ineffective antiseptic in the form of silver proteinates on account of the slow release of silver ion. Other metals like copper have been included in preparations to enhance ionisation and the bacterial lethality of silver. Soluble silver salts exert some selectivity in their antibiotic action against common pathogens, but Gram-negative organisms tend to be more sensitive to silver ion than Gram-positive bacteria (Table 6.1).[11]

The greatest advances in water purification in recent times are associated with silver–copper technology. Like silver, the astringent and antibiotic properties of copper have been appreciated for many years.[13] In water purification, copper is markedly toxic to fungi and algal infections, and active against coliform bacteria. In 1915, Thomas described the use of copper in purification of swimming pools.[13] Copper is an essential trace metal in the human body, but

Table 6.1 Bacterial sensitivities to soluble silver compounds.[11]

	Staphylococcus aureus (Gram + ve)	*Salmonella typhimurium* (Gram − ve)
Silver Compound		
Nitrate	1 in 80[a]	1 in 1,250
Citrate	1 in 600	1 in 4,500
Lactate	1 in 110	1 in 690
Colloidal Silver Proteinates		
Strong	1 in 15	1 in 175
Mild	1 in <12	1 in 70

[a]dilutions

at high doses (20 g), soluble copper salts are toxic and can evoke haemorrhage, gastroenteritis, nephritis and coma. The copper content of food is expected to be in the range 0.1 to 444 mg kg^{-1}.[14]

It is unclear when silver–copper technology was first introduced for water purification but it came to prominence in the early 1980s as a means of controlling outbreaks of *Legionella pneumophila,* the causative agent of Legionnaire's disease.[15] Current views are that silver and copper ions liberated into water systems with or without electrolytic means act synergistically to kill sensitive bacteria. Thus copper cations avidly bind anionic moieties on the bacterial cell envelope/membrane, thereby denaturing it and increasing its porosity to silver ion.[7] Increased penetration of silver into the bacterial cell leading to denaturation of DNA/RNAases, and other enzymic proteins involved in respiratory and metabolic process leads to bacterial cell death. *In vitro* studies have demonstrated that the silver–copper combination is highly efficacious in eliminating planktonic bacteria in hot and cold water systems, but is less effective in hard water areas where ionisation of silver is inhibited by build up of scale on electrodes. In addition, silver ion binds readily to chloride in water to precipitate as relatively insoluble silver chloride.

Numerous reports have been published validating the value of silver–copper filtration as a means of eliminating troublesome *Legionella* sp. and other pathogenic bacteria in hospital water systems (Figure 6.1). In one such controlled evaluation, silver: copper balances of >0.04 and >0.4, respectively, were found to be efficacious in reducing bacterial colonisation; the study also found that the methodology was cost-effective and could be easily installed and maintained.[16,17] Studies conducted by Nigel Pavey of the BSIRA[ii] laboratories in Bracknell, UK, have demonstrated that silver–copper ionisation systems (40 µg L^{-1} silver, 400 µg L^{-1} copper) are highly effective in eliminating *Legionella* sp. in cold and hot water systems involving soft water.[7]

Whilst the efficacy of silver–copper filtration is seen to be acceptable in the short-term in controlling *Legionella* sp., some authors have expressed reservations as to its value over long periods.[18,19] On the other hand, there now seems to be overwhelming evidence from multicentre trials and re-evaluation of

[ii]Building Services Research and Information Association.

the technology that silver-copper filtration is the only disinfection modality to fulfil the following criteria:

- Efficacy against *Legionella* sp. in laboratory trials;
- Prevents Legionnaire's disease in individual hospitals;
- Prevents Legionnaire's disease in controlled trials;
- Validation of observations and confirmatory reports from multiple hospitals over a prolonged period (5–11 years).[8]

Recent recommendations are that the system is safe providing that silver and copper ion concentrations are monitored regularly and maintained within internationally recognised guidelines.[20] Review of the experimental and clinical studies between 1997 and 2007 revealed that silver–copper ionisation is an effective method for controlling *Legionella* sp. bearing in mind that eradication cannot be achieved by any method in isolation. Maintaining water at high temperatures has been shown to maximise the efficacy of the system.

So far limited evidence exists to show that *Legionella pneumophila* develop a resistance to silver–copper in the long term.[21,22] However, as far as I am aware, no studies have been seen to confirm that bacterial resistance to silver–copper filtration is genetically based or that organisms exhibit characteristic plasmids and *Sil* proteins as delineated by Gupta and Silver in isolates of *Salmonella typhimurium* (pMGH100) in the Massachusetts General Hospital as discussed in Chapter 5.[23]

Although experimental studies suggest that silver zeolite coatings on stainless steel surfaces may be suitable in controlling *Legionella pneumophila* in air ducts, condensation pans and exhaust vents, the full benefits of this technology remain to be seen.[24]

6.5 Regulatory Requirements for Silver in Water Purification

European legislation does not regulate the use of silver concentrations in water. The Drinking Water Directive (80/778/EEC) identified water quality indicators and classified silver in Group C as being an indicator which in large amounts "might be toxic" without specifying thresholds.

In the United States, current guidelines specify maximal levels of silver in drinking water at $0.1 \, mg \, L^{-1}$ in long-term exposures set by the Environmental Protection Agency (EPA) and laid down in the National Secondary Drinking Water Regulations 2002. This replaces earlier regulation specifying maximal silver concentration in drinking water at $0.05 \, mg \, L^{-1}$ in 1985 and $0.09 \, mg \, L^{-1}$ in 1989.[25]

At present, the World Health Organization (WHO) does not regulate silver in drinking water as, in the view of their expert committees, available information is "insufficient to identify a health standard".[26] Similar views were taken by HSE in the UK (A. Lloyd, personal communication, 2001);[27] its view in 1980 was that the $10 \,\mu g\, L^{-1}$ standard published in the Drinking Water Directive (which forms the basis of German and all other EU drinking water standards) has no toxicological basis. The conditions of approval limit use to no more than 90 days in one year and a maximum dose of $80 \,\mu g\, L^{-1}$.

Colloidal silver products, which were used earlier to improve water quality, are not legally permitted in the European Union, United States and many other countries for reasons of safety and lack of efficacy.

References

1. A. R. Flegal, C. L. Brown, S. Squire, J. R. M. Ross, G. M. Scelfo and S. Hibdon, Spatial and temporal variations in silver contamination and toxicity in San Francisco Bay, *Environ. Res.*, 2007, **105**, 34.
2. R. W. Boyle, *Geochemistry of Silver and its Deposits, with Notes on Geochemical Prospecting for the Element*, Department of Energy, Mines and Resources, Ottawa, 1968, Geological Survey of Canada Bulletin 160.
3. K. Scow, M. Goyer and L. Nelken, *Exposure and Risk Assessment of Silver Final Report*, US Environmental Protection Agency, Office of Water Regulation and Standards, Washington DC, 1981, EPA Report No. 440481017.
4. W. P. Eckel and T. A. Jacob, Ambient levels of 24 dissolved metals in US surface and ground waters, *Proceedings of 196th American Chemical Society Division of Environmental Chemistry*, Los Angeles, 1988, **28**, pp. 371–372.
5. J. C. Monica, Nano-Silver EHS Backgrounder, Nanotechnology Law Report, posted 17 July 2008, www.nanolawreport.com/2008/07/articles/nanosilver-ehs-backgrounder/, accessed 20 November 2009.
6. Health and Safety Executive, *Legionnaire's Disease. The Control of Legionella Bacteria in Water Systems. Approved Code of Practice and Guidance*, HSE Books, Sudbury, 3rd edn, 2000, www.hse.gov.uk/pubns/books/l8.htm, accessed 20 November 2009.
7. A. Hambidge, Reviewing the efficacy of alternative water treatment techniques, *Health Estate*, 2001, **55**, 23.
8. J. E. Stout and V. L. Yu, Experiences of the first hospitals using copper–silver ionisation for Legionella control: implications for the evaluation of other disinfection modalities, *Infect. Control Hosp. Epidemiol.*, 2003, **24**, 563.
9. N. Sylvestry Rodrigues, K. R. Bright, D. R. Uhlmann, D. C. Slack and C. P. Gerba, Inactivation of *Pseudomonas aeruginosa* by silver in tap water, *J. Environ. Sci. Health, Part A.*, 2007, **42**, 1579.
10. V. Goetsch, Colloidal silver, a fresh look, in *Colloidal silver knowledge base*, SynerGenesis Inc., Pengilly, MN, 2006, www.colloidalsilver.org/colloidal_silver_knowledge_base.htm, accessed 20 November 2009.
11. G. Sykes, *Disinfection and Sterilization: Theory and Practice*, E. & F. N. Spon, London, 1958, Spon's General and Industrial Chemistry Series.
12. B. D. Davis, 'Principles of sterilisation' and 'Principles of Chemotherapy', in *Bacterial and Mycotic Infections of Man*, ed. R. J. Dubois, Lippincott, Philadelphia, 1952, ch. 34, 35, pp. 707–775.

13. T. Sollemann, *A Manual of Pharmacology and its Applications to Therapeutics and Toxicology*, Saunders, Philadelphia, 1942, pp. 1111–1114.
14. C. W. Lindow, C. A. Elvehjem and W. H. Peterson, Copper, *J. Biol. Chem.*, 1929, **82**, 465.
15. M. Best, V. L. Yu, J. Stout, A. Goetz, R. R. Muder and F. Taylor, *Legionellacae* in the hospital water supply. Epidemiological link with disease and evaluation of a method for control of nosocomial legionnaire's disease and Pittsburgh pneumonia, *Lancet*, 1983, **2**, 307.
16. L. K. Landeen, M. T. Yahya and C. P. Gerba, Efficacy of copper and silver ions and reduced levels of free chlorine in inactivation of *Legionella pneumophilia*, *Appl. Environ. Microbiol.*, 1989, **55**, 3045.
17. Z. Liu, J. E. Stout, L. Tedesco, M. Boldin, C. Hwang, W. F. Viven and V. L. Yu, Controlled evaluation of copper-silver ionisation in eradicating *Legionella pneumophilia* from hospital water distribution systems, *J. Infect. Dis.*, 1984, **169**, 919.
18. A. Goerz and V. L. Yu, Copper-silver ionisation: cautious optimism for Legionella disinfection and implications for environmental culturing, *Am. J. Infect. Control*, 1997, **25**, 449.
19. Y. S. Lin, J. E. Stout, V. L. Yu and R. D. Vidic, Disinfection of water distribution systems for *Legionella*, *Semin. Respir. Infect.*, 1998, **13**, 147.
20. S. P. Cachafiero, I. M. Naviera and I. G. Garcia, Is copper-silver ionisation safe and effective in controlling *Legionella*? *J. Hosp. Infect.*, 2007, **67**, 209.
21. U. Rohr, M. Senger, F. Selenga, R. Turley and M. Wilhelm, Four years of experience with silver-copper ionisation for control of *Legionella* in a German university hospital hot water plumbing system, *Clan. Infect. Dis.*, 1999, **29**, 1507.
22. M. G. Hwang, H. Katayama and S. Ohgaki, Inactivation of *Legionella pneumophilia* and *Pseudomonas aeruginosa*: evaluation of the bactericidal activity of silver cations, *Water Res.*, 2007, **41**, 4097.
23. A. Gupta, K. Matsui, J.-F. Lo and S. Silver, Molecular basis for resistance to silver cations in Salmonella, *Nat. Med.*, 1999, **5**, 183.
24. P. Rusin, K. Bright and C. Gerba, Rapid reduction of *Legionella pneumophilia* on stainless steel with zeolite coatings containing silver and zinc ions, *Lett. Appl. Microbiol.*, 2003, **36**, 69.
25. W. L. Roper, *Toxicological Profile for Silver*, Agency for Toxic Substances and Disease Registry, US Public Health Service, Washington DC, 1990.
26. World Health Organization, Guidelines for Drinking-Water Quality, WHO, Geneva, 3rd edn, 2008, www.who.int/water_sanitation_health/dwq/gdwq3rev/en/, accessed 20 November 2009.
27. The Drinking Water (Undertakings) (England and Wales) Regulations 2000, Statutory Instrument 2000, No. 1297.

CHAPTER 7

Silver Technology and Antibiotic Textiles

7.1 General Features

Increasing standards of personal hygiene and awareness of antisocial and embarrassing odours, exudates and skin discomfort due to infections in the skin and its appendages have led to a recent surge in the development of antibiotic or biofunctional textiles. In addition, research has indicated an urgent need for the production of clothing to be used in elderly and immuno-compromised patients at risk of methicillin-resistant *Staphylococcus aureus* (MRSA) and similar life threatening infections (P. Nicholson, personal communication).[1]

Silver is the oldest and most efficacious of the antibiotics known; it is also the safest.[2] In the past 40 years, silver has been incorporated into or coated on a wide range of natural or synthetic materials for wound care and other medical devices.[3] Silver-coated textiles have a special value in reducing risks of pathogenic infection such as *Staphylococcus aureus* in patients with atopic eczema.[4,5] Numerous patent applications have been filed recently claiming new technology for "silvering" textile fibres with activity against antibacterial or antifungal infections. In each case, the textile fibres serve as a vehicle for delivery of bioactive silver ion (with or without promotion by a direct electric current) to eliminate or otherwise protect against bacterial imbalances in the skin, microbial over growth accompanied by excessive odour or local discomfort.[6–8] Silver is increasingly used as an antibiotic in hygiene clothing, sports and leisure wear, underclothing and medical textiles.

Hygiene considerations in the design and manufacture of antibiotic garments and production of biofunctional textiles include:

- the type and source of the fibres to be used;
- their influence on the normal physiology and protective function of the skin at expected sites of contact;
- the duration of antibiotic action required;
- expected patterns of use of the finished garment.

Issues in Toxicology No. 6
Silver in Healthcare: Its Antimicrobial Efficacy and Safety in Use
By Alan B. G. Lansdown
© Alan B. G. Lansdown 2010
Published by the Royal Society of Chemistry, www.rsc.org

Technological processes can be expensive, hence the cost:benefit ratio will be of importance to manufacturers.

The principal considerations in silver-containing antibiotic textiles are:

1. **Fibre type:** Preliminary investigations examine the durability of fibres, their suitability for long-term contact with human skin, their durability in the presence of skin secretions and excretions, resistance to degradation by bacterial or fungal infections and safety (hypoallergenicity, lack of ability to evoke irritancy on sensitive areas of skin, minimal risk of causing asthma). Fibres should not degrade to release toxic by-products.
2. **Silver-coating technology:** Techniques for silver coating or impregnation of textile fibres will achieve a sustained rate of silver ion release; the treatment will result in a homogeneous distribution of silver over the surface of fibres or through the interstices of individual threads. The silver-fibre binding should be stable and ionic release long-lasting and sufficient to withstand everyday wear-and-tear, exposure to ultraviolet (UV) light and other environmental factors.
3. **Safety *vs.* efficacy:** Silver ion release in the presence of skin moisture or exudates will be sufficient for therapeutic or prophylactic action against expected and unusual pathogenic organisms, but insufficient to evoke an imbalance in skin infections. Release of silver ion should be sufficient to allow for the binding of some free ion to the surface layers of epidermal keratin, anions on the skin surface (notably Cl^-) and within the finished textile. Only "free" Ag^+ is of antibiotic value.

In each case, antibiotic textiles containing silver or another agent will comply with nationally and internationally recognised standards of safety and good manufacturing practice. Occupational safety and exposure to such technology as silver vaporisation, slurry dipping and production of nanoparticulate silver forms an essential part of any manufacturing process.

7.2 Natural and Synthetic Fibres in Textile Production

Considerable experience has been gained in the past 50 years in using silver-treated fabrics in production of wound care dressings but, until recently, biocidal textiles for clothing, domestic use and hygiene products have not been widely available. Hydrocolloids, hydrogels, alginates, polyurethane foams (*e.g.* Lyofoam), silicone-coated polyurethane and sodium carboxymethyl cellulose have variously been found beneficial in managing wound exudates in wound bed preparation; with added metallic silver or soluble silver compound, dressings to control the bacterial balance and associated odours and patient discomfort.[3,9,10,11]

Natural fibres commonly used in the textile industry include cellulose, cotton, linen, lyocell-derived from wood pulp, silk and wool.[1] Man-made fibres include acetate, acrylic, Lycra®, nylon, polyesters, polyurethane, polypropylene, polyethylene terephthalate (PET), rayon and viscose. They vary greatly in their physico-chemical properties, responses to water (including skin moisture and secretions), durability and capacity to bind organic colourings and antibiotic materials such as silver, copper, triclosan and organic protein denaturants, which differ in their fibre-bonding capacity and mechanisms of action against micro-organisms. A silver–nylon cloth developed for use in wound care was claimed to show theoretical advantages over other silver-coated textiles in its ability to continuously release silver ion for as long as the dressing is in contact with the skin, but silver ionisation was enhanced by applying a small direct electric current.[6]

Many illustrations demonstrate that the efficacy of antibiotic textiles depends on levels of attachment of biocidal molecules to fibre surfaces or their incorporation into the interstices between fibres by adsorption or covalent binding. In the case of silver, the efficacy of the antibiotic textile is proportional to the amount of silver bound to the fibre surfaces and its potential for ionisation in contact with the skin. The molecular mechanics of antibiotic textile production are complex but can be expected to involve bioactive moieties available on fibre surfaces and the kinetics of ion release in the presence of skin exudates, moisture, or metabolic products of infection (including enzymes and toxins).[12]

Metallic silver, colloidal silver, silver alloy, silver zeolite complexes, silver adsorbed to charcoal and silver sulfadiazine have variously been used in coating natural and synthetic materials for medical devices, sutures and textiles for use in clothing and hygiene products. In each case, the chemistry is well known, but the recent introduction of tetrasilver tetroxide (Ag_4O_4, which contains two monovalent and two trivalent silver ions in each molecular crystal) as an antibiotic component of textiles is less well documented.[13] This application claimed that woven and non-woven hydrophilic fabrics treated with tetrasilver tetroxide are resistant to sunlight and maintain excellent antimicrobial properties with a wide spectrum of activity against Gram-positive and Gram-negative organisms, yeasts and viral infections. Nanosilver particles with their high surface area to volume ratio are increasingly used in textile preparation; they are efficacious as antibiotics and are safe. Importantly, minute particles of silver bound to the surface of natural or synthetic textile fibres do not influence the performance, functionality, durability or physical properties of treated fibres. Nanoparticles introduced into cotton, polymers and cellulosic fibres during initial manufacture/synthesis or during the finishing process by processes ranging from simple immersion to silver ion spray methods are claimed to produce a more homogeneous and controlled silver coating or impregnation.[14,15]

A cellulose and seaweed based fibre has recently been introduced for use in textiles and hygiene products such as compression stockings and bandages for burns and skin wounds.[16,17] The fibre produced by the German companies, Zimmer AG and SeaCell GmbH, in a lyocell process has proved to be of particular value in textiles used in proximity of skin exposed to or at risk of fungal infections including *Tinea pedis* (athlete's foot) and *Candida* sp. (ringworm). The new cellulosic fibre matrix is stable and unaffected by prolonged contact with human skin. It is tolerant to washing.[16] SeaCell fibres are manufactured by adding finely ground algal extracts to a spinning solution to which cellulose solution is added. Following successive processes of pulp treatment, cellulose suspension, spinning and filtration, dilute silver nitrate is added to produce the antibiotic SeaCell Active® textile.[16,17] Silver ion readily binds carbonyl, carboxy and hydroxyl moieties on the cellulose fibres to form a resilient product for incorporation into antibiotic socks and hygiene clothing. The technology developed for product manufacture ensures homogeneous distribution of metallic silver deposition throughout the cellulosic fibre without altering its durability or physic-chemical properties.[17,18]

7.3   Infections in the Human Skin and its Appendages

The human skin contains a balanced population of micro-organisms which varies greatly from one region of the body to another and according to age, sex, race and geographic location, the health of the body, and standards of personal hygiene. Most bacteria and yeasts normally present in the skin are commensal organisms, but some become opportunist pathogens in event of skin injury, atopic eczema, abnormal physiological or metabolic states, or in immuno-compromised individuals. The classic definition of infection does not always aid the clinician in determining whether a patient is in the process of developing a life-threatening condition.[19] The Geneva-based Association for Professionals in Inspection Control and Epidemiology (APIC) has published guidelines for hand hygiene in healthcare settings and documented a comprehensive survey of the normal population of micro-organisms in human skin and their transmissibility through skin contact.[20]

Skin infections take the form of:

• Resident populations;
• Transitory populations.

Resident organisms including anaerobic bacteria such as *Propionobacteria* sp. are found more commonly in deeper parts of hair follicles and are not usually influenced by topical antiseptics or hand-washing. They will be transferred in inter-personal contact and contact with common user facilities. Transient flora including *Staphylococcus aureus, Staphylococcus epidermidis, Proteus mirabilis,* β-haemolytic streptococci and coagulase-positive staphylococci are commonly seen as transitory bacteria colonising superficial regions of the skin; they are more amenable to transfer or removal by routine washing than resident organisms, which locate in deeper regions of hair follicles. Transitory infections are acquired by inter-personal contact, contaminated clothing or environmental surfaces and are most frequently associated with healthcare-associated infections. In nurses and healthcare workers, skin infections (notably on the hands) are commonly acquired through contact with patients and hospital facilities, indicating an urgent need for stringent hygiene protocols; *Staphylococci* sp. (ten species have been identified) and yeasts are commonly involved.

The human skin becomes colonised by bacteria from birth, but as an individual ages, so the resident and transitory populations change. In adults

Micrococci sp. and staphylococci are predominant transitory species, but *Propionobacteria* sp. become more established as skin lipid secretion increases. In normal individuals, bacterial and yeast populations are highest in the region of the nose, perineal skin, axilla and clefts between the toes where moisture is greater and the skin softer. However, staphylococcal infections can occur as transitory infections on more exposed areas of skin, particularly in cracks and crevices and between keratinised cells in superficial aspects of the stratum corneum. *Micrococci* sp. are also found on healthy skin but tend to be more prevalent in childhood. Of the resident bacteria, anaerobic species of *Propionobacteria* are commonly associated with hair follicles and may be associated with acneform lesions (*P. acnes*). Males are reported to carry more aerobic staphylococcal bacteria in their skin than females.[21]

Pseudomonas aeruginosa is an aerobic Gram-negative bacterium occurring as a transient infection in moist areas of the anogenital, axilliary and external ear regions.[22] This infection is of principal concern in patients with open wounds and burns where its pathogenicity may have fatal consequences. Silver sulfadiazine is clinically effective in treating *P. aeruginosa* infections associated with swimming pool rashes, and where infection is associated with erythema, purulence and foul smell.

In hygiene practice, antiseptics applied to the skin surface are generally effective in reducing the transient population, although it has been claimed that treatment with agents which specifically inhibit Gram-positive organisms may be associated with increases in Gram-negatives, *i.e.* cause an imbalance in the microflora.[23] Silver ion released from wound dressings, medical devices and textiles will inhibit a wide range of Gram-positive and Gram-negative organisms in culture and in clinical practice.

Dermatophyte infections include species of fungi that cause skin changes, irritation and discomfort. They include three genera of fungi causing ringworm and athlete's foot (*Microsporum, Trichophyton* and *Epidermophyton*).[24–26] Ringworm is commonly associated with inflammatory changes of the toe clefts, hair and nails possibly extending to a chronic dermatitis with vascular changes and pustular reactions. Athlete's foot infections, which are commonly transferred through personal contact, use of public facilities (swimming baths, *etc.*), coal miner's shower rooms and contaminated footwear, are attributable to *Tinea pedis* infections common in Europe and North America in adults greater than 15 years of age.[27] Pathological manifestations of *Tinea pedis* infections include peeling of superficial tissues of the stratum corneum, maceration and fissuring with patient discomfort.

7.4   Influence of Textiles on the Human Skin

The interaction between textiles and the skin is complex and ill-defined. Discussion necessarily requires appreciation of normal skin functions as a primary protective tissue separating individuals from their environment, and contact with xenobiotic fibres of natural or synthetic origin. Textiles applied to the skin separate the tissue from its environment and to some extent supplement or moderate its protective function. Transitory or longer lasting alterations in the physiological and anatomical profile in the skin can be expected, albeit with low personal awareness or discomfort.

The protective functions of the skin are variously achieved by the horny stratum corneum, hair, eccrine and apocrine glands, nerve endings, melanocytes, nails and vascular patterns. Exposed areas of skin are fashioned by such environmental factors as pressures and contact with the environment, desiccating factors (temperature, wind speed, humidity), friction, warmth and light intensity (UV light). The skin serves to control water loss from the body through transepidermal "leakage" and sweat gland activity. Evolutionary changes over many millennia have lead to marked regional differences in hair cover, glandular distribution patterns, thermo- and tactile receptors, skin thickness and states of hydration. Areas of pressure keratinisation characteristic of the soles of the feet or palms of the hands reflect prolonged contact with hard surfaces, whereas the soft, moister and sensitive areas of the axilla, anogenital regions, nasal and external ear regions are associated with low friction and environmental impact. The skin is considered to be in a perpetual state of dynamic equilibrium with its surrounding environment. If that environment is modified in some way by a textile or substances released by it, the skin may be expected to change in its colour, state of hydration, and thickness (Table 7.1).

Although the skin is composed of cells of epidermal, mesodermal and neuro-epidermal origin, changes in temperature, the state of hydration, pressures, light energy, *etc.* can influence different cell types differently, though in normal skin the responses are complementary. Thus, thick woollen fabrics warm the

Table 7.1 Textile-induced changes in the human skin.

Influence	Physical effect	Response in skin or appendages
Warmth	Increased sweating Vascular changes	Increased hydration of tissues, Skin blush,
Decreased UV exposure	Depressed melanocyte activity	Skin pales
Abrasion though rough texture of textile	Increased loss of cells from skin surface, hair loss	Increased cell proliferation in basal epidermis, reduced hair growth
Tightness of textile to skin surface	Reduced air circulation, depressed vitamin D_3 synthesis, increased temperature , hydration	Skin is softer, moister and prone to colonisation with commensal bacteria and yeasts
Release of chemicals from fabric	Local concentrations of xenobiotic materials in surface layers of the epidermis, penetration of hair, hair follicles or sweat glands	1. Percutaneous absorption, systemic absorption and distribution. 2. Interaction with physiological or metabolic events 3. Allergenic response

tissue, induce increased sweating and sebum secretion as a cooling measure, and enhance transepidermal water loss. The resulting increase in the hydration of the tissues is conducive to increased percutaneous absorption of water-soluble or lipophilic materials in contact with the skin surface, and proliferative activity and colonisation of transient commensal bacteria. Moisture together with epidermal cell debris is conducive to growth of micro-organisms. Increased moisture promotes the ionisation of any silver compound used as antibiotic in textiles and facilitates its binding to sulfydryl groups on epidermal keratin. Experience with sustained silver-release wound dressings has shown that only a small proportion of silver released from a high silver-content wound dressing is absorbed percutaneously.[28]

Toxic responses attributable to contact between textiles and the skin is a large and complex subject. Virtually all natural and man-made fibres are allergenic in some people and such general terms as shirt dye dermatitis, photo-toxic dermatitis and shoe dermatitis are documented. Storrs recorded that untreated natural and synthetic fabrics and fibres used in clothing manufacture rarely cause skin problems, but that chemicals added to them in manufacturing processes are a potential cause of irritancy and allergic dermatitis in predisposed persons.[29] Of the 3000 chemicals used in clothing and shoes in 1978, about 1200 were dyes and pigments, others being softeners, detergents, preservatives, flame retardants and conditioning agents.[30,31] Silver ion is a rare sensitiser but has not been reported to cause allergy when used as an antibiotic in clothing.[9,32] Further studies are required.

7.5 Silver Technology in Antibiotic Textiles

Silver has been introduced as an antibiotic into natural or man-made fibres with the objective of producing textiles with:

- an homogeneous coating or impregnation;
- a propensity to liberate bioactive Ag^+ for bactericidal and fungicidal action;
- silver ion release consistent with current safety standards and sufficiently durable to provide therapeutic and prophylactic effect over the expected lifetime of the clothing, footwear or other hygiene product;
- antibiotic action that is not diminished by successive garment washes or exposure to UV light or other environmental influences.

The "silvering" treatment of textile fibres may be achieved by coating or embedding in ionisable silver compounds during the initial spinning process or during the finishing process of yarn or fabrics.[33] Pretreatment of textile fibres with resins or other modification may be desirable to promote attachment of the silver particles to prolong the antibiotic viability of the finished product. A large number of patent applications for new technological processes for silver coating of polyamide, polyurethane, cotton and other fibres are documented. Whereas silver coating in the spinning phase is more costly, the treatment is more permanent and better controlled than silvering as an after treatment.

Treatments of textile fibres with ionisable silver compounds range from simple immersion of fibres or "doping" in silver nitrate solutions to the more costly and advanced technology involving plasma-coating, ultrasound sono-chemical techniques, dielectric barrier discharge and silver ion sorption, and exposure of fibres to high energy silver-ion electron beam assisted deposition (IBAD).[14,34,35] Clearly, the more advanced technologies achieve a more consistent silver coating or impregnation and hence improved antibiotic efficacy with longer-lasting antibiotic efficacy.

An example of immersion coating is seen in the production of antibiotic sutures for wound closure and tissue engineering with a silver glass complex (Bioglass®) as discussed in Chapter 5.[36,37] Stable and homogeneous coatings were achieved on silk or Vicryl (polyglactin 910) resorbable sutures using an optimised aqueous slurry dipping technique in which the fibres were exposed to the glass under controlled environmental conditions. The silver–Bioglass coating was stable and did not affect the dynamic or mechanical properties of the fibres and commendable antibacterial and clinical effects were demonstrated *in vitro* and *in vivo* (see Figure 5.11).[38,39] In an alternative technology based on immersion, tetrasilver tetroxide was introduced into textile fibres by initial immersion in ammoniacal silver nitrate followed by treatment with an oxidising agent capable of depositing the novel silver antibiotic within the threads of the textile.[13,40] This novel silver precipitate was claimed to be sparingly soluble in water but sufficient to provide a sustained antibiotic action in woven and non-woven textiles.

Plasma technology has been developed for the silver coating of textile fibres and yarns.[41,42] Textile fibres modified by exposure to silver atom beam become progressively coated with metallic silver for weaving and use in hygiene clothing. Swiss Materials Science and Technology (St Gallen, Switzerland) claimed that the technology enhanced the biocompatibility of the textiles and that silver coating imparts conductivity to fibres enabling their use in textile based electrodes and electro-stimulation products.[41,42] The "sputter" process patented for silvering three-dimensional knitted fabrics also used an electric current to aid silver deposition. High energy plasma discharge across two electrodes in an argon atmosphere is used to ionise silver and to coat textile fibres without altering their tensile strength or durability.[33] Bio-Gate Innovative Technologies in association with the Bremen-based Fraunhofer Institute for Manufacturing Engineering and Applied Materials Research developed a technology introduction of elemental silver particles into polymers.[14] They developed a vacuum vaporisation and re-condensation process on a continuous liquid film to produce silver particles of 5–10 μm for incorporation into textile fibres. The technology has been applied to antibiotic catheter production where polymer coatings of 10^{12} to 10^{13} activated silver nanoparticles per gram in polyurethane and silicone materials have provided efficacious antimicrobial action in *in vitro* studies.[43]

Silver modification of PET textiles for use in heart valve sewing cuffs for prevention of prosthetic valve related endocarditis involved an advanced silver coating technology developed by the Spire Corporation (Bedford, MA).[35,44-46] PET polyester fabric, uncrimped and heat set was coated with metallic silver (3000 Å thick) using a high energy ion beam assisted deposition (IBAD) technology. Silver vapour is formed by an electron beam and used to bombard the surface of materials to be coated. Although the technology was initially developed for imparting antibiotic properties to in-dwelling catheters, it was subsequently adapted for use in silvering a wide range of fibres, polymers and prosthetic materials.

Manufacture of the cellulosic fibres with antibiotic properties in the Lyocell process involves an initial grinding and preparation of brown and blue algae, suspension and isolation of pure cellulose fibres, and spinning.[16,47] The purified cellulosic fibres are exposed to and readily bind and become impregnated with silver, zinc and copper to produce a "loaded" and antibiotically efficacious SeaCell® fibre for use in hygiene textiles and footwear. Kinetic analysis has demonstrated that cellulosic fibres become loaded with silver after exposure to 0.01 or 0.1N silver nitrate in 1–3 minutes. Production of the finished Sea Cell Active® cellulosic fibres was complemented by laboratory tests establishing their antibacterial durability for 60 standard washing cycles, compatibility with human skin and predictable safety in use.

Spacer fabrics have been developed for use in footwear and products which require sufficient cushioning from the skin to allow thermoregulation, moisture control, pressure distribution and elasticity for comfort. Silver has been incorporated into three-dimensional knitted fabrics comprising two outer layers linked with pile threads and enclosing an air layer providing the desired breathability for the finished product. Heide noted that the functionality of space fabrics depends upon the materials selected and their stability in the presence of moisture, perspiration and continued wear-and-tear.[33] Sputter and plasma discharge methods using an atmosphere of argon are preferred to vaporisation methods for coating fabrics as the durability of the resulting coating is claimed to be more durable and efficacious as an antibiotic. Experience with silver-coated spacer fabrics including thread materials such as polyester or polyamide provide excellent antimicrobial efficacy, long-lasting effect and safety in use.[33] Many new products have been introduced in recent years.

7.6 Summary

The use of silver in the production of antibiotic textiles for hygiene clothing, garments for hospital patients and clinical staff, and for personal and domestic products, is a relatively new science. New technology with advanced development in atomic silver coating, sputter and plasma treatments has enabled production of natural and synthetic fibres with even distribution of nanocrystalline silver for use in hygiene clothing, footwear, medical sutures and prosthetic cardiac devices. The various methods of silver doping, immersion and atomic silver spray technologies have ensured durable and long-lasting antibiotic protection of textiles for human skin contact, efficacy against expected skin pathogens and safety in use. However, a large proportion of the information currently available is still based on laboratory studies, with evaluation of antibiotic efficacy in *in vitro* tests using type strains of known human pathogens. There is an urgent need for wide-ranging, accurately monitored and statistically validated product-in-use surveys to substantiate the commercial and clinical value of antibiotic textiles.

7.7 Further Reading

U.-C. Hipler and P. Elsner, *Biofunctional Textiles and the Skin,* Karger, Freiberg, 2006, Current Problems in Dermatology, Vol. **33**.

References

1. G. Ricci, A. Patrizi, F. Bellini and M. Medri, Use of textiles in atopic dermatitis: care of atopic dermatitis, *Curr. Probl. Dermatol.*, 2006, **33**, 127.
2. A. B. G. Lansdown, Silver in health care: antimicrobial effects and safety in use, *Curr. Probl. Dermatol.*, 2006, **33**, 17.
3. A. B. G. Lansdown, A. Williams, S. Chandler and S. Benfield, Silver absorption and antibacterial efficacy of silver dressings, *J. Wound Care*, 2005, **12**, 113.
4. A. Gauger, Silver-coated textiles in the therapy of atopic eczema, *Curr. Probl. Dermatol.*, 2006, **33**, 152.
5. A. Gauger, M. Mempel, A. SchekatzM, T. Schäfer, J. Ring and D. Abeck, Silver-coated textiles reduce *Staphylococcus aureus* colonisation in patients with atopic eczema, *Dermatology*, 2003, **207**, 15.
6. E. A. Deitch, A. A. Marino, V. Malakanok and J. A. Albright, Silver nylon cloth: *in vitro* and *in vivo* evaluation of antimicrobial activity, *J. Trauma*, 1987, **27**, 301.
7. C. S. Chu, A. T. McManus, B. A. Pruitt and A. D. Mason, The therapeutic effects of silver nylon dressings with weak direct current on *Pseudomonas aeruginosa* infected burn wounds, *J. Trauma*, 1988, **28**, 1488.
8. J. A. Spadaro, S. E. Chase and D. A. Webster, Bacterial inhibition by electrical activation of percutaneous silver implants, *J. Biomed. Mater. Res.*, 1986, **20**, 565.
9. A. B. G. Lansdown and A. Williams, How safe is silver in wound care, *J. Wound Care*, 2004, **13**, 131.
10. S. Bale, Cost-effective wound management in the community, *Prof. Nurse*, 1989, **12**, 589.
11. D. A. Morgan, Wound management products on the drug tariff, *Pharm. J.*, 1999, **263**, 820.

12. U. Wollina, M. B. Abdel-Naser and S. Verma, Skin physiology and textiles, *Curr. Probl. Dermatol.*, 2006, **33**, 1.
13. M. S. Antelman, High performance silver (I,III) oxide antimicrobial textile articles, *US Pat.*, 6436420, 2002.
14. T. Bechert and P. Steinrücke, Infective surfaces—nanoparticles attacking micro-organisms, *Fraunhofer Magazine*, 2003, **1**, 68.
15. S. Gupta, Keeping textiles fresh for longer, *J. Asia Text. Apparel*, 2008, February, 1.
16. S. Zikeli, Production process of a new cellulosic fibre with antimicrobial properties, *Curr. Probl. Dermatol*, 2006, **33**, 110.
17. U.-C. Hipler, P. Elsner and J. W. Fluhr, A new silver-loaded cellulosic fibre with antifungal and antibacterial properties, *Curr. Probl. Dermatol.*, 2006, **33**, 165.
18. U.-C. Hipler, P. Elsner and J. W. Fluhr, Antifungal and antibacterial properties of a silver-loaded cellulosic fabric, *J. Biomed. Mater. Res. B. Appl. Biomater.*, 2005, **77B**, 156.
19. P. D. Thompson, What is infection? *Am. J. Surg.*, 1994, **167**(Suppl. 1A), 7S.
20. E. L. Larsen, APIC guideline for hand washing and antisepsis in healthcare settings, *Am. J. Infect. Control*, 1995, **23**, 251.
21. W. C. Noble, *Microbiology of the Human Skin*, Lloyd-Lukje Medical Books, London, 1981.
22. J. H. Hall, J. L. Callaway and J. P. Tindall, *Pseudomonas aeruginosa* in dermatology, *Arch. Dermatol.*, 1968, **97**, 312.
23. N. H. Shehadeh and A. M. Kligman, The effect of topical antibacterial agents on the bacterial flora of the axilla, *J. Invest. Dermatol.*, 1963, **40**, 61.
24. Y. M. Clayton, Scalp ringworm (*Tinea capitis*), in *Superficial Fungal Infections*, ed. J. Verbov, MTP Press, Manchester, 1984, pp. 1–8.
25. C. De Vroey, Epidemiology or ringworm (dermatophytosis), *Semin. Dermatol.*, 1985, **4**, 185.
26. S. Rothman, G. Knox and D. Windhorst, *Tinea pedis* as a source of infection in the family, *Arch. Dermatol.*, 1957, **75**, 270.
27. J. J. Leyden and A. M. Kligman, Inter-digital athlete's foot: the interaction of dermatophytes and residual bacteria, *Arch. Dermatol.*, 1978, **114**, 1466.
28. T. Karlsmark, R. H. Agerslev, S. H. Bendz, J. Larsen, J. Roed Petersen and K. E. Andersen, Clinical performance of a new silver dressing, Contreet Foam, for chronic exudating venous leg ulcers, *J. Wound Care*, 2003, **12**, 351.
29. F. J. Storrs, Dermatitis from clothing and shoes, in *Contact Dermatitis*, ed. A. A. Fisher, Lea and Febiger, Philadelphia, 1987, pp. 283–337.
30. M. Braitman, Dermatitis and fabrics, *J. Med. Soc.*, 1955, **52**, 757.
31. E. Cronin, Clothing and textiles, in *Contact Dermatitis*, Churchill Livingstone, Edinburgh, 1980.
32. C. Trimmer, Disinfectants, in *The Irritant Contact Dermatitis Syndrome*, eds. P. G. M. Van der Valk and H. I. Maibach, CRC Press, Boca Raton, FL, 1996, pp. 77–86.

33. M. Heide, U. Möhring R. Hänsel, M. Stoll, U. Wollina and B. Heinig, Antimicrobial textile three dimensional structures, *Curr. Probl. Dermatol.*, 2006, **33**, 179.

34. M. Kostiæ, N. Radic, B. M. Obradoviæ, S. Dimitrijeviæ, K. M. Kuraica and P. Škundric, Antimicrobial textile preparation; silver deposition on dielectric barrier discharge treated cotton/polyester fabric, *Chem. Ind. Chem. Eng. Quart.*, 2008, **14**, 219.

35. E. Tobin and R. Bambauer, Silver coating of dialysis catheters to reduce bacterial colonisation and infection, *Ther. Apher. Dial.*, 2003, **7**, 504.

36. J. J. Blaker, S. N. Nahzat and A. R. Boccacini, Development and characterisation of silver-doped bioactive glass-coated sutures for tissue engineering and wound healing applications, *Biomaterials*, 2003, **25**, 1319.

37. J. J. Blaker, A. R. Boccaccini and S. N. Nazhat, Thermal characterisations of silver-containing bioactive glass-coated sutures, *J. Biomater. Appl.*, 2005, **20**, 81.

38. A. B. G. Lansdown, J. J. Blaker, I. Thompson, P. Pavan, A. R. Boccaccini and L. L. Hench, Bioglass® and Silver-Bioglass®—a novel therapy using bioactive scaffold and silver "doped" sutures in acute wounds, *Wound Rep. Regen*, 2003, **11**, P.063.

39. A. B. G. Lansdown, Bioactive dressings: old ideas, new technology, *Br. J. Nurs.*, 2007, Tissue Viability. Suppl., **16**, S3.

40. S. T. Dubas, P. Lumlangdudsana and P. Potiyaraj, Layer-by-layer deposition of antimicrobial silver nanoparticles on textile fibres, *Colloids Surf. A*, 2006, **289**, 105.

41. NanoEurope, Plasma-coated yarns for medical applications, Media Release, 19 July 2007.

42. Medical Device Link, Technology news surfactant treatment—plasma-coating process produces multifunctional medical textiles, European Medical Device Manufacture, Canon Communications, 2007.

43. U. Samuel and J. P. Guggenbichler, Prevention of catheter related infections: the potential of a new nano-silver impregnated catheter, *Int. J. Antimicrob. Agents*, 2004, **23S1**, S75.

44. K. S. Tweeden, J. D. Cameron, A. J. Razzouk, R. W. Bianco, W. R. Holmberg, R. J. Bricault, J. E. Barr and E. Tobin, Silver modification of polyethylene terephthalate textiles for antimicrobial protection, *Am. Soc. Art. Int. Org. J.*, 1997, **43**, M475.

45. K. S. Tweeden, J. D. Cameon, A. J. Razzouk, W. R Holmberg and S. J. Kelly, Biocompatibility of silver-modified polyester for antimicrobial protection of prosthetic valves, *J. Heart Valve Dis.*, 1997, **6**, 553.

46. D. Langaki, M. F. Ogle, J. D. Cameron, R. A. Lirtzman, R. F. Schroeder and M. W. Mirsch, Evaluation of a novel bioprosthetic heart valve incorporating anti-calcification and antimicrobial technology in a sheep model, *J. Heart Valve Dis.*, 1998, **7**, 633.

47. S. Zikeli, Lyocell fibers with health-promoting effect through incorporation of seaweed, *Chem. Fibers Int.*, 2001, **51**, 272.

CHAPTER 8

The Toxicology of Silver

8.1 Introduction

Silver is ubiquitous in the human environment, being found in the air we breathe, drinking water, plant and animal foods, and in the soil. The vast majority of people have small amounts of silver in their bodies without any obvious clinical signs or metabolic disturbances.[1] There is no tangible evidence that silver functions as a trace metal in the human body or fulfils a physio-logical or biochemical role in any tissue. However, silver metal and most of its inorganic salts readily ionise to some extent in the presence of moisture, body fluids and secretions to release the biologically active Ag^+ which interacts with body tissues and is absorbed into the body by oral ingestion, inhalation or percutaneous routes to be deposited in soft and hard tissues. In most people, the amount of silver accumulating in the body through every day exposures is very low ($< 2.3\,\mu g\,L^{-1}$), but higher concentrations can be expected in workers engaged in silver industries, photographic work, X-radiography and silver plating.

Recent advances in biotechnology and the wider appreciation of the anti-biotic properties of silver have led to its use in a wide variety of dressings for the management of infected wounds, medical devices, textiles and domestic appliances. As noted in Chapter 3, there is limited evidence to show that silver uptake in these situations will significantly alter the blood levels (argyraemia) or tissue content of silver or produce recognisable signs of toxicity. However, increasing use of silver as a coating or impregnate in materials used in "implantable" or "in-dwelling" devices—intravascular and intraperitoneal catheters, bone cements and prostheses, dental materials, cardiovascular valve sewing cuffs, stents and intracerebral cannulae—illustrate situations where metallic silver or Ag^+ are introduced directly into the circulation with the possibility of higher levels of argyraemia and tissue silver concentrations. This chapter discusses the implications of transient rises in argyraemia and silver excretion.

Issues in Toxicology No. 6
Silver in Healthcare: Its Antimicrobial Efficacy and Safety in Use
By Alan B. G. Lansdown
© Alan B. G. Lansdown 2010
Published by the Royal Society of Chemistry, www.rsc.org

8.2 Argyria, Argyrosis

Argyria is the principle condition associated with excessive silver exposures; it is not a disease and there is no evidence that argyria *per se* is life-threatening. Case studies are reported where patients with profound argyria have died through pre-existing diseases or causes unrelated to silver-induced pathology.[2,3] Argyria may be defined as a permanent or long-lasting grey or blue-grey discolouration of the skin attributable to prolonged exposure to metallic silver or ionisable silver salts (Figure 8.1). It can occur in a wide variety of situations and be localised as in the case of implantation of silver-loaded acupuncture needles, or generalised throughout the body with discolouration of the eyes (argyrosis) and internal organs.[4–9] It is most noticeable in skin areas exposed to solar radiation. Fair skinned people with low levels of ultraviolet (UV) energy absorbing melanin are more vulnerable to argyria.[8] At most, argyria should be considered as a cosmetically undesirable occupational risk or complication associated with consumption of unregulated colloidal silver products as recommended remedies for treating immunodeficiency virus (AIDS), acquired immune deficiency syndrome, cancer, tuberculosis, shingles, tonsillitis, bubonic plague, recurrent boils, acne and chronic fatigue.[9–13] Argyria may be a cause of severe disfigurement.

Argyria (also listed as argyriasis, argyrism and argyrosis) is predominantly a condition affecting the skin and its appendages attributable to minute particles of metallic silver, silver sulfide or silver selenide containing granules in the connective tissues of the dermis in the region of sweat ducts, hair follicles, blood vessels and nerves.[4,5,7,14] Hill and Pillsbury did not have access to modern-day microanalytical equipment and recorded that the silver-laden deposits were

Figure 8.1 Argyria: a case of grey skin discoloration attributable to chronic ingestion of a colloidal silver preparation. (*By courtesy of B. A. Bouts, Ohio, USA*).

either metallic silver or in "combined form" as albuminate, chloride, sulfide or sulfate, and that "some type of albuminate and the chloride must play a role".[9] The fingernails, cornea, conjunctival membranes and buccal mucosa may also discoloured.[15] Argyria is distinguished from argyrosis which is more specifically applied to deposition of silver sulfide in the cornea or conjunctiva of the eye and which on rare occasions may lead to semi-blindness or impaired night vision through obstruction of light rays reaching the retina[16–18] (see Chapter 9). The higher incidence of argyrosis in workers exposed to silver nitrate or silver oxide suggests that deposition of silver precipitates in the eye precedes frank manifestations of argyria.[19] Slit lamp examinations and electro-physiological and psycho-physiological tests reveal no functional defects in affected patients and that argyrosis is a benign side effect.[16,17]

Argyria and argyrosis are closely related to the amount of silver in the systemic circulation, but the minimal level of argyraemia and silver intake leading to frank manifestations of the condition are equivocal and possibly depend on the rate and route of silver exposure and its duration.[9] Estimates for the minimum exposure period, blood concentrations and total body burden of silver necessary to evoke argyria and argyrosis vary greatly according to the date of the study and the type of evaluation conducted, and the sensitivity of the analytical technology available.[1] Study of individual case studies can be misleading, as illustrated by Hill and Pillsbury who tabulated 357 cases of argyria reported up until 1939 mainly following administration of silver nitrate or colloidal silver preparations (Argyrol, Collargol, Neo-Silvol, Cryptagol, Protargol), with onset of the condition being noted in children below the age of ten.[9] They quoted a statement attributable to a Dr Spiegel in which he stated that: "I have had hundreds of cases in which as much as 12 g of silver arsphenamine have been given without any unpleasant sequali". At the time argyrias were classified under:

- General therapeutic argyria;
- Local therapeutic argyria;
- Therapeutic argyria;
- Industrial argyrosis;
- General industrial argyria.

Dosage regimens and manifestations of argyria varied greatly and none gave accurate indications of the minimal amounts of silver uptake necessary to evoke skin discolouration. Routes of administration ranged from oral, intranasal, gargle and throat swab, throat painting and lip painting over periods ranging for a few days to several months. Instillation of silver nitrate or Collargol into the urethra, vagina or bladder to alleviate venereal disease were listed causes of generalised argyria.

Gaul and Staud were perhaps first to attempt a quantitative and qualitative evaluation of silver deposition in the human body and to estimate concentrations of silver necessary to evoke argyria using bio-spectrophotometric analyses.[20,21] In 1935, they were aware of the increasing incidence of argyria, particularly in young children dosed intranasally with Argyrol or Neo-Silvol for respiratory infections and reported that:

> "*Argyrol solutions were available at that time in any strength (25–50% solution once daily) to be used in the throat, nose or ears as often as the physician desires, with perfect safety and freedom from irritation*".[20,21]

Colloidal silver atomisers were also available. Gaud and Staud also evaluated the argyrogenic potential of the antisyphilitic drug, silver arsphenamine (presumed to contain > 19% arsenic and 12–14% silver). Its efficacy was attributed to "two chemotherapeutic agents, arsenic and silver and possibly the catalytic action of silver on the arsphenamine molecule".[21] In summarising reports of ten argyric patients published between 1914 and 1934, Gaul and Staud concluded that silver arsphenamine could not be given at a dose exceeding 8 g because of the risk of argyria and that administration of 18 g or more is apt to be followed by argyria.[21] They noted that, "in some patients", a discolouration developed after total intravenous doses of 4, 7, or 8 g of the drug whereas, in others, doses of 10, 15 or 20 g led to symptoms of argyria. In their own research, they examined the so-called silver line (biospectrogram) in the skin of people not known to have been exposed to silver occupationally or therapeutically, and those treated with silver arsphenamine.[21] On the basis of their observations in four patients and using a microphotometer to calibrate the density of the silver line, they estimated that normal skin would contain equivalent to 0.1–0.5 g of silver arsphenamine and that light to very deep discolourations represented injection of 8–20 g of the drug. In a study of 500 biospectrograms of skin biopsies from children and adults, the density of the silver line in 5% was judged to be equivalent to a patient dosed with 2–5 g of the drug; however, there were wide discrepancies, possibly attributable to age-related differences in body silver retention and patient lifestyle. Gaul and Staud indicated that the silver content of the body was directly proportional to the concentration administered but seemed to be oblivious of the fact that silver is eliminated from the body in the bile and urine.[1,9] In four cases, they claimed total doses of silver arsphenamine (intravenous) of 1.45, 2.05, 4.9 and 5.775 g led to skin silver concentrations of $0.01–7.0 \times 10^{-6}$ g, but it is unlikely that the quantitative methods available at the time were sufficiently accurate to justify these estimations[21] (see Chapter 2). In summary, whilst these workers provide a wide insight into the problems associated with silver ingestion or therapeutic use, their recommendations should be viewed with caution in view of more competent work conducted more recently. Gaul and Staud's claims that "argyria becomes clinically apparent after a silver retention approximating to an equivalent of 8 g of silver arsphenamine based on biospectrophotometric analysis" is regarded as inadequate statistically and technically flawed. The validity of the observations is also questioned on the inconsistency of the 70 case studies reported and unsound assumptions. It is noteworthy that several of the reported case studies were possibly complicated by exposures to other metals such as bismuth, mercury and arsenic, all of which show evidence of skin toxicity. As discussed below, the molecular formula of silver arsphenamine is not yet chemically resolved such that the actual amount of silver administered cannot be known.

In 1981, the US Environmental Protection Agency (USEPA) reasoned that a total body burden of at least 1 g silver was consistent with argyrosis,[22] though other estimates suggest levels ranging from 1–30 g for soluble silver salts;[8,23,24] however, it is generally recognised that, in overt argyria, the total amount of silver found in the skin is increased by 80–100 fold, or in one exceptional case 8000 fold.[23–27] Argyria is rarely encountered these days on account of tighter regulations in occupational health conditions and mandates for consumer safety; as such, modern day scientists are precluded from quantifying silver concentrations in discoloured tissues using accurate mass spectrophotometry or microanalytical means. Argyria is also attributable to occupational exposures to metallic silver, silver oxide and silver nitrate fumes or aerobic particles; exposure to silver in jewellery or use of silver salts for irrigation of the nasal tract or mucosal membranes of the urinogenital tract. Topical treatment with silver sulfadiazine in burn wound therapy, and excessive use of silver acetate antismoking remedies have been implicated.[26–30] A rare case was reported of a patient who developed argyria after injecting silver nitrate intravenously as an antisclerosant to alleviate varicose veins.[31] Rarely in these cases are the details of silver exposure given or the actual blood silver levels reported.

The histopathological features of true argyria and the chemical profile of silver deposition have been amplified by use of electron microscopy and X-ray microanalysis.[4] Argyria takes the form of small black–brown refractile granules deposited in the dermis in the region of the basal epidermal lamina and hair follicles, small blood vessels and sweat ducts, but not in epidermal structures. The electron-dense granules of silver sulfide and/or silver selenide ranged in size from 30–100 nm and were most numerous in relation to arterioles, nerve cells (Schwann cells), dermal elastic and collagen fibres and macrophages in the papillary layer of the dermis.[3,4] Quantitative studies in patients exhibiting occupational argyria showed melanocytes to be normal in size and shape, and showing the characteristic reaction for dihydroxyphenylalanine (DOPA) consistent with melanogenesis. Substantive evidence to show that silver overload stimulates melanogenesis has not been seen.[32] As discussed later, macrophages have a central role in the metabolism of silver in soft tissues like the liver and kidney.[33] In the skin, silver granules were enclosed within lysosomal vacuoles. Electron microscopy has confirmed that silver-laden granules were absent from epidermal structures and that no evidence is available to support earlier theories that silver is excreted or transferred from dermis to epidermis through an intact epidermal basement membrane in normal human skin.[4,34] Chemical analysis of the silver-containing granules revealed the presence of high X-ray absorption peaks consistent for silver, sulfur and selenium, but evidence was also provided for the presence of lead, osmium, copper, mercury, titanium and iron;[7] metals other than silver were rarely seen in electron-dense granules in macrophage lysosomes and may have been inhaled from fumes in the factory atmosphere. Bleehan *et al.* considered that neutron activation analyses confirmed that silver accumulated preferentially in the skin, bone and liver and may be derived from such extreme sources as cooking vessels, as well as the wide diversity of products containing silver for antibiotic purposes.[4,25]

The mechanism for silver precipitation in the skin as silver sulfide, silver selenide or as minute particles of metallic silver in the mid-dermis is still unclear. Aasath *et al.* referred to the importance of selenium in the skin as a potential factor in argyria and estimated that the selenium content in the skin of patients with argyria was tenfold higher than normal; silver precipitates as the highly insoluble selenide, which contributes to the long-lasting or permanent manifestations of argyria.[7,35,36] Solar radiation undoubtedly has a role in reducing silver complexes in the peripheral circulation and it is well known that Ag^+ avidly precipitates as silver sulfide, also a highly insoluble blackish discolouration. Buckley and Terhaar suggested that an equilibrium exists in argyric skin between the relatively insoluble precipitates of silver sulfide and silver selenide in the dermis and soluble silver mercaptides which ionise above pH7 to release free Ag^+.[34] These workers argued that an increase in soluble silver leads to precipitation of more insoluble silver by a reductive process as illustrated in the following formula:

$$Ag^o \underset{+e}{\overset{-e}{\rightleftarrows}} Ag^+$$

The process is associated with lysosomes, where the molar concentration of silver was estimated by the number of granules to be 1×10^{-5}. Buckley and Terhaar argue that molar concentration of silver is "similar to the minimum amount required to maintain oxidative stability in colloidal dispersions like Argyrol or mild silver proteins".[34] Furthermore, it seems that a critical amount of soluble silver exists below which any lysosomal reductase enzyme is inactive and silver remains solubilised. Sulfur on the other hand is determined by X-ray microanalysis and the ratio of silver to sulfide in argyric tissues is similar to that of silver sulfide (Ag_2S) in an inorganic standard.[33]

Argyrosis or deposition of silver sulfide in the cornea is pathologically distinct from several reported instances of silver damage in the eye through inappropriate use of silver nitrate or other silver therapy for ocular cysts, conjunctivitis or periocular infections.[37] Wadhera and Fung referred to argyrosis as the most common form of localised argyria characterised by a permanent dirty grey to brownish discolouration of the eye and contiguous parts.[13] Whereas ocular damage arising from the use or abuse of silver nitrate or colloidal silver preparations may be expected to cause severe discomfort through the astringent properties of nitric acid, argyrosis arising as part of a generalised argyria is not a cause of discomfort. Some ocular difficulties may be presented by accumulation of black silver sulfide deposits in the cornea and possibly the lens, retina and around the optic nerve as discussed in Chapter 9.

Case studies illustrate that argyrosis may develop following use of silver in cosmetics, antiseptic eye drops, mild silver protein (Argyrol), occupational exposure to silver in soldering and accidents in silver plants.[13,18,38–41] Histopathologically, the lesions revealed silver-containing granules in cornea, upper eyelid, eyelid margin, Descemet's membrane and basement membrane, and superficial substantia propria of the conjunctiva.[40,42] Slit lamp investigations, confocal microscopy, light and electron microscopy with energy dispersive X-ray microanalysis have demonstrated reflective silver-containing granules anterior to and within the corneal basement epithelium, the anterior stroma, Bowman's layer.[38,39] Lysosomally bound silver-containing granules were confirmed also within degenerative keratinocytes and extracellularly in association with collagen fibres and cellular debris.

Argyroses are rare these days in view of improved standards of occupational safety and hygiene, but they may occur indirectly as a result of workers rubbing their eyes and eyelids with silver-contaminated cloths or gloves. Indiscriminate use of silver-containing eyelash cosmetics and tints over a long period will lead to argyrosis and a permanent "black eye syndrome".[40] Silver compounds may ionise within lachrymal secretions to be absorbed locally into the eye and surrounding membranes.[41] As with more generalised argyrias, argyrosis may occur in individuals consuming colloidal silver products for gastrointestinal diseases of rhino-laryngeal infections.[12,13] On occasions, they may present in the form of conjunctival melanomas, but a patient's clinical history and biopsies with appropriate microscopy are available to confirm the true nature of these benign silver granules.[42]

8.3   Toxicology of Silver in Cell Cultures

In vitro toxicology is the newest branch of the science of safety evaluation and has become increasingly important in product development and regulation in recent years as an alternative to *in vivo* testing using live animal models.[43] The recent upsurge of interest in conducting preliminary toxicity studies using bacteria, fungi, protozoa or human cell lines in culture is largely attributable to anxieties over the excessive use of live animals in medical experiments and acknowledgement of the relevance of the financial and temporal restraints in research programmes.[44,45]

Models of *in vitro* toxicology have been developed over the past 40 years principally as a predictive tool for assessing mutagenic and carcinogenic potential of drugs, food and food additives, chemicals and processes in industrial plant, occupational exposures and environmental contaminants. Guidelines set out by the International Agency for Research on Cancer (IARC) defined tests in prokaryotic cells (bacteria), mammalian cell lines and *Drosophila melanogaster* to identify substances capable of evoking:

- chromosomal and DNA damage;
- cell transformation;

- mitotic cross over and gene conversion or expression;
- cytological changes as evidence of carcinogenic potential.[46]

These tests have been re-appraised and updated as newer information has been presented and their predictive capacity defined.[43] The battery of *in vitro* tests now includes cellular and subcellular studies designed to identify:

- the sensitivity of specific isolated cell types or tissues to xenobiotic damage;
- target sites on cell membranes and intracellular sites;
- molecular mechanisms of cellular damage;
- cytoprotective mechanisms;
- intracellular management and metabolism of toxic materials.

In each case, cell lines derived from human or mammalian tissues are exposed to test materials under defined environmental conditions and with clear end points.[47] Such tests are reproducible, relatively inexpensive in relation to preclinical studies in live animals, and readily available in most laboratories. *In vitro* studies in isolated cells allow investigation of macromolecular mechanisms of toxicity and the role of mitochondria, lysosomal systems and intracellular enzymes in detoxification.

8.3.1 *In vitro* Cytogenicity

Cell culture systems have been developed in recent years to examine the intracellular metabolism of xenobiotic materials and mechanisms of toxicity.[48] The tests have been specifically adapted according to the objective of the investigations and the expected target organ/tissue in the human body (*e.g.* liver, kidney, fibroblast, *etc.*). Whereas isolated hepatocytes have provided useful information regarding phase I and phase II transformation of xenobiotic materials and the role of metabolising enzymes, serial cultivations of human epidermal keratinocytes have provided information on the cytotoxicity of silver ion eluted from wound dressings, medical devices and textiles.[49–52] Disaggregated epidermal cells, sheets of epidermis and explants of partial or full thickness human skin have been grown successfully in cell, organ and explants cultures to provide reproducible and informative models for predictive toxicology. Thus serial cultivations of strains of human diploid epidermal keratinocytes leading to keratinising colonies from single cells have formed the basis for many experiments designed to investigate the role of silver ion in impairing wound re-epithelialisation and repair following injury.[49,50] These experiments demonstrated the importance of fibroblasts in modulating colony formation in keratinocytes and maturation leading to the expression of stratified keratinised epithelium. Such models provide the opportunity to investigate molecular patterns of toxicity and intracellular target sites for xenobiotics like silver, which is known to bind avidly with the sulfydryl residues of cysteine thereby impairing keratohyalin synthesis.[51] Further, specific culture conditions can be created to investigate certain phenotypic traits and capacity for proliferation and maturation.[50] However, like human diploid fibroblasts, cultured keratinocytes have a finite lifetime in culture—possibly a maximum of 50 cell generations. The plating efficiency of human keratinocytes declines with advancing age of the donor.

More specific *in vitro* toxicity studies have evaluated the mechanisms and potential risks of silver ion as a cause of leukopenia, as recorded in clinical studies.[53,54] Low levels of silver ion released from metallic silver as in dental devices or silver sulfadiazine were inhibitory and cytotoxic to cultured monocytes, lymphocytes and neutrophils in culture media.[55] Hansbrough *et al.* demonstrated that silver suppressed the oxidative burst in human peripheral blood neutrophils stimulated by N-formyl-L-methionyl-L-leucyl-L-phenylalanine (mitogen) at sub-clinical concentrations, and inhibited neutrophil and lymphocyte activity.[56] These effects were partially reversed when cells were washed and resuspended in medium before exposure on silver ion.[57] Human diploid fibroblasts and fresh human donor dermal fibroblasts were inhibited by short-term exposure to silver sulfadiazine and impaired proliferation was associated with marked changes in cell morphology, cytoplasmic deterioration and degeneration of nuclei and cell organelles.[58] Further studies with fibroblast cell lines in culture have revealed details of "protective systems".[59] Thus growth factors such as platelet-derived growth factor (PDGF), epidermal growth factor (EGF) and basic fibroblast growth factor (β-FGF), which all play instrumental roles in cell motivation, migration and functional maturation in epidermal repair following injury, have been shown to cytoprotect dermal fibroblasts from injury by silver sulfadiazine. This suggests that cells activated by growth factors are more resistant to the toxic effects of silver antibiotic, but this is unconfirmed by clinical or *in vivo* studies.

In vitro cell systems designed to evaluate the cytotoxicity of Ag^+ released from nanocrystalline silver, silver nitrate, silver sulfadiazine and a wide range of silver-containing medical devices have routinely sought to monitor loss of cell viability in relation to ionic concentration. However, in view of the wide diversity of silver-containing products and variations in the ionising potential of the silver compounds and complexes used, it is impractical to compare patterns of cytogenicity against any cell type in culture other than in terms of free silver ion.[60] A proportion of the ion released from any product will precipitate as silver chloride, silver phosphate or silver proteinates with constituents of culture media, and hence be inactivated.[60] The cytolethal concentration is defined as the level of Ag^+ that binds to and inactivates sulfydryl groups or other protein residues on cell membranes, and that which is absorbed intracellularly to interact with and disable key metabolic enzymes including DNA/RNA-ases.[61]

Cellular and subcellular changes in isolated cell lines to silver are similar to those reported above for bacterial cells though eukaryotic cells differ in their sensitivities to Ag^+. Hildago *et al.,* confirmed the very low concentrations of silver nitrate (> 7–550×10^{-5}%) required to kill *Staphylococcus aureus, Citrobacter freundii* and *Pseudomonas* aeruginosa.[62] In contrast, when human fibroblasts were treated with silver nitrate in the range of 7×10^{-5}% with foetal calf serum, high cytotoxity (71 ± 19%) was observed within 2 h of incubation, with silver ion affecting mitochondrial function. Fibroblast sensitivity was time and silver concentration dependant. Cytotoxicity was enhanced causing 76% growth inhibition at concentrations of 14×10^{-5}% silver nitrate. Foetal calf serum was shown to exert a cytoprotective effect on cultured fibroblasts. Fibroblasts tend to be more sensitive that keratinocytes, although when both cell types were grown in the same medium, sensitivities to Ag^+ released from nanocrystalline silver, silver nitrate or commercially available wound dressings were similar.[63] Colorimetric assays using spectrophotometric analysis of the blue formazan product of 3-(4,5-dimethylthiazol-2-yl)-2,5-diphenyltetrazolium bromide at wavelengths of 540–595 nm have proved a sensitive index of cell death,[64] but such methods as flow cytometry, analysis of growth patterns and migration have been complimentary.[65] Recent cytokinetic and morphological studies in cultured human lung fibroblast and glioma cell lines exposed to nanocrystalline silver suggest that cells are more vulnerable during G_2/M phase of the cell cycle and that cell injury is principally due to the action of the Ag^+ on mitochondrial cell function, ATP synthesis, reduced oxygen reactive species and oxidative stress.[66] The implications of these findings on the use of silver in wound healing and cancer therapies are speculative at present.

8.3.2 Screening Assays for Carcinogenicity and Mutagenicity

The *in vitro* tests required for full preliminary screening for mutagenicity and carcinogenicity as set out in the IARC guidelines[46] have not been completed for silver. Nevertheless available knowledge provided by *in vitro* cytotoxicity tests and *in vivo* experience indicates unequivocally that silver is not carcinogenic in any tissue and should be placed in the "No Risk" category, or Class D.[67,68]

Whilst metallic silver is inert in the presence of biological materials, most silver-containing products ionise to some extent in the presence of moisture or body fluids to release biologically active Ag^+ to interact with inorganic anions, protein residues expressing sulfydryl, carbonyl, carboxy, imidazole and other moieties on prokaryotic or eukaryotic cell membranes or intracellularly. These interactions leading to denaturation, inhibition and destruction of key metabolic enzymes such as glucose oxidase, glutathione peroxidise, alkaline phosphatase, DNAases and other enzymes involved in lipid peroxidation and oxidative mechanisms are contributory to cell death.[57,62,63,69–71] Silver as silver sulfadiazine complex interacts with and denatures isolated DNA.[72] Silver ion is noted as a potent inhibitor of trypsin and chymotrypsin, with specific binding between the *Asp* 102 and the δ-nitrogen atom of *His* 57 resulting in impairment in the acetylation rate constant in serine proteases.[73]

Specific *in vitro* test results quoted in support of the non-carcinogenicity/mutagenicity of silver in regulatory appraisals by the USEPA include two older studies showing silver nitrate to be ineffective in backcross mutations in *Escherichia coli* from streptomycin-resistance to non-dependance,[74] and recombinant assays where almost insoluble silver chloride was shown incapable of damaging DNA in a strain of *Bacillus subtilis* at sublethal concentrations.[75] Silver sulfadiazine lacked mutagenic activity when tested in assays against *Salmonella typhimurium*.[76] More recently, silver (Argion) as an antibiotic component of dental cements was shown to be non-mutagenic in four strains of *Salmonella* in the microsome test as defined by Ames test in the presence or absence of S9 fraction from rat liver, whereas other materials used in dental cements such as zinc phosphate, polycarboxylate and glass ionomer cements may have mutagenic properties.[77]

8.4   Systemic Toxicity

8.4.1   General Aspects

Silver is absorbed into the human body mainly through the diet, drinking water and by inhalation; a small amount may enter the systemic circulation by per-cutaneous route, but this is unlikely to be of toxicological significance.[78,79] Uptake is proportional to the ionisation rate of the silver source in the presence of moisture and body fluids, the route of exposure and its duration (as discussed in Chapter 3). Silver is absorbed into tissues and into the circulation in the form of protein complexes (albumins < macroglobulins < other plasma proteins) for distribution to soft tissues (other than the brain and nervous system) and bone in all parts of the body.[1,80,81] The liver and kidney are major organs in the metabolism and elimination of silver from the body, but neither is a target organ for silver toxicity.

Argyria and argyrosis are commonly recognised as signs of the systemic "toxicity" of silver in humans, but there is minimal evidence that even in severe cases of either condition that overdose of silver is fatal; where death has been recorded it has been attributed to pre-existing health problems unrelated to silver.[20] Other signs of silver exposure include silver lines on the gums (especially in the region of teeth with silver-containing dental amalgams), discolouration of the nail beds and darkening of the buccal mucosae. Reports of bleeding gums as a consequence of silver ingestion may result from the cor-rosive effects of silver nitrate rather than to the toxicity of silver, or alter-natively, may be attributable to other toxic factors such as mercury, lead or copper—all of which have been associated with gingivitis.

Evaluation of the true systemic toxicity of silver through acute or chronic exposure in humans or laboratory animal models is extremely difficult in view of the lack of reliable and consistent clinical or experimental data, and the erroneous use of argyria as a criterion for toxicity. Hill and Pillsbury docu-mented a vast amount of experimental and clinical data relating to adminis-tration of silver nitrate, colloidal silver preparations and silver arsphenamine derived from very early publications until 1939, but the reliability of the information is contradictory and of questionable value.[9] They recognised that silver is metabolised to most tissues in the body with the possible exception of the nervous tissue and muscles, but then state that administration of inorganic silver compounds, intravenously and subcutaneously, produced effects that are most marked in the central nervous system with rigidity of the legs, loss of voluntary movements and interference in the motor nerve supply of the heart. The lack of these effects has been substantiated in more recent work (see Chapter 9).

In clinical studies, discolourations of the skin and eyes are cardinal outward signs of exposure to silver but, on occasions, long-term exposure of workers in silver industries fail to show overt signs of argyria, yet silver deposits have been identified in their internal organs. Gaul and Staud maintained that, regardless of the cutaneous site of silver deposition, the quantity of silver in the biopsies was directly proportional to the total dosage of silver administered intravenously in silver arsphenamine administered (range 1.45 g to 5.775 g),[20] but this has not been validated using less toxic silver compounds or by routes of expected silver absorption these days. They predicted that a person of 50 years of age would have a mean silver retention of 0.23–0.48 g of silver equivalent to 1–2 g of silver arsphenamine.

It is unfortunate that many documents for regulatory purposes are based on predictions from clinical case studies or trials involving small numbers of patients/volunteers exposed to silver nitrate, silver sulfadiazine, colloidal silver, silver cyanide and silver arsphenamine where the anions can be expected to present evidence of toxicity. The most reliable evidence of exposure to silver or ionisable silver compounds and alloys available to practitioners nowadays is through evaluation of raised silver levels in blood or urine. As noted in Chapter 3, faecal silver levels are a poor guide to silver exposures by oral route as up to 90% of silver ingested is not absorbed gastrointestinally. Using accurate mass spectrophotometric assays, levels of silver in body fluids and tissues may be reliably monitored for occupational safety assessment and reference value.[1]

Current opinion is that the acute toxicity of silver and most silver compounds is very low and correlates well with the solubility and ionisation of the silver compound under consideration. Thus, the LD_{50} of the relatively insoluble silver oxide in mice is $2820 \, \text{mg kg}^{-1}$, whereas the more soluble compounds like silver nitrate, colloidal silver and silver cyanide show LD_{50} values of 129, 100 and $125 \, \text{mg kg}^{-1}$, respectively.[82] The USEAP and World Health Organization International Programme on Chemical Safety (IPCS) have reviewed published work dating from the 1930s and designated an oral reference dose (RfD) of $5 \, \mu\text{g kg}^{-1}$ body weight for silver as an estimate of daily exposure of humans likely to be without appreciable health risk during a person's lifetime.[83,84] (Interestingly, Gaul and Staud speculated that, while the daily intake of silver through contact or exposure to silver-plated appliances was likely to be infinitesimal, the total increment from chemical action and mechanical friction over a period of many years or decades becomes a sizeable amount.) The exposure criteria for inhalation limits of silver are not available at present, but the American Conference of Governmental Industrial Hygienists (ACGIH) recommended a time-weighted average of $0.01 \, \text{mg m}^{-3}$ threshold limit value for occupational exposure to metallic silver dust.[85] Ag^+ is absorbed more readily from soluble salts than from metallic silver, silver chloride or silver alloys. An

RfD for dermal silver is not practicable on account of the exceedingly low percutaneous penetration of the metal through intact skin.[86] Burn wounds are more sensitive to silver penetration, but other than in clinical management, these are unlikely to be exposed directly to silver.[87] Experience in human patients and animal models shows that even in extensive skin wounds, a large part of the free Ag^+ binds to the proteins of wound exudates or cell debris and is unavailable for absorption. Nevertheless, Coombs *et al.* recorded blood silver levels as high as $310\,\mu g\,L^{-1}$ without evidence of argyria or irreversible damage in soft tissues.[1,87]

An exceptional case study is reported of a chronically ill lady who had consumed approximately 30 mg of silver nitrate capsules daily for alternate periods of two weeks and which led to blood silver levels of $0.5\,mg\,L^{-1}$ by one week after treatment; argyraemia declined slowly in three months.[88] The argyraemia was remarkably high in this patient and accompanied by high levels of silver excretion in saliva, urine and faeces, but generalised argyria was milder than expected and systemic toxicity to silver of minimal toxicological significance.

In experimental animal studies, the systemic toxicity of silver by oral route is not readily appreciated.[89] Dogs dosed with radiolabelled (^{110m}Ag) silver nitrate in drinking water absorbed approximately 10% of the ingested silver whereas humans, monkeys rats and mice absorbed less than 1%.[81] As in humans, silver uptake is influenced greatly by the age and state of health of individuals, and the composition of their diet.[81] Although it is commonly accepted these days that silver entering the human body will be mostly excreted *via* the urine or liver,[1,19,87] a misleading report quoted that:

"the major route of excretion of silver is via the gastrointestinal tract, predominantly through desquamation of silver-containing cells of the alimentary tract, and that urinary excretion has not been reported to occur even after intravenous injection".[89]

Acute toxicity studies in laboratory animals often quoted in evaluating risks associated with oral consumption of silver have used absurdly high and unrealistic levels of silver to evoke fatalities. Thus, a single dose of 500 mg of colloidal silver was shown to evoke severe weight loss, muscle weakness, pulmonary failure and fatality in dogs within 12 h;[90] $50\,mg\,kg^{-1}$ body weight of silver nitrate led to 50% fatality in mice within 14 days[91] and an intraperitoneal administration of $0.2\,g\,kg^{-1}$ body weight of silver nitrate led to fatality in six out of ten guinea pigs after seven days.[86] Intraperitoneal administration of $20\,mg\,kg^{-1}$ silver nitrate in rabbits was also fatal and associated with severe liver and kidney failure with silver granules being identified in both organs.[92]

In subacute experimental studies, adult rats have been shown to tolerate 0.25% silver nitrate in drinking water for up to 16 months (equivalent to 75% lifespan).[93] As in humans, the oral toxicity of silver nitrate or silver acetate is influenced by the quality of the diet. Vitamin E and selenium have been shown to protect against the toxicity of higher oral doses (1500 parts per million) of these silver salts administered in diet or drinking water.[94] Petering (1976) considered that the influence of silver consumption in man is not well understood because of the low gastrointestinal absorption of the metal, but it has been assumed that a single oral dose of 10 g of silver nitrate is fatal.[61,95] This is possibly attributable to the corrosive influence of the nitrate anion and its gastrointestinal irritancy. However, when human volunteers were given 50 mg of metallic silver for 20 consecutive days, they exhibited minimal metabolic changes and survived well.[96] The patients exhibited transitory changes in plasma enzymes and total lipids, but their urine chemistry was normal.

Clinical and experimental studies indicate that uptake of metallic silver (including nanoparticulate forms) by any route can be toxic and that dietary silver intake of $10–20 \, \mu\text{g day}^{-1}$ may lead to the measurable amounts of silver in the circulation $(<2.3 \, \mu\text{g L}^{-1})^{1}$ and accumulation of silver in soft tissues and bone. Profound argyria through precipitation of insoluble silver salts in the skin can be severely disfiguring but is not fatal. A total body burden of 9 mg silver might accumulate over a 50-year period through expected intake, absorption and excretion values.[61] USEPA estimated that as much as 10% of ingested silver is absorbed, but it is unclear whether these older estimates were based on metallic silver intake or use of colloidal silver preparations that have been commonly implicated in argyria symptoms.[83] Experiments in laboratory animals suggest that very little silver is absorbed gastrointestinally and that up to 99% is excreted. Estimates suggest that, in the absence of specific data, about 13% of airborne silver particles would enter the lungs based upon airborne particle dynamics in the human respiratory tract and be available for ionisation and absorption through pulmonary mucosae.[83] The role of mucociliary process in clearing silver particles is appreciated. Percutaneous absorption of metallic silver, ionisable silver compounds or silver alloys through intact or broken skin is very low and of minimal interest in evaluating the systemic toxicity of silver through acute or chronic exposures.[78–79,98]

In view of its widespread use as an antibacterial in wound care and in antibiotic medical devices, more detailed and scientifically valid studies have been conducted on silver sulfadiazine than other silver compounds and silver-related industrial processes. Silver sulfadiazine is less soluble that silver nitrate and lacks its corrosive anion. It is rarely associated with argyria-like symptoms and is less toxic.[87,99–101] Silver sulfadiazine as a 1% cream has been investigated extensively in burn wound therapy where percutaneous absorption is proportional to the extent of wounds.[87,102,103] Silver was not detectably absorbed through the intact arm skin of two volunteers.[87] Silver absorbed through burned skin in rats was bound to plasma proteins of a similar molecular size to those of the transferring molecule.[104] Clinical studies in burned patients show unequivocally that silver is excreted from the body in the urine and that a "threshold" of blood silver excretion exists at about $100 \, \mu\text{g L}^{-1}$, since above this level, increases in urinary excretion are more variable.[87] In the opinion of some authors, this variability makes monitoring percutaneous silver uptake by plasma silver concentrations alone unreliable.[87,105]

Experimental studies in which silver sulfadiazine was applied to injured skin showed that percutaneous absorption of silver was fourfold higher than in intact skin and that the silver absorption from 1% silver nitrate in the same model was approximately threefold higher.[106] In animals exposed to silver nitrate or silver sulfadiazine, there was a strong tendency for silver to be deposited in wound debris and wound exudates proteins. Hoekstra *et al.* examined silver sulfadiazine therapy in incisional wounds in pigs and reported that the antibiotic preserved viable epidermal tissue and acted as a wound healing promoter.[107] Systemic blood silver levels increased to a peak within 7–21 d post-operatively and declined as wounds re-epithelialised. This contrasts with earlier work where less than 1% silver was absorbed from silver sulfadiazine cream in scalded pig skin in 48 h following injury.[108]

In introducing his new burn wound therapy, Fox (1968) intended that his ointment containing 30 mM sulfadiazine per kg would be applied topically.[99] However, subsequent patent applications have been examined for silver sulfadiazine in aqueous suspension to be administered orally or subcutaneously as a non-toxic antibiotic against *Plasmodium berghei* and systemic infections.[109] Experimental studies in mice showed it to be minimally toxic in mice at 1050 mg kg^{-1} day^{-1} for 30 d, but this dose caused minimal local granulomatous reactions when injected subcutaneously. Acute doses of silver sulfadiazine were toxic when administered intraperitoneally at >550 mg kg^{-1}.

In recent years, silver sulfadiazine and nanocrystalline silver coatings have been developed for use in the production of antiseptic catheters for renal dialysis, central venous intubation, intracerebral implantation and intra-urethral insertion.[110–112] The first three of these involve direct exposure of the antibiotic as a catheter coating or impregnate to the blood system and plasma proteins that bind the free Ag$^+$ released. Signs of systemic toxicity have not been seen following use of any silver product employed as an antibiotic in in-dwelling catheters. Commonly, silver sulfadiazine is used with chlorhexidine gluconate or other antibiotics to improve the antibiotic spectrum of catheters which are prone to infection with a wide range of bacteria and fungi and which give rise to biofilm formations.[110] In an experimental study in which plastic catheters were coated with bioactive silver nanoparticles, silver release monitored by ^{110m}Ag radiolabel over 10 d was efficacious against Gram-negative and Gram-positive infections but non-toxic.[112] Approximately 15% of the silver coating (0.1–30 μg thick) eluted from catheters within 10 days; 8% of this was voided in faeces and 3% deposited at implantation sites with no detectable organ accumulation. A lack of toxic change was observed in polyurethane endotrachial tubes coated with silver sulfadiazine in a clinical study, although some reduction in mucosal secretions were noted in 46 patients incubated for 24 h.[113] The low toxicity of silver sulfadiazine is attributed to the cleavage of the molecule/complex to free Ag$^+$ and sulfonamide.[100] The Ag$^+$ binds to plasma proteins or is precipitated as silver chloride or silver phosphate, whilst the sulfonamide residue is readily excreted in urine.[99,100]

8.4.1.1 Silver Arsphenamine (Silver Salvarsan)

Silver arsphenamine was introduced as an efficacious antisyphilitic drug in 1918, and according to Gaul and Staud, its effectiveness was attributed to the combined action of the two metal ions released, Ag^+ and As^{+++} and the possible catalytic action of the silver ion.[21] These ions have been discussed much in older literature as a cause of argyria and argyrosis. Arsphenamines are inorganic arsenic compounds with a high level of action against protozoan parasites, trypanosomes, spirochaetes, amoebic dysentery, plasmodia, *etc.* They were introduced by Paul Ehrlich (1909) and were in use at least until 1942.[114] Arsphenamines oxidise readily following injection with the rate of oxidation being dependent upon temperature, acidity and the chemical nature of the compound. The arsenic released from arsphenamines is excreted slowly in the bile. Sollemann noted that silver arsphenamine possibly contains silver as an admixed colloidal metal,[114] but its constitution is not established (Figure 8.2). The arsphenamines are toxic in the liver, kidneys, skin and many other tissues. The IARC reported that there is sufficient evidence for the carcinogenicity of inorganic arsenic compounds in humans and that cases of skin cancer have been reported in people exposed to arsenic through medical treatment.[115] The carcinogenicity of organic arsenicals is less clear.[116] The formula of arsphenamine (Salvarsan) originally introduced by Paul Ehrlich in 1909 is shown as "A" in Figure 8.2, but more recent reports consider arsphenamine to be a complex of trimers and pentamers "B" and "C" respectively. Gaul and Staud considered that the silver content of silver arsphenamine to be approximately 14% with arsenic at 19%, but these analyses may be misleading in view of more recent research.[21]

8.4.2 Organ Toxicity

Silver has no nutritional value in the mammalian body but it is regarded by some as a "normal" body component. The toxicological significance of the minute concentrations found in most tissues of the human body is unclear.[1] The liver, kidney, skin and eye have been repeatedly identified as prime target tissues for silver toxicity, but humans exposed occupationally or receiving silver sulfadiazine burn wound therapy show no unequivocal evidence that silver (as Ag^+ or a protein complex) is a cause of irreversible damage or pathological change. (Arsenic released from silver arsphenamine is far more likely to evoke toxic damage in many organs.[114,117]) More equivocal information is provided by case studies of individuals who have consumed silver nitrate or colloidal silver products (against medical advice), or used silver containing preparations in an inappropriate manner.[10–11,37,68,118,119]

8.4.2.1 Liver

The liver in all mammals is a large multifunctional organ comprising 2–4% of the total body weight. Hepatocytes forming at least 90% of the liver mass have a central role in the metabolism and transformation of xenobiotic materials entering the body and excretion of hydrophobic compounds and products of catabolism *via* the biliary pathway.[48,120] Kupffer cells derived from reticuloendothelial precursors function as phagocytic macrophages; they reside within

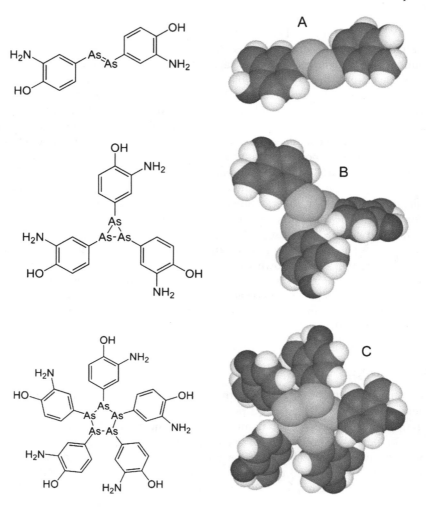

Figure 8.2   Presumed structure of arsphenamine.

the sinusoidal epithelium where they remove particulate matter including metallic silver and insoluble silver complexes from the blood. (As discussed in Chapter 9, cells of the blood–brain barrier function in a similar manner.) Experimental studies, mainly in rats, have demonstrated that the size of the liver and its metabolic capacity for xenobiotic materials varies according to age, the composition of the diet and state of health, but that changes in hepatocellular enzyme levels are a valuable guide to the metabolic state of the tissue and can provide valuable evidence for early onset of clinical or toxicological damage.[48]

Clinical studies on suspected cases of silver toxicity and experimental studies in laboratory animals routinely identify the liver as a target organ for metals including lead, cadmium, mercury and copper, but irreversible pathological change attributable to silver exposure by any route is hard to find—even in situations where argyric individuals are exposed chronically to high levels in diet, drinking water, occupationally or therapeutically. Nevertheless, silver does accumulate in the liver from the blood stream in which it is transported in the form of macroglobulin complexes (77%), albuminates (15%) or otherwise bound to plasma proteins.[121–123] Tracer studies indicate that silver is cleared rapidly from the circulation and accumulates in liver prior to elimination *via* the biliary tract to be excreted in faeces.

Although the liver is not a recognised storage organ for silver, tracer studies suggest that its removal from the organ may be delayed.[100] Thus, a cancer patient treated intravenously with 0.1 mg radiolabelled silver nitrate, tracer seen initially in erythrocytes declined by 70% within seven minutes, but post-mortem analysis after 195 days showed high residual levels of silver in his liver.[121] Faecal excretion of silver in this patient averaged 2.8% per day in the first two days but levelled to approximately 1.3% per day in the following 21 days. Urinary silver excretion was low at all times ($<0.3\%$ per day). A second case study recorded that more than 50% of radiolabelled silver was retained in the liver of a 29-year-old silver worker 16 days following inadvertently inhaling [110m]Ag-silver and [65]Zn in a factory accident, and that 90% of silver was eliminated in faeces.[122] Other post-mortem analyses in patients dying through causes unrelated to silver exposure show high levels of silver in liver following accidental exposures to silver nitrate, silver therapies for severe burn wounds and consumption of colloidal silver preparations or silver proteinates for respiratory complaints.[37,124] Where histopathological reports have been published, silver deposits identified mostly as blackish pigmentations in connective tissues, macrophages (Kupffer cells) and vascular endocardial cells in liver biopsies. The famous Blue Man of Barnum and Bailey's Circus was supposed to have a total body silver burden of 90–100 g, of which only 0.7% was estimated in liver (0.24% in kidney, 21% in bone, 0.16% in muscle, 0.15% in heart and 0.011% in brain), but the reliability of the data is questionable in view of the low accuracy of analytical techniques available at the time.[125] Nevertheless, Gettler *et al.* noted a lack of hepatotoxic effects at these concentrations.[125]

From the fragmentary clinical information available, it is not possible to calculate accurately the expected half-life of silver in the human liver. It may be as long as 52 d as suggested by Newton and Holmes on the basis of their observations in a case of accidental inhalation of silver (and where silver was detected in faeces after 300 days),[122] or 15 days as calculated by the International Commission for Radiological Protection,[126] but accurate estimations based upon modern analytical methods are urgently required.

Hepatocellular management of silver is illustrated by reference to burn wound patients treated with silver sulfadiazine. Such studies show that blood silver levels increase in patients with $>5\%$ total body surface area burns (TBSA) and that rises in key liver function enzymes such as γ-glutamyl transpeptidase, aspartate aminotransferase and alanine aminotransferase occur.[1,87] Enzymic changes occurring 4–35 days following injury might be construed as early evidence of hepatocellular damage and dysfunction, but this was not confirmed in biopsies.[87] A pronounced hepatocellular deposition of silver ($14\,\mu g\,g^{-1}$ wet weight), as reported in a fatal case, was entirely contained within lysosomal vacuoles and was not associated with cytoplasmic pathology. A severely burned patient treated with silver sulfadiazine for three months showed $9.5\,\mu g\,g^{-1}$ silver deposition in his liver with no evidence of liver damage. Other cases of burns wounds treated with silver nitrate "soaks", resulting in $14\,\mu g\,g^{-1}$ silver in liver biopsies, were also without pathology.[123] Marshall and Sneider described a case of a 46-year-old lady who had self-prescribed silver nitrate "sticks" to treat bleeding gums resulting from poor fitting dentures.[127] It is not known how much silver she consumed over 2.5 years, but she developed a local argyria in her nasal region and a liver biopsy taken to investigate a painful cyst revealed dense black silver deposits in hepatocytes in the vicinity of portal veins. No other pathology was reported and the silver content in the liver of this lady is not known.

Risks of liver damage in workers exposed to silver dust, silver nitrate or other silver compound occupationally seem to be low.[16,17] Pifer *et al.* examined 27 men exposed long term to insoluble silver compounds and reported seeing no appreciable evidence of argyria; blood silver was raised but not associated with pathological changes in internal organs.[128] Elsewhere, reports of patients exposed to silver claim that high levels of the metal seen in post-mortem or biopsy samples provide evidence of toxicity, but this is not established even where blood liver function enzymes are significantly higher than normal.[37,123,124,129] This is well-illustrated by a controlled study of 30 healthy volunteers given daily oral doses of 50 mg of silver leaf for 20 days.[96] Statistically significant hypophospholipidaemia, hypoglyceridaemia, hypoglycaemia and increases in HDL cholesterol accompanied by a lesser falls in total lipids, alkaline phosphatise, glutamate oxaloacetate transaminases, glutamate pyruvate transaminase, creatinine phosphokinase, γ-glutamyl transpeptidase and lactic dehydrogenase were not associated with urinary pathology or abnormalities in serum protein or albumin levels.[96] The authors concluded that, in the absence of demonstrable toxicity, silver might be used beneficially in the treatment of such conditions as diabetes mellitus, obesity and atherosclerosis. This has not been confirmed, but further work is indicated to examine whether the propensity of Ag^+ to combine with sulfydryl residues and be excreted by a glutathione-dependant pathway is a complication.[130]

Evidence for the intracellular management of silver is provided by *in vitro* studies where hepatocytes are exposed to soluble silver compounds.[71] Studies using freshly isolated hepatocytes exposed to 30–70 mM silver nitrate demonstrated the strong capacity of silver to bind to and impair the homeostasis of intracellular thiols as a central feature of its cytotoxicity. The cytotoxic effect of silver was enhanced by pretreatment of the hepatocytes with diethylmaleate, which reduced glutathione availability, but was protected by alpha-tocopherol which acted against the onset of silver-induced lipid peroxidation. The ion chelator, deferoxamine, failed to prevent loss of cell viability in the presence of Ag^+.

Experimental studies in animal models provide further evidence for the low hepatotoxic risks associated with silver uptake by any route. Silver nitrate or

Table 8.1 Biliary excretion patterns of silver in laboratory animals.[83]

Species	Silver Nitrate Administration	Silver Excretion in Bile
Sprague Dawley Rat (male)	0.01 mL iv over 2 h	0.25 μg/min/kg
New Zealand White Rabbit (male)	0.01 mL iv over 2 h	0.05 μg/min/kg
Dog (mongrel male)	0.01 mL iv over 2 h	0.005 μg/min/kg

silver sulfadiazine applied to experimental burns led to an increase in silver uptake in the liver but no evidence of hepatocellular damage.[131–133] An unconfirmed observation by Sugawara and Sugawara demonstrated that subcutaneous injection of silver nitrate in rats impaired copper uptake and metabolism by inhibiting synthesis of the copper-binding protein caeruloplasmin.[134] Silver binding in the liver was unaffected. This study showed that serum silver was located mostly in the albumin fraction with some bound to caeruloplasmin. The authors claimed that silver uptake by the rat liver, at least, is not dependent on copper concentrations, but that biliary excretion may be influenced by the copper bound in the tissue and that metallothionein carrier proteins are involved. An unseen Russian study cited by the National Institutes of Health (CAS Registry) claimed that injection of $2\,mg\,kg^{-1}$ silver nitrate in rats resulted in a twofold increase in hepatic metallothioneins.[135] Silver excretion patterns in bile have been confirmed in rats, rabbits, dogs following oral administration or acute intravenous injection, but silver mobilisation from liver deposits and patterns of excretion vary greatly between species (Table 8.1).[83,119] Greater concentrations of silver accumulated in the dog liver ($2.9\,\mu g\,Ag\,g^{-1}$ tissue weight) compared to rat ($1.24\,\mu g\,g^{-1}$) or rabbit ($2.12\,\mu g\,g^{-1}$). Radioactive tracer studies indicate in the rat, at least, silver excreted in the bile is largely bound in a low molecular weight complex with glutathione.[130,136]

8.4.2.2 Kidney

The kidney is a site for temporary deposition of silver in cases of argyria and following chronic exposure to silver sulfadiazine in burns patients, but it is not a target organ for silver toxicity.[5,37,126,137] Silver is excreted in the urine, but experience with silver sulfadiazine in moderate to severe burn wound patients shows that urinary silver excretion is inferior to the biliary route.[1,87] Urinary excretion of silver is low at blood levels of $<100\,\mu g\,L^{-1}$, but a threshold for serum silver excretion would appear to exist at serum silver concentrations of about $100\,\mu g\,L^{-1}$. Above this level significant rises in urine silver occur, but they tend to be more variable.[87] This observation suggests that measure of urinary silver is a poor means of quantifying silver absorption. Proteinuria was not shown to influence urinary silver excretion.

Silver deposits in renal epithelia, basal membranes of the glomeruli, blood vessel walls and macrophages occur as granular deposits in lysosomal vacuoles following high doses of silver nitrate or therapy with silver sulfadiazine for burns of at least 5% (TBSA).[3,87,138] There is no clear correlation between levels of argyraemia and renal damage. Rarely have cases been seen where silver deposition has been implicated as a cause of renal dysfunction in human patients, even at blood silver concentrations of $> 300\,\mu g\,L^{-1}$ and where urinary excretion was $100-400\,\mu g\,L^{-1}$ (normal range $< 1\,g\,L^{-1}$).[100] Also there are inconsistencies in the relationship between the extent of burn wounds and treatment with silver sulfadiazine and the amount of silver deposited in the kidney. Although a patient dying from severe burns showed silver of $0.18\,\mu g\,L^{-1}$ in renal biopsies and electron microscopic evidence of silver granules in tubular epithelia, these were unrelated to cellular damage.[87] Elsewhere, no silver was seen in renal biopsies of patients treated with silver sulfadiazine and dying with 80–90% TBSA burns.[105] Similar discrepancies are illustrated in occupational health studies where 37 workers in a silver smelting plant showed $< 0.005\,\mu g\,g^{-1}$ urinary silver ($11\,\mu g\,L^{-1}$ in blood, $15\,\mu g\,g^{-1}$ faeces).[19] No renal impairment was noted, but the authors calculated that human exposure to metallic silver equivalent to a threshold limit value of $0.1\,mg\,m^{-3}$ is expected to lead to a faecal excretion rate of $1\,mg$ silver per day. Faecal silver might be a more accurate measure of body silver than urinary excretion in some situations, but it is greatly influenced by the quality of the diet and other human factors (age, health state, *etc.*). The half-life of silver in the human kidney has been estimated as ten days,[26,126] but this requires revision using more accurate analyses than were available in 1959.

A nephrotic syndrome was reported in a 62-year-old lady who had been treated with silver sulfadiazine for eight weeks for 78% TBSA burns.[139] Biopsy showed membranous proliferation in renal tubules and interstitial glomerulonephritis. Maitre *et al.* reported two contrasting cases of silver sulfadiazine therapy for burns; one was a diabetic, 65-year-old lady treated with $100\,g$ weekly for 18 months and who developed renal dysfunction but low blood silver ($38\,\mu g\,L^{-1}$), the second a 19-year-old lady who sustained thermal cutaneous burns on her legs.[140] The second patient exhibiting very high blood silver ($440\,\mu g\,L^{-1}$) and urinary silver ($12\,\mu g\,L^{-1}$) showed normal renal and hepatic function tests.

Consumption of about $35\,g$ of an undefined colloidal silver preparation for treatment of a duodenal ulcer in a 52-year-old patient who died of coronary heart failure led to pronounced argyria.[3] Blood silver levels were not recorded, but the patient showed dense silver deposits in the walls of most blood vessels, in the basal membranes of renal glomeruli, and macrophages, but only rarely in epithelial basement membranes. Aaseth *et al.* examined a patient treated with silver nitrate for several months for gingival erosions and measured high levels of biologically inert silver in biopsies.[35] They emphasised that silver was deposited in the form of silver selenide and not as silver sulfide as is commonly assumed. They concluded that silver selenide is non-toxic since renal function was completely normal and that negligible "reactive" changes were present in renal biopsies.

Cases of renal and urinary tract pathology have been recorded in patients treated with silver nitrate instilled into the renal pelvic cavity or into the upper urinary tract to alleviate haemorrhages, obstructions and infectious lesions in oriental medicine.[37] It is unclear how frequently this therapy is practised these days, but retrograde instillation of 0.5% silver nitrate is seemingly still used in south-eastern Asian countries to alleviate infections of the filarial worm (*Wucheria bancrofti*) which is spread by mosquito vectors.[141–144] Pathological information on most cases is fragmentary, but it is expected that even 0.5% silver nitrate solutions will prove irritant to the delicate renal and urothelial membranes through the corrosive properties of nitric acid, but more severe complications can be expected in the rare cases where concentrations of 2.5 or 3.0% have been administered. In one such case, a 16-year-old man was treated with 2.5% silver nitrate by intra-urethral stent instillation twice at intervals of two months for haematuria developed a severe and progressive unilateral hydronephrosis with a narrowing of ureter over the following five months.[141] The affected ureter was surgically excised and pathological analysis revealed marked inflammatory changes, vascularisation and epithelial damage but deposition of silver was not reported. Anuria occurred in a patient treated with 3% silver nitrate to treat chyluria, and both kidneys were involved.[145] In a 69-year-old lady referred to hospital for pyuria and renal dysfunction 20 days following retrograde instillation of silver nitrate to quench renal haemorrhage, renal function tests showed elevated blood urea nitrogen and serum creatinine levels with red and white cells in her urine.[146] X-radiography and computerised tomography scans revealed a mineralisation in her right kidney affecting collecting ducts and parenchymal tissues. This "argyrosis of the upper urinary tract" syndrome diminished following unilateral nephectomy and blood urea nitrogen levels and creatinine normalised. In patients examined following instillation of silver nitrate into the bladder to combat haemorrhages or infections, silver deposits were identified on epithelial surfaces and renal papillary membranes, but despite inflammatory and oedematous changes due to nitric acid, pathological changes attributable to silver deposition have not been identified.

Whilst some believe instillation of 0.5% silver nitrate into the renal pelvic cavity to alleviate painful symptoms of chyluria is without serious risk, complications including acute necrotising ureteritis with obstructive uropathy have been reported.[143] Thus a 38-year-old lady developed a severe haematuria two days following therapy and renal function tests showing increased blood urea nitrogen, raised creatinine and electrolytes as signs of renal damage.[147] X-radiography confirmed bilateral radio-opacity (presumably due to silver deposition) throughout her renal pelvis and ureters, and evidence of papillary necrosis but her renal function tests normalised within one month following haemodialysis. In these representative cases, there is no clear evidence that silver deposited as silver sulfide or other form is detrimental either in terms of renal function or urinary tract damage, and that where transitory changes have occurred, they are predictably due to the acidity of the nitric acid released.

Experimental studies conducted in laboratory animals confirm that silver metal, silver nitrate and silver sulfadiazine are unlikely causes of irreversible renal damage or changes in the urinary tract, but that where changes do occur, they result from administration of metallic silver or silver compounds at concentrations greatly in excess of those experienced by humans through occupational exposure or antibiotic therapies.[148,149] Available studies illustrate further wide interspecies variations in urinary silver excretion and deposition of silver granules in renal tissues. Rats exposed to chronic silver nitrate (12 mM) administration in drinking water for 81 weeks showed a progressive accumulation of silver in glomerular basement membranes but no obvious cytological damage;[132] rabbits showed minimal silver uptake following 10% silver nitrate therapy for burns. Moffat and Creasey demonstrated that silver nitrate administered to rats and rabbits in drinking water ($1500 \, mg \, L^{-1}$ for 4–20 weeks) influenced the permeability of medullary vessels; they also confirmed that basal glomerular epithelia were targets for silver deposition in the rat and that most silver was deposited in proximal aspects of the Loops of Henle where it tended to be asymmetrical and highest in the region of collecting ducts.[150] Their observation that higher numbers of renal parenchymal cells appeared degenerate in rats exposed to silver nitrate has not been confirmed. Other studies confirm inter-strain variations in urinary excretion rates, and silver retention in the kidney and liver in different strains of rat. However, inconsistencies in experimental details and dosing regimens make direct comparison of the different published studies difficult to evaluate.[148–151]

8.4.2.3   Skin

The human skin is exposed to silver through occupational exposures (silversmiths and jewellery workers, industrial processes, photography, *etc.*), medical devices coated or impregnated with silver as antibiotic protection, and coinage. Argyria as discussed above is the principal complication of silver exposure by all routes, although true and long-lasting argyria occurs less commonly through topical exposure than through inhalation or oral intake.

Argyria-like discolourations arising through dermal contact with ionising silver products usually involve a deposition of silver sulfide or other insoluble silver complex in superficial cells of the epidermis and in hair shafts and nail. This is rarely long-lasting as these cells are lost by natural desquamation and through everyday wear and tear. In contrast, argyria resulting from inhalation or oral ingestion of silver is invariably dermal and located in the region of basal membranes of sweat and sebaceous glands, and in and around blood vessels and connective tissue fibres in the papillary region.[4,5,14,34] Penetration of Ag^+ or silver complexes from epidermis to dermis is very low. Percutaneous excretion of silver does not occur.

Epidermal keratinocytes and dermal fibroblasts are vulnerable to the toxic effects of silver in tissue culture[63] and predict erroneously that silver may be a cause of:

- cytotoxicity and reduced viability in exposed tissue;
- impaired growth factor regulation and recognition at target sites;

- impaired cell migration patterns in normal and damaged tissue, and functional maturation;
- impaired wound healing.

Experience using animal models and clinical studies of patients treated with silver nitrate, silver sulfadiazine and sustained silver release wound dressings show that silver (as Ag^+) exhibits a minimal toxicity and that inherent protective mechanisms within the skin and its appendages are usually sufficient to offset the detrimental effects of this xenobiotic metal. Secondly, silver exerts a beneficial influence in advancing the healing process in open wounds and burns.[106]

Appraisal of toxic effects of metals on the skin will include consideration of:

1. Local tissue damage following topical contact (including corrosive effect, astringency, cytotoxicity);
2. Impairment in melanogenesis and melanocyte function;
3. Immunological changes leading to delayed hypersensitivity and allergic reactions;
4. Carcinogenicity.

The extent of tissue injury and the depth of penetration of any material in contact with the skin surface is influenced by the manner in which it is presented (metallic form, ionisable compound), acidity, temperature, vehicle/solvent (water soluble, lipophilic or amphiphilic) and the state of hydration of the tissue.[152,153] Toxic changes may be measured in terms of frank morphological change, functional or physiological impairment, and degenerative changes. The skin, unlike most other tissues in the body, exists in a state of dynamic equilibrium with its surrounding environment and is subject to perpetual mitotic activity in the epidermal basement membrane. Essential homeostatic mechanisms exist for maintaining functional cell mass in all areas throughout an individual's lifetime.[154] Minor changes in cytological character or intracellular or pericellular environment are liable to influence the homeostatic pathways in the epidermis, possibly leading to changes which are initially invisible to the naked eye. The influence of xenobiotic materials on such subtle features as cell cycle time, keratinisation patterns, epidermal cell turnover and physico-chemical properties have been well researched using kinetic analysis, histo-chemical techniques and *in vitro* systems.

In the "normal skin" of non-pressure areas (*e.g.* forearm, back), post-mitotic cells migrate in a columnar fashion and undergo a well-defined sequence of changes consistent with functional maturation and ultimately death in the outer stratum corneum as mature keratinocytes. In view of the more rapid rates of epidermal cell proliferation in areas of pressure keratinisation (palms of hands, soles of feet, calluses, *etc.*), migration patterns tend to lose the columnar patterns and cells reaching the skin surface frequently show incomplete keratinisation before desquamation from the skin surface. In mammalian skin, the epidermis comprises about 10% of the total skin thickness and mass, but provides essential protective functions against the toxic action of xenobiotic materials and conserving the water balance in deeper (living) tissues. The ratio of epidermis to dermis is specific for every area of skin in the human and animal body, but changes in relation to age, nutritional status and the health of the individual.[155]

Repair systems in the skin and epidermal appendages are well-defined and dependent upon appropriate endocrine and paracrine hormonal balances, genetic up-regulation of growth factors and cytokines. Nutrition and trace metal ion gradients are critical in maintaining epidermal homeostasis and repair following injury. The age and state of health of an individual are important in controlling the structural integrity of the skin and its response to injury. As a number of wound dressings are currently available, special discussion is desirable to explain current view on the role of Ag^+ in wound bed preparation, which for many years has been a fundamental clause in wound management.[156–160]

8.4.2.3.1 Topical Action and Influence of Silver on Homeostatic Mechanisms. Metallic silver is inert in contact with the skin, but the biologically active Ag^+ released in the presence of moisture and secretions on the skin surface is potentially capable of influencing the character of the tissue and inherent homeostatic mechanisms in a similar manner to other metallic cations. Some of the Ag^+ released through ionisation of metallic silver (including nanosilver particles and silver alloys) and inorganic and organic silver compounds provides an antimicrobial function through oligodynamic mechanisms, but much of the remainder binds strongly to sulfydryl groups and other anionic residues on superficial keratinocytes, hair shafts and exposed surfaces of nail.[127] The action of keratin in binding xenobiotic ions is an essential feature of the cytoprotective function of the stratum corneum in all areas of skin. The keratin-Ag^+ complexing and silver precipitation as silver sulfide or silver selenide is influenced by:

- the concentration of Ag^+ released;
- solar energy, temperature and pH;
- the character of the skin–hair cover, epidermal thickness, surface moisture and state of hydration, occlusion, keratinisation type (pressure or non-pressure areas);
- duration of contact;
- the maturity of the keratin molecule.

Available evidence indicates that the penetration of silver (as Ag^+ or other form) into intact or damaged skin is very low and that percutaneous absorption is negligible; most free ion binds strongly to sulfydryl, imidazole and carboxyl residues in epidermal keratin.[78,79,161,162] No decrease in radioactivity was noted in human forearm skin treated topically with ^{110m}Ag after 24 or 48 h and maximum absorption was calculated at <4%. Karlsmark *et al.* confirmed the low percutaneous penetration of silver in five patients with leg ulcers (mean area 15.6 cm^2) treated with a high silver content wound dressing (Contreet Foam, Coloplast, Denmark).[98] After 28 days therapy, blood silver levels increased by 4–80 nMol L^{-1}; these are well within the reference levels of

1–93 nMol L^{-1} calculated for humans not knowingly exposed to silver therapeutically or occupationally.[163] Karlsmark *et al.* also noticed a narrow band of blackish residue in peri-ulcer areas, suggesting that some of the Ag$^+$ released was precipitating as silver sulfide or metallic silver deposit at dressing changes and when skin was exposed to daylight.[98] Deposition of silver in wound debris and exudates is appreciated histologically and in smears, but may not be recognised by visual inspection. This is not regarded as a toxic change.

Evidence of toxic change is seen in skin areas treated with corrosive and highly astringent concentrations of silver nitrate where local irritancy and tissue damage are largely attributable to the acidity of the nitrate anion, concentration and duration of exposure rather than the silver cation.[114] Silver nitrate applied topically for antiseptic purposes turns the skin white, then grey and eventually black through precipitation of metallic silver or insoluble silver sulfide and keratin binding Although silver nitrate is mildly astringent to some areas of skin at concentrations of 1%, the caustic effects of concentrations of 10% and higher are beneficially used in the form of lunar caustic, pencils and sticks (sometimes mitigated with potassium nitrate) for removing troublesome calluses, granulations, warts and other disfigurements.[114,164] Permanent changes have not been recorded and skin wounds have repaired normally. A patient using silver nitrate sticks to treat a lesion on her peri-orbital folds experienced some discomfort, which passed uneventfully when the treatment was discontinued.[165] On rare occasions, 1% silver sulfadiazine cream (Flamazine) has been reported to cause a transitory hardening of the skin peripheral to ulcers and burn wounds (personal observations.)

Experimental studies in rodents have shown that silver nitrate (0.5%) or silver sulfadiazine (1% as Flamazine) applied subacutely to intact or surgically induced skin wounds are non-toxic.[106] Ag$^+$ released from silver nitrate precipitated as silver sulfide to turn the hair and skin shafts brown-black, whereas the ion released from silver sulfadiazine rarely produced discolouration. Silver deposits were demonstrated histologically along the wound margins in wound debris and proteins of wound exudates (Figure 8.3). No evidence was seen of deep penetration of silver into the epidermis or lateral migration from the wound site. Histochemical techniques using immunocytochemistry demonstrated that Ag$^+$ was absorbed by intact viable cells of the wound margin where it induced and bound metallothioneins I and II.[166] Metallothioneins are cysteine rich ($\approx 30\%$) proteins with a capacity to bind certain trace and xenobiotic metals.[167] Metallothioneins are found in all living tissues of the human body and serve a cytoprotective function against the toxic effects of certain xenobiotic metals (zinc, copper, silver, gold, cadmium and other metal elements).[168–170] Silver–metallothionein complexes are stable and the silver ion is inactivated. Other experiments have shown that only when the protective role of metallothioneins is saturated are signs of cell damage and degeneration appreciated histologically.[166] Metallothioneins bind zinc and copper ions as co-factors in enzymes required in cell proliferation; they exhibit mitogenic properties and have been shown to participate in molecular and subcellular mechanisms of cell replication in damaged tissue.

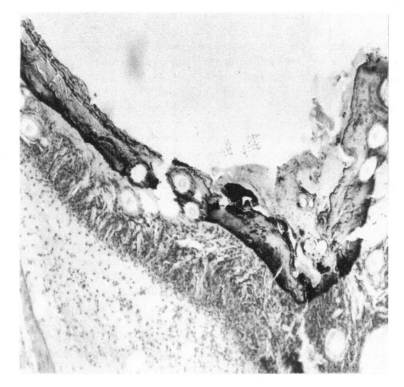

Figure 8.3 Silver sulfide precipitates in wound debris in an experimental rat wound
treated with 0.5% silver nitrate for two days. The black precipitates are
located mainly in the proteinaceous exudates ($\times$ 10 objective).

True toxic changes attributable to the topical action of silver in mammalian
skin are few. At the moment, there is no good evidence to show that silver
binding in the outer layers of the stratum corneum in human or animal skin
influences the homeostatic pathways of normal or damaged skin, or alters the
epidermal barrier function. In argyria, silver sulfide or silver selenide granules
enclosed within lysosomal vesicles in macrophages and dermal fibroblasts, or
loose within the connective tissue, are inactive and of no toxicological significance.[7,171] Elevated concentrations of silver occurring in the skin of workers
exposed occupationally to silver are without adverse effects other than the
marked disfiguration attributable to argyria.[3–5,19] The threshold level of blood
silver consistent with overt argyria is not known, but estimates suggest that
total body silver of 4–5 g to 20–40 g can produce the clinical picture of
argyria.[23] Other estimates suggest that oral doses of 6 g of silver nitrate or 6.3 g
silver arsphenamine given intravenously, corresponding to 3.8 and 0.9 g of
elemental silver respectively, will evoke argyria with no evidence of "toxic"
damage or subtle effects.[5] Skin rashes recorded in 2.3% of patients treated with
silver sulfadiazine cream were possibly attributable to delayed hypersensitivity
to silver, sulfadiazine or another constituent of the emollient.[24,100,172,173] The

concentration of silver in full thickness biopsies of patients with argyria was estimated using neutron activation analysis to be 80–100 times higher than seen in individuals not knowingly exposed to silver.[24,25]

The implications of silver-containing products in wound management are complex. More than 30 different formulations of sustained silver release wound dressings are currently marketed, but most scientific publications and product evaluations are directed more towards an evaluation of their capacity to control wound infections and exudates, and their efficacy in alleviating wound pain and patient discomfort. In general, products such as silver sulfadiazine, Silvadene hydrocolloid and most sustained silver-release wound dressings are recognised for aiding wound healing and wound bed preparation with no visible signs of toxic effect.[100,160,174–178] However, some authors contend that silver sulfadiazine therapy retards re-epithelialisation and healing in chronic ulcers and burns, or has no noticeable effect.[179–182] In an experimental study of porcine wounds, Hoekstra *et al.* claimed that silver sulfadiazine "preserved viable epidermal tissue" and promoted granulation tissue, but retarded re-epithelialisation and was mildly irritant.[182] They reported that newly formed epidermis appeared parakeratotic, spongy and pseudocarcinomatous with changes suggesting that the silver was impairing wound contraction as a manifestation of toxicity. These studies do not account for clinical studies in which silver sulfadiazine was shown to promote expression of key growth factors involved in wound repair and where silver sulfadiazine with epidermal growth factor advanced re-epithelialisation in human wounds by 25–50%.[183]

8.4.2.3.2 Melanogenesis and Melanocyte Function. Argyria is the principle form of discolouration in the skin following exposure to silver by ingestion, inhalation and occasionally by topical contact. However, other sources of discolouration in the skin may include silver protein complexes (unidentified) or increased melanin production. Melita *et al.* reported that, when a large amount of silver is present in the skin, photo-activated reduction of the metal leads to bluish-grey discolourations and causes a generalised stimulation in melanin production.[32] Others have theorised that Ag^+ in the presence of light energy stimulates melanogenesis.[184,185] Clinical studies of biopsies taken from patients with profound argyria through occupational silver exposure, the chronic consumption of silver acetate antismoking remedies or ingestion of colloidal silver products have failed to confirm this hypothesis using electron microscopy and histochemical techniques.[4,5,26] Bleehan *et al.* specifically noted that epidermal melanocytes in biopsies from argyric patients were entirely normal in size and shape, and exhibited characteristic DOPA-oxidase reaction for tyrosinase.[4]

8.4.2.3.3 Delayed Hypersensitivity and Allergic Reactions. Dermatoses and discolourations of the skin attributable to exposure to metals, metal alloys and metallic salts are common, but whilst many people tolerate metals in their solid state, they acquire allergic dermatitis when exposed to ions in solution.[25] Thus, whilst silver in jewellery and wire is largely inert, Ag^+ released in the presence of moisture and silver nitrate solution have been shown to be allergenic and irritant following occupational and therapeutic exposures.[186,187]

Identification of allergies to metals is frequently complicated by the presence of contamination with other allergenic metals such as nickel, chromium and cobalt but cross-sensitisation is rare. Hollinger considered that the human immune system was particularly sensitive to silver in view of experimental observations showing that B and T-lymphocytes and monocytes were vulnerable to silver in culture.[188] He cited earlier studies in which mice exposed to 0.05% silver nitrate exhibited selective induction of anti-nucleolar anti-fibrillarin antibodies after five weeks.[189,190]

The first evidence of silver allergy in humans related to the use of the colloidal protein Argyrol. Howard reported a case of a young lady treated with nasal drops twice weekly for chronic purulent otitis media who developed swelling of her face, generalised urticaria and a near collapse state, which cleared when the therapy was withdrawn.[191] Later, Criep reported the case of a 26-year-old man who had used Argyrol nasal sprays for pharyngitis and hay fever.[192] This patient developed severe asthma and increased discomfort, which receded when the nasal spray was discontinued. Standard patch test and intradermal injections of 1% Argyrol produced skin reactions which confirmed allergy and the existence of specific serum antibodies. Fisher considered that allergic reactions to Argyrol should be classified as "immediate type hypersensitivities".[25] Fundamental research on allergies to silver nitrate established that allergy and skin reactions were directly proportional to Ag^+ concentration and that freshly prepared solutions and solutions kept in the dark were less allergenic.[186,187] Gaul and Underwood reported a case of silver allergy in a patient who developed a vesicular reaction following exposure to 10% silver nitrate as used in marking out patch test sites.[187] In another patient, silver nitrate applied to an area of eczema on his heel led to increased pain and dermatitis.[193] Patch tests revealed allergy to 5 or 10% "aged" (more ionised) silver nitrate solutions, but not freshly prepared reagent.

Silver allergies are reported where silver nitrate, colloidal silver or silver sulfadiazine have been used therapeutically, or where individuals have been exposed to inorganic silver compounds occupationally. In the therapeutic application of 1% silver sulfadiazine cream (Flamazine), Dollery noted that 2.3% of patients supposed to have developed allergy to the Ag^+ released from the antibiotic in wound exudates;[100] however, patch testing revealed allergies to components of the emollient cream (cetyl alcohol and propylene glycol) and not silver.[194,195] The true extent of allergies to silver in silver sulfadiazine preparations is not known, although reports are published detailing patch testing procedures in patients exhibiting febrile reactions, hyperpigmentation and dermatitis.[196–199] It is unfortunate that when patients develop discomfort through use of silver sulfadiazine or other silver-containing products used in wound care, patch tests are not routinely carried out for financial reasons or shortage of time. It is a case of the "silver doesn't suit you, we'll try something else" situation.

Allergies to silver in industrial exposures can occur under a wide variety of situations, but many are incompletely investigated. Fisher recorded that metallic silver in jewellery is not a cause of dermatitis, even though allergic reactions are seen occasionally following exposure to metallic silver and silver in coinage (silver is rarely used in coinage these days).[25] Occupational allergic contact dermatitis was confirmed in an atopic patient exposed to silver and other heavy metals in the jewellery trade.[200] This 30-year-old man had been exposed to metals and precious stones for many years but developed multiple skin lesions. He was subjected to tests consistent with European standards and revealed positive reactions to a fragrance mix (undefined), colophonium and 0.5% silver nitrate but not to tests for other metals. No cases of silver allergic contact dermatitis were confirmed in a survey of 93 workers.[201] Nickel allergy is a common problem associated with inexpensive jewellery, but contact sensitisation occurs with metals such as palladium, cobalt, beryllium, rhodium, vanadium, gold, platinum and mercury—in that order of frequency.[25]

Technicians exposed to silver, silver coating, radiography, explosives manufacture and general silver work are subject to the risks of silver allergies. However, the true incidence of contact allergic dermatoses in these professions is not known as only a small of those investigated are published. Heyl described a patient who developed dermatitis following electroplating work in which he was exposed to "silver coat" containing silver cyanide with small quantities of sodium cyanide, sodium carbonate and sodium nitrite.[202] He showed positive patch tests to a 1% solution of the silver coat preparation but did not respond to metallic silver. "Silver worker's finger" is an unusual occupational hazard in silver working, but in one case the overt signs of allergy resembled melanocytic lesions.[203] Another form of occupational hazard was reported in workers exposed to the "white salt" (silver fulminate) used in the explosives industry.[204] Although silver fulminate may not be used much these days, the associated irritant and allergic reaction known in the trade as "fulminate itch" are occupational hazards in the production of explosive snaps. Workers occasionally experienced stomatitis as an additional complication. Silver dyes containing up to 5% silver nitrate used in cosmetic parlours for colouring eyebrows and eyelashes have been implicated in silver allergies.[25] In each case, the delayed hypersensitivity reactions were triggered not by metallic silver but Ag^+. Patients showed skin reactions when challenged with 1% silver nitrate or solutions of the suspected silver salt (silver cyanide in the case of silver coat allergy).

A rare case of silver allergy in a lady radiographer exposed to silver chloride illustrates that Ag^+ may be released from this virtually insoluble compound through formation of a soluble complex with sodium thiosulfate.[205] She presented with a papular type form of eczema beneath her watch strap (nine carat gold) following exposure to silver chloride complexed with sodium thiosulfate and sodium nitrate. Patch tests confirmed sensitisation to 1% silver chloride complexed with sodium thiosulfate, 1% silver nitrate and the fixing fluid. The reaction was possibly enhanced by the moisture under the watch strap, which increased the state of hydration of her skin and sensitivity to silver as a mild allergen.

Metal allergies can be expected more frequently in professions like dentistry through exposure to amalgams,[206] but allergies are more commonly attributable to nickel, cobalt and chromium exposure rather than silver.[207,208]

8.4.2.3.4 Bone. Bone toxicity is not widely reported in pre-clinical and clinical evaluations of silver in medical devices or in industrial toxicology, but there are strong indications that not only does silver deposit in bone (probably in a hydroxyapatite complex),[209] but that Ag^+ interacts with calcium (Ca^{2+}) and magnesium (Mg^{2+}) in reconstituted systems and in *in vitro* models.[210–212] Tupling and Green further suggest that silver ions induce calcium release from the sarcoplasmic reticulum in skeletal muscle by acting on the calcium release channels and the calcium pump mechanism, presumably through oxidising sulfydryl groups.[213] This suggests that bone and possibly cartilage may be susceptible to toxic changes following long-term exposure to silver-release devices and silver-containing biomaterials used in orthopaedic surgery, but this has not been established. Greater attention in published work has been turned to evaluation of antibiotic efficacy of the silver (Ag^+) released from medical devices and its capacity to protect against biofilm formation.[214–216]

Silver deposition in bone has been demonstrated in workers exposed occupationally in refineries in northern Sweden using atomic absorption spectrometry, neutron activation analysis and X-ray emission analysis.[217] The cytotoxicity of silver in bone has been investigated using *in vitro* cultures of human bone fibroblasts/osteoblasts and clinical observation of implants following surgery. Alt *et al.* examined the cytotoxicity of a methacrylate bone cement loaded with metallic silver particles (5–50 nm) (NanoSilver).[214,215] They demonstrated that incubation of osteoblasts in the presence of NanoSilver did not accentuate release of the marker enzyme lactate dehydrogenase within 48 hours and that the total protein content of cultured cells was unaltered. Osteoblast growth patterns retained normal viability and growth in the presence of Ag^+. In a second *in vitro* cytotoxicity study, the biocompatibility of a silver-loaded orthodontic silver alloy was evaluated in the presence of osteoblast-like cells.[218]

Dental alloys are subject to a low progressive release of ions such as silver, gold, palladium, cobalt and titanium in the course of many years implantation.[206,207] In an attempt to evaluate the long-term safety of silver and alloy commonly used in dentistry, metals and metal complexes were exposed to osteosarcoma-derived cells in culture.[219] After 48 hours incubation, silver wire failed to produce a statistically insignificant reduction in cell growth and reduction in mitotic index as signs of toxicity, but this effect was greatly enhanced in the presence of copper. Alkaline phosphatase activity was markedly depressed in the presence of individual metals as a mark of toxicity.

In their experience, Spadaro *et al.* claimed that addition of silver ions to polymethylmethacrylate bone cements not only provides an efficacious antibiotic faculty but might be an attractive alternative to the conventional organic antibiotics in providing maximal effectiveness at low concentrations with "low risk".[220] They demonstrated that four silver salts (oxide, chloride, sulfate or phosphate) were entirely biocompatible with surrounding tissue and that changes were consistent with a mild foreign body reaction. With the exception of the oxide, all seemed to maintain the compressive strength of rabbit bone, following implantation in para-spinal muscles. Also compatible with bony and peri-osseous tissue are the silver-coated orthopaedic fixing pins designed for external fixation.[221]

Clinical studies in patients implanted with bone cements and silver-containing orthopaedic materials have mostly failed to provide evidence of toxicity even in cases of profound argyria and high blood silver concentrations. In one exceptional case, a patient fitted with a revisional Christensen total hip arthroplasty developed a 1000-fold increase in blood silver following use of a silver-impregnated bone cement.[222] Hip joint fluid and biopsies from her acetabular cavity exhibited very high silver concentrations ($103.3\,\mu g\,L^{-1}$) which regressed in the two years following removal of the silver cement. Blood silver concentrations declined from $6.3\,\mu g\,L^{-1}$ to $2.2\,\mu g\,L^{-1}$ in this period. These high systemic and local silver concentrations were not associated with osteological damage in this patient, but the observation of neurological damage and its association with silver toxicity is debatable.[223]

Recent advances in biomedical materials have seen introduction of silver loaded composites as bone scaffolds.[224] BioGlass® silver comprising addition of a suitable silver compound (*e.g.* silver oxide) into a stable bioactive glass has been developed for use in fracture repair and orthopaedic surgery. The Bio-Glass scaffold is claimed to show a hierarchical porous structure similar to that of cancellous bone. It degrades in the presence of peri-osseous macrophages to release silicon and calcium ions to promote bone repair, and up to 2% ionised silver is available for antibiotic action. Preliminary studies have demonstrated silver–bone biocompatibility and beneficial therapeutic effect. Further clinical studies are awaited.

8.4.2.3.5 Miscellaneous Soft Tissues. In clinical cases of generalised argyria, silver is metabolised to most tissues of the human body where it is deposited in the form of silver sulfide, silver selenide or as insoluble protein complex. These deposits are probably inert and confined to lysosomal vesicles or intercellular spaces as demonstrated in the liver and kidney,[87] and hence of minimal toxicological significance. Frank pathological changes attributable to silver deposition are rarely encountered and hormonal patterns do not reflect silver-related damage in the pituitary, adrenal or thyroid glands.

Questionable pathology relates to the ability of silver sulfadiazine or silver nitrate to evoke changes in white blood cell populations (WBC) following therapy in burn wounds. Thus, low WBC of $\leq 5000\,mm^{-3}$ were recorded in 47.5% of patients treated with silver sulfadiazine for severe burns, and in 13 of 30 patients (43.3%) patients receiving silver nitrate.[225] However, Fuller and Engler suggested that factors other than silver absorption, including stress, might be implicated in post-burn leukopenias.[226] In their extensive survey of 101 burn wound clinics in North America they concluded that, whatever its causation, post-burn leukopenia is of small risk, but that where WBC fall to less than $2000\,mm^{-3}$, it is prudent to withdraw silver sulfadiazine therapy as a precaution. There is sufficient experience to demonstrate that burn wound therapy with silver sulfadiazine is safe in general practice and that, where WBC changes have been recorded, leukopenia and neutropenia have normalised following withdrawal of the silver therapy without adverse sequalae.[53]

However, cases of fever and leukopenia associated with silver sulfadiazine were reported at the University of Wisconsin Burns Centre;[53,54] blood WBC of $2860 \, mm^{-3}$ were associated with neutropenia with increased numbers of immature band forms in peripheral blood following discontinuation of silver therapy and with continued treatment. The authors concluded that a small number of their patients may have been subject to an undiagnosed silver allergy, but that leukopenia occurring secondary to silver sulfadiazine is probably an innocuous and a self-limiting phenomenon. Agranulocytosis may be a further complication of silver sulfadiazine, but as in other reported cases, these regress when treatment is discontinued.[227]

Occupational risks associated with inhaling low levels of silver dust or silver nitrate are not well documented, but in cases of generalised argyria, silver sulfide or other insoluble complexes have been identified in nasal membranes and in the proximal respiratory tract with occasional evidence of phagocytosis in macrophages. Squamous metaplasia may be a complication, but the prevalence of this and other changes in the respiratory tract among silver workers these days is likely to be minimal as regulations for health and safety at work are enforced.[10,22,67,228] Other cases of acute inhalation of silver nitrate dust and droplets have been associated with respiratory failure, falls in blood pressure, spasms, paralysis through effects on phrenic muscles, gastrointestinal irritancy and fatality, but the true aetiology of the conditions is unclear. It expected that the underlying pathology is more likely to be attributable to the corrosive effects of the nitric acid released and not silver or Ag^+ as inferred in some published reports.[82,229,230]

Soft tissue pathology has been observed in a small number of patients treated with silver nitrate by unusual and possibly illegal routes. In these cases, local tissue damage and fatalities are also almost certainly attributable to the corrosive effects of the nitric acid and not silver uptake or toxicity. Rare cases are recorded where silver nitrate has been instilled intravaginally in an attempt to induce abortion and where patients have died through severe abdominal pain, acute circulatory failure and collapse within three hours. A German lady treated with 7% silver nitrate as an abortifacient agent exhibited extensive corrosive injuries in her genital tract; her foetus and placenta were aborted, but she died of congestive changes in her lungs, brain and kidneys.[231] Other cases where dilute silver nitrate has been used as a urinogenital tract disinfectant have recorded silver deposits in mucosal linings but no obvious pathology. Silver precipitates have been identified within and between epithelial cells of cervical tissue by light and electron microscopy following use of silver nitrate solution for cauterisation of tumours of the cervix.[232] Rare cases are recorded where small quantities silver nitrate as 10–15% paste were instilled into the fallopian tubes through a cannula to induce sterility.[233] It is not known to what extent this technique is practised these days, but tubal blockage through the astringency and irritancy of the nitrate anion (15% silver nitrate paste) was a common occurrence.

The literature on the use and abuse of silver nitrate is vast, but used under appropriate conditions and fused with potassium nitrate as a diluent, it is a safe and efficacious agent for general cautery.[114,164] Being highly soluble in water, silver nitrate solutions are readily applied to desired areas and soak into warts or verrucous tissues. Brown discolourations of the tissues are usual and without toxic risk, but precautions are recommended against the accidental ingestion of products which can lead to severe pain, sialorrhea, diarrhoea, vomiting, coma and convulsions. Caustic pencils and applicators of silver nitrate for topical use contain 75–95% silver nitrate.

8.5 Carcinogenicity

In 1971, the International Agency for Research on Cancer (IARC) began a programme to prepare monographs on the ability of environmental chemicals to cause cancer in humans.[46] IARC evaluated evidence compiled from individual case reports and epidemiological studies of cancer associated with specific chemicals such as silver. Three criteria were adopted in determining a causal relationship between exposure to a given chemical and human cancer:

1. The relationship between exposure and evidence of carcinogenic change is unequivocal and not subject to bias.
2. That carcinogenicity is not a "chance observation" or attributable to uncontrolled variables.
3. Pathological changes are progressive, possibly tissue specific and consistent.

Carcinogenicity studies have been conducted in live animal models for many years, but extrapolation of observations in terms of human risk is still questionable. In the absence of unequivocal evidence of carcinogenicity for a chemical from human studies, provisional predictions have been made on the basis of available human epidemiology and results of animal studies.[46] *In vitro* tests using bacteria, yeasts, human cells and *Drosophila* are available as a guide to mutagenicity and may provide a preliminary screen for carcinogenicity, but discrepancies have been reported.[46]

> The US Department of Health and Human Resources (DHHS) has monitored the toxicology and carcinogenicity of many thousands of individual chemicals and chemical processes on a regular basis, but has so far failed to present unequivocal evidence that silver or silver-related processes present a risk of causing cancer in any tissue on the basis of human or animal data.[234]

Silver is classified by the US Environmental Protection Agency under Category D as "Not classifiable as a Human Carcinogen".[83,119] It is not shown to be mutagenic in the limited number of tests so far conducted and no human cancers have so far been reported following exposure to any form of silver, its alloys or compounds. Carcinogenicity to metallic silver, silver nitrate and colloidal silver has been demonstrated in some animal models but published data are inconsistent, incomplete and insufficiently controlled for valid judgement on predictable human risk to be made.

An early study documented a 32% incidence of fibrosarcomas in Wistar rats implanted subcutaneously with 1.5 cm discs of metallic silver foil within 275 and 625 days post-operatively.[235] Other tumours were not reported in these rats and no placebo animals were included. A later study involving intramuscular injection of a suspension of silver particles in trioctanoin in Fischer rats produced no observable tumours at injection sites in 24 months (80% lifetime), whereas animals injected with the known carcinogenic metal cadmium developed fibrocarcinomas within 16.5 months.[236] Furst later discussed the clinical implications of subcutaneous sarcoma and considered that the vast majority of rodents develop these responses at sites of metal implantation and that the extent of the biological reaction was determined by the physicochemical features of the implanted material—a smooth surface being a major requirement.[237] The so-called "Oppenheimer Effect" produced by implanting silver foil discs was similar to that observed following implantation of plastic discs. Other work suggested that the subcutaneous response to implanted silver foil varied according to the anatomical site,[238] but this remains to be ratified. Colloidal silver of unknown ionisable silver content produced fibrosarcomas at sites of intravenous or subcutaneous injection in a BD-strain of rats, but the time of appearance and the true incidence of the tumours are not known accurately.[239]

In summary, the carcinogenicity of silver or silver compounds in humans is not known; no study in animals or in human populations exposed to the metal by any route, has substantiated beyond all resonable doubt that silver is carcinogenic. Further detailed clinical studies on workers exposed to high levels of silver occupationally over long periods are urgently required. Other studies should include pathological evaluation of patients subject to long-term implantation with silver-containing antibiotic prostheses, cardiovascular devices, bone cements and dental materials, each of which release some Ag^+ on a continual basis.

8.6   Teratogenicity

Limited evidence is available to show that administration of silver, silver nitrate, silver sulfadiazine or other silver compounds in pregnancy is a cause of infertility, impaired foetal growth or abnormal development in any species. Dubin *et al.* administered 1% silver nitrate by intrauterine injection to 13 cynomolgus monkeys between 27 and 43 days of pregnancy, and noted vaginal bleeding after 1–2 days and termination of pregnancy.[240] However, two out of seven animals re-mated, became pregnant again and delivered healthy offspring, indicating that silver did not permanently damage the reproductive system. The silver content of the newborn is not known.

Experiments conducted specifically to demonstrate the pathogenicity of silver in the central nervous system have shown that foetal fats exposed to silver on days 18 or 19 of gestation showed precipitates of insoluble silver in lysosomes of neurons and astroglial cells, and basal cell membranes of the blood–brain barrier.[241] Volumetric analysis of the developing foetuses revealed a reduction in hippocampal pyramidal cells and that the perikarya of pyramidal cells were target sites for silver toxicity.[242,243] No behavioural studies were reported and the pathological influence of the silver deposits is not known.

8.7 Further Reading

C. K. Chan, F. Jarrett and J. A. Moylen, Acute leukopenia as an allergic reaction to silver sulfadiazine in burn patients, *J. Trauma*, 1976, **16**, 395.

K. Norlind, Further studies on the ability of different metal salts to influence the DNA synthesis in human lymphoid cells, *Int. Arch. Allergy Appl. Immunol.*, 1986, **79**, 83.

References

1. A. T. Wan, R. A. J. Conyers, C. J. Coombs and J. P. Masterton, Determination of silver in blood, urine and tissues of volunteers and burn patients, *Clin. Chem.*, 1987, **37**, 1683.
2. H. W. Dietl, A. P. Anzil and P. Mehraein, Brain involvement in generalised argyria, *Clin. Neuropathol.*, 1984, **3**, 32.
3. H. Steininger, E. Langer and P. Stommer, Generalised argyrosis, *Deutsch. Med. Wochenschr.*, 1990, **115**, 657.
4. S. S. Bleehan, D. J. Gould, C. I. Harrington, T. E. Durrant, D. N. Slater and J. C. E. Underwood, Occupational argyria; light and electron microscopic studies and X-ray microanalysis, *Br. J. Dermatol.*, 1981, **104**, 19.
5. R. J. Pariser, Generalised argyria: clinicopathologic features and histological studies, *Arch. Dermatol.*, 1978, **114**, 373.
6. Y. Tanita, T. Kato, K. Hanada and H. Tagami, Blue macules of localised argyria caused by implanted acupuncture needles, *Arch. Dermatol.*, 1985, **121**, 1550.
7. S. Sato, H. Sueki and A. Nishimura, Two unusual cases of argyria: the application of an improved tissue processing method for X-ray microanalysis of selenium and sulphur in silver-laden granules, *Br. J. Dermatol.*, 1999, **140**, 158.
8. G. F. Nordberg and L. Gerhardsson, Silver, in *Handbook on Toxicity of Inorganic Compounds*, ed. H. G. Seiler and H. Sigel, Marcel Dekker, New York, 1988, pp. 619–623.
9. W. R. Hill and D. M. Pillsbury, *Argyria: The pharmacology of silver*, Williams and Wilkins, Baltimore, MD, 1939, pp. 69–127.
10. Department of Health and Human Services, US Food and Drug Administration, Over-the-counter drug products containing colloidal silver ingredients or silver salts, *Fed. Regist.*, 1996, **61**, No. 200 (October 15, 1996), 53685.
11. B. A. Bouts, Images in clinical medicine, argyria, *N. Engl. J. Med.*, 1999, **340**, 1554.
12. N. S. Tomi, B. Kränke and W. Aberer, A silver man, *Lancet*, 2004, **363**, 532.

13. A. Wadhera and M. Fung, Systemic argyria associated with ingestion of colloidal silver, *Dermatol., Online J.*, 2005, **11**, 12.

14. W. R. Buckley, C. F. Oster and D. W. Fassett, Localised argyria: the chemical nature of the silver containing particles, *Arch. Dermatol.*, 1965, **92**, 697.

15. Y. M. Sue, J. Y. Lee, M. C. Wang, T. K. Lin, J. M. Sung and J. J. Huang, Generalised argyria in two haemodialysis patients, *Am. J. Kidney Dis.*, 2001, **37**, 1048.

16. K. D. Rosenmann, A. Moss and S. Kon, Argyria: clinical complications of exposure to silver nitrate and silver oxide, *J. Occup. Med.*, 1979, **21**, 430.

17. A. P. Moss, A. Sugar, N. A. Hargett, A. Atkin, M. Wolkstein and K. Rosenmann, The ocular manifestations and functional effects of occupational argyrosis, *Arch. Ophthalmol.*, 1979, **7**, 906.

18. M. W. Scroggs, J. S. Lewis and A. D. Proia, Corneal argyrosis associated with silver soldering, *Cornea*, 1992, **11**, 264.

19. G. D. DiVincenzo, C. J. Gordiano and L. S. Schriever, Biologic monitoring of workers exposed to silver, *Int. Arch. Environ. Health*, 1985, **56**, 207.

20. L. E. Gaul and A. H. Staud, Clinical spectroscopy: distribution of silver in the body or its physiopathological retention as a reciprocal of the capillary system, *Arch. Dermatol. Syphilol.*, 1935, **52**, 775.

21. L. E. Gaul and A. H. Staud, Clinical spectroscopy: seventy cases of generalised argyria following organic and colloidal silver medication, including a biometrospectrometric analysis of ten cases, *J. Am. Med. Assoc.*, 1935, **104**, 1387.

22. US Environmental Protection Agency, *An Exposure and Risk Assessment for Silver*, USEPA Office of Water, Washington DC, 1981, EPA Report Number 440-4-81017.

23. J. Seimund and A. Stolp, Argyrose, *Z. Haut. Geschlechtskr.*, 1968, **43**, 71.

24. A. B. Molokhia, B. Portnoy and A. Dyer, Neutron activation analysis of trace elements in the skin, *Br. J. Dermatol.*, 1979, **101**, 565.

25. A. A. Fisher, *Contact Dermatitis*, Lea and Febiger, Philadelphia, 1987, pp. 710–744.

26. B. W. East, K. Boddy, E. D. Williams, D. MacIntyre and D. A. McLay, Silver retention, total body silver and tissue silver concentrations in argyria associated with exposure to an anti-smoking remedy containing silver acetate, *Clin. Exp. Dermatol.*, 1980, **5**, 305.

27. Y. Ohbo, H. Fukuzako, K. Takeuchi and M. Takigawa, Argyria and convulsive seizures by ingestion of silver in a patient with schizophrenia, *Psychiatry Clin. Neurosci.*, 1996, **50**, 89.

28. S. M. Lee and S. H. Lee, Generalised argyria after habitual use of $AgNO_3$, *J. Dermatol.*, 1994, **21**, 50.

29. P. Sugden, S. Azad and S. Erdmann, Argyria caused by an earring, *Br. J. Plast. Surg.*, 2001, **54**, 252.

30. N. M. Fisher, E. Marsh and R. Lazova, Scar-localised argyria secondary to silver sulphadiazine cream, *J. Am. Acad. Dermatol.*, 2003, **49**, 730.

31. W. B. Shelley, E. D. Shelley and V. Burmeiser, X-ray microprobe and electron microscopic study was made of the remarkable blue-black pigmentation that sunlight elicits in patients with argyria, *J. Am. Acad. Dermatol.*, 1987, **16**, 211.

32. A. C. Melita, K. Butterworth and M. A. Woodhouse, Argyria-electron microscopic study of a case, *Br. J. Dermatol.*, 1966, **78**, 175.

33. M. Westhofen and H. Schafer, Generalised argyrosis in man: neuratological, ultrastructural and X-ray micro-analytical findings, *Arch. Otolaryngol.*, 1986, **243**, 260.

34. W. R. Buckley and C. J. Terhaar, The skin as an excretory organ in argyria, *Trans. St John's Hosp. Dermatol. Soc*, 1973, **59**, 33.

35. J. Aaseth, A. Olse and J. Halse, Argyria: tissue deposition of silver as selenide, *Scand., J. Clin. Lab. Chem.*, 1981, **41**, 247.

36. J. Aaseth, J. Halse and J. Falch, Chelation of silver in argyria, *Acta Pharmacol., Toxicol.*, 1986, **59**, 471.

37. S. D. M. Humphreys and P. A. Routledge, The toxicology of silver nitrate, *Adverse Drug React. Toxicol. Rev.*, 1978, **17**, 115.

38. U. Schlötzer-Schrehardt, L. M. Holbach, C. Hofmann-Rummeltt and G. O. Naumann, Multifocal corneal argyrosis after explosion injury, *Cornea*, 2001, **20**, 553.

39. V. Sánchez-Huerta, G. De Wit-Carter, E. Hernández Quintela and R. Narajo-Tackman, Occupational corneal argyrosis in art silver solderers, *Cornea*, 2003, **22**, 604.

40. M. J. Gallardo, J. B. Randleman, K. M. Price, D. A. Johnson, S. Acosta, H. E. Grossniklaus and R. D. Stuttling, Ocular argyrosis after long-term self-application of eyelash tint, *Am. J. Ophthalmol.*, 2006, **141**, 198.

41. K. U. Loeffler and W. R. Lee, Argyrosis and the lachrymal sac, *Graefes Arch. Clin. Exp. Ophthalmol.*, 2005, **255**, 146.

42. L. Zografos, S. Uffer and L. Chamot, Unilateral conjunctival-corneal argyrosis simulating conjunctival melanoma, *Arch. Ophthalmol.*, 2003, **121**, 1483.

43. S. C. Gad, Alternative to *in vivo* studies in toxicology, in *General and Applied Toxicology*, ed. B. Ballantyne, T. C. Marrs and T. Sylversen, MacMillan, London, 1999, Vol. **1**, pp. 401–424.

44. M. Balls and S. A. Horner, The FRAME inter-laboratory programme on *in vitro* cytotoxicology, *Food Chem. Toxicol.*, 1985, **23**, 205.

45. S. C. Gad, Recent developments in replacing, reducing and refining animal use in toxicologic research and testing, *Fundam. Appl. Toxicol.*, 1990, **15**, 8.

46. International Agency for Research on Cancer, *Long and short term screening assays for carcinogens: a critical appraisal*, World Health Organization, Lyon, 1980, IARC Monographs on the Evaluation of the Carcinogenic Risk of Chemicals to Humans Supplement 2.

47. Interagency Coordination Committee on the Validation of Alternative Methods (ICCVAM), ICCVAM Test Method Evaluation Report: *in vitro* Cytotoxicity Test Methods for Estimating Starting Doses for Acute Oral Systemic Toxicity Testing, National Toxicology Program, Research Triangle Park, NC, 2006, NIH Publication No. 07-4519, http://iccvam.niehs.nih.gov/methods/acutetox/inv_nru_tmer.htm, accessed 23 November 2009.

48. S. D. Gangolli and J. C. Phillips, The metabolism and disposition of xenobiotics, in *Experimental Toxicology: The Basic Issues*, ed. S. M. Anderson and D. M. Conning, Royal Society of Chemistry, Cambridge, 1993, pp. 181–201.

49. J. G. Rheinwald and H. Green, Serial cultivation of strains of human epidermal keratinocytes: the formation of keratinising colonies from single cells, *Cell*, 1975, **6**, 331.

50. K. A. Holbrook and H. Hennings, Phenotypic expression of epidermal cells *in vitro*: a review, *J. Invest. Dermatol.*, 1983, **81**(Suppl. 1), 2s.

51. M. Pruniéras, M. Régnier and D. Woodley, Methods for the cultivation of keratinocytes with an air-liquid surface, *J. Invest. Dermatol.*, 1983, **81**(Suppl. 1), 28s.

52. S. K. Bhatia and A. B. Yetter, Correlation of visual *in vitro* cytogenicity ratings of binomials with quantitative *in vitro* cell viability measurements, *Cell Biol. Toxicol.*, 2007, **10**, 315.

53. R. L. Gamelli, T. P. Paxton and M. O'Reilly, Bone marrow toxicity by silver sulphadiazine, *Surg. Gynecol. Obstet.*, 1993, **177**, 115.

54. F. Jarrett, S. Ellerbe and R. Demling, Acute leucopenia during burn therapy with silver sulphadiazine, *Am. J. Surg.*, 1978, **135**, 818.

55. J. C. Wataha, P. E. Lockwood and A. Schedle, Effect of silver mercury and nickel ions on cellular proliferation during extended low dose exposures, *J. Biomed. Mater. Res.*, 2000, **2**, 360.

56. J. F. Hansbrough, R. L. Zirpata-Sirvant and M. L. Cooper, Effects of topical antibacterial agents on the human neutrophil respiratory burst, *Arch. Surg.*, 1991, **126**, 603.

57. R. L. Zapata-Sirvant and J. F. Hansbrough, Cytotoxicity to human leukocytes by topical antimicrobial agents used in burn care, *J. Burn Care Rehabil.*, 1993, **14**, 132.

58. R. L. McCauley, H. A. Linares, V. Pelligrini, D. N. Herndon, M. C. Robson and J. P. Heggers, *in vitro* toxicity of topical antimicrobial agents to human fibroblasts, *J. Surg. Res.*, 1989, **46**, 267.

59. R. L. McCauley, Y. Y. Li, V. Chopra, D. N. Herndon and M. C. Robson, Cytoprotection of human dermal fibroblasts against silver sulphadiazine using recombinant growth factors, *J. Surg. Res.*, 1994, **56**, 378.

60. R. E. Burrell, Scientific perspectives on the use of topical silver preparations, *Ostomy Wound Manage.*, 2003, **49**(Suppl. 5A), 19.

61. H. G. Petering and C. J. McClain, Silver, in *Metals and their Compounds in the Environment; occurrence, analysis and biological relevance*, ed. E. Merian, VCH Weinheim, 1991, pp. 1191–1202.

62. E. Hidago, R. Bartolome, C. Barroso, A. Moreno and C. Dominguez, Silver nitrate antimicrobial activity related to cytotoxicity in cultured human fibroblasts, *Skin Pharmacol. Appl. Skin Physiol.*, 1998, **11**, 140.

63. V. K. Poon and A. Burd, *In vitro* cytotoxicity of silver implication for clinical wound care, *Burns*, 2004, **30**, 140–147.

64. T. Mosmann, Rapid colorimetric assay for cellular growth and survival: application to proliferation and cytotoxicity assays, *J. Immunol. Methods*, 1983, **65**, 55.

65. E. C. Smoot, J. O. Kucan, A. Roth, N. Mody and N. Debs, *in vitro* testing for anti-bacterials against human keratinocytes, *Plast. Reconstruct. Surg.*, 1991, **87**, 917.

66. P. V. AshaRani, G. L. K. Mun, M. P. Hande and S. Valiyveettil, Cytotoxicity and genotoxicity of silver nanoparticles in human cells, *ACS Nano*, 2008, **19**, 1021.

67. US Environmental Protection Agency, *Silver*, Integrated Risk Information System (IRIS), Environmental Criteria and Assessment Office, Office of Health and Environmental Assessment, Cincinnati, OH, 1992, www.epa. gov/oppsrrd1/REDs/old_reds/silver.pdf, accessed 23 November 2009.

68. W. L. Roper, *Toxicological Profile for Silver*, Agency for Toxic Substances and Disease Registry, US Public Health Service, Atlanta, GA, 1990.

69. S. Nakamura and Y. Ogura, Mode of inhibition of glucose oxidase by metal ions, *J. Biochem.*, 1968, **64**, 439.

70. S. J. Stohs and D. Bagchi, Oxidative mechanisms in the toxicity of metal ions, *Free Radic. Biol. Med.*, 1995, **18**, 315.

71. C. Baldi, C. Minoia, A. Di Nucci, E. Capodaglio and L. Manzo, Effects of silver on rat hepatocytes, *Toxicol. Lett.*, 1988, **41**, 261.

72. H. S. Rosenkranz and S. Rosenkranz, Silver sulfadiazine: interaction with isolated deoxyribonucleic acid, *Antimicrob. Agents Chemother.*, 1972, **2**, 373.

73. J. L. Chambers, C. G. Kreiger, M. Krieger, L. Kay and R. M. Stroud, Silver ion inhibition of serine proteases: crystallographic study of silver trypsin, *Biochem. Biophys. Res. Comm.*, 1974, **59**, 70.

74. M. Demerec, G. Bertani and J. Flint, A survey of chemicals for mutagenic action on *E. coli*, *Am. Nat.*, 1951, **85**, 119.

75. H. Nishioka, Mutagenic activities of metal compounds in bacteria, *Mutat. Res.*, 1975, **31**, 185.

76. E. C. McCoy and H. S. Rosenkranz, Silver sulphadiazine: lack of mutagenicity, *Chemotherapy*, 1978, **24**, 87.

77. C. Kaplan, N. Diril, S. Sahin and M. C. Cehreli, Mutagenic potentials of dental cements as detected by Salmonella/microsome test, *Biomaterials*, 2004, **25**, 4019.

78. J. E. Wahlberg, Percutaneous toxicity of metal compounds, *Arch. Environ. Health*, 1965, **11**, 210.

79. J. J. Hostýnek, R. S. Hinz, C. R. Lorence, M. Price and R. H. Guy, Metals and the skin, *CRC Crit. Rev. Toxicol.*, 1993, **23**, 171.

80. J. Dequidt, P. Vasseur and J. Gromez-Potentier, Experimental toxicological study of some silver derivatives, *Bull. Soc. Pharm. Lille*, 1974, **1**, 23–25.

81. J. E. Furchner, C. R. Richmond and G. A. Drake, Comparative metabolism of radionuclids in mammals IV. Retention of silver-^{110m}Ag in the mouse, rat, monkey and dog, *Health Phys.*, 1968, **15**, 505.

82. B. Venugopal and T. D. Luckey, in *Metal Toxicity in Mammals-2: Chemical Toxicity of Metals and Metalloids*, Plenum Press, New York, 1978, pp. 32–36.

83. US Environmental Protection Agency, *Silver (CASRN-7440-22-4)*, Integrated Risk Information System (IRIS), Environmental Criteria and Assessment Office, Office of Health and Environmental Assessment, Cincinnati, OH, www.epa.gov/iris/subst/0099.htm, accessed 23 November 2009.

84. International Programme on Chemical Safety, *Summary of Toxicological Data of Certain Food Additives: Silver*, IPCS, Geneva, 1977, WHO Food Additives Series No. 12.

85. American Conference of Governmental Industrial Hygienists, *Documentation of the Threshold Limit Values and Biological Exposure Limits*, ACGIH, Cincinnati, OH, 4th edn, 1980.

86. J. E. Wahlberg, Percutaneous toxicity of metal compounds, *Arch. Environ. Health*, 1965, **11**, 201.

87. C. J. Coombs, A. T. Wan, J. P. Masterton, R. A. J. Conyers, J. Pedersen and Y. T. Chia, Do burn wound patients have a silver lining?, *Burns*, 1992, **18**, 179.

88. H. Blumberg and T. N. Carey, Argyraemia: Detection of unsuspected and obscure argyria by the spectrographic demonstration of high blood silver, *J. Am. Med. Assoc.*, 1934, **103**, 1521.

89. L. J. Casarett and J. Doull, *Toxicology: the Basic Science of Poisons*, MacMillan, New York, 1975, pp. 967–969.

90. S. S. Shouse and G. H. Whipple, I. Effects of the intravenous injection of colloidal silver upon the haematopoietic system in dogs, *J. Exp. Med.*, 1931, **53**, 413.

91. A. A. Goldberg, M. Shapero and E. Wilder, Antibacterial colloidal electrolytes: the potentiating of the activities of mercuric-phenylmercuric-, and silver ions by colloidal sulphonic anion, *J. Pharm. Pharmacol.*, 1949, **2**, 20.

92. F. La Torraca, Anatomic, histopathological and histochemical aspects of acute experimental intoxication with silver salts, *Folia med. (Naples)*, 1972, **45**, 1065.

93. K. N. Ham and J. D. Tange, Silver deposition in rat glomerular basement membrane, *Aust. J. Exp. Biol. Med. Sci.*, 1972, **50**, 423.

94. A. T. Diplock, J. Green, J. Bunyan, D. MacHale and I. R. Muthy, Vitamin E and stress. 3 The metabolism of δ-alpha-tocopherol in the rat under dietary stress with silver, *Br. J. Nutr.*, 1967, **21**, 115.

95. C. F. Cooper and W. C. Jolly, Ecological effects of silver iodide and other weather modification agents; a review, *Water Resour. Res.*, 1970, **6**, 88.

96. D. C. Sharma, P. Sharma and S. Sharma, Effect of silver on circulating lipids and cardiac and hepatic enzymes, *Ind. J. Physiol. Pharmacol.*, 1997, **41**, 285.
97. USEPA Drinking water criteria Document for Silver, EPA Number 600X85040, 97p, 1985.
98. T. Karlsmark, R. H. Agerslev, S. H. Benz, J. R. Larsen, J. Roed-Petersen and K. E. Andersen, Clinical performance of a new silver dressing, Contreet Foam for chronic exuding venous leg ulcers, *J. Wound Care*, 2003, **12**, 205.
99. C. L. Fox, Silver sulphadiazine—a new topical therapy of Pseudomonas aeruginosa, *Arch. Surg.*, 1968, **96**, 184.
100. C. Dollery, Silver sulphadiazine, in *Therapeutic Drugs*, Churchill Livingstone, Edinburgh, 1991, Vol. **2**.
101. R. J. White and R. Cooper, Silver sulphadiazine: review of the evidence, *Wounds UK*, 2005, **1**, 51.
102. K. Hirakawa, Determination of silver and cerium in the liver and kidney from a severely burned infant treated with silver sulphadiazine and cerium nitrate, *Radioisotopes*, 1983, **32**, 59.
103. X. W. Wang, N. Z. Wang, O. Z. Zhang, R. L. Zirpata-Sirvant and J. W. L. Davies, Tissue deposition of silver following topical use of silver sulphadiazine in extensive burns, *Burns*, 1985, **11**, 197.
104. Y. Matuk, Distribution of radioactive silver in the subcellular fractions of various tissues of the rat and its binding to low molecular weight proteins, *Can. J. Physiol. Pharmacol.*, 1983, **61**, 1391.
105. M. G. Boosalis, J. T. McCall, D. H. Ahrenholz, L. D. Solem and C. J. McClain, Serum and urinary silver levels in thermal injury patients, *Surgery*, 1985, **101**, 40.
106. A. B. G. Lansdown, B. Sampson, P. Laupattarakasem and A. Vuttivirojana, Silver aids healing in the sterile skin wound: experimental studies in the laboratory rat, *Br. J. Dermatol.*, 1997, **137**, 728.
107. M. J. Hoekstra, P. Hupkens and P. Dutrieux, A comparative burn wound model in the New Yorkshire pig for the histopathological evaluation of local therapeutic regimens: silver sulphadiazine cream as a standard, *Br. J. Plast. Surg.*, 1993, **46**, 585.
108. R. Lazare, P. A. Watson and G. D. Winter, Distribution and excretion of silver sulphadiazine applied to scalds in the pig, *Burns*, 1974, **1**, 5.
109. M. S. Wysor, Orally administered silver sulphadiazine: chemotherapy and toxicology in CF-mice; *Plasmodium berghei* (malaria) and *Pseudomonas aeruginosa*, *Chemotherapy*, 1975, **21**, 302.
110. J. I. Greenfield, L. Sampath, S. J. Popilskis, S. R. Brunner, S. Stylianos and S. Modak, Decreased bacterial adherence and biofilm formation on chlorhexidine and silver sulphadiazine-impregnated central venous catheters implanted in swine, *Crit. Care. Med.*, 1995, **23**, 894.
111. T. S. Elliott, Role of antimicrobial central venous catheters for the prevention of associated infections, *J. Antimicrob. Chemother.*, 1999, **43**, 441.

112. D. Roe, B. Karanandikar, N. Bonn-Savage, B. Gibbins and J.-B. Roullet, Antimicrobial surface functionalization of plastic catheters by silver nanoparticles, *J. Antimicrob. Chemother.*, 2008, **61**, 869.

113. L. Berra, T. Kolobow, P. Laquerriere, B. Pitts, S. Bramati, J. Pohlmann, C. Marelli, M. Panzeri, P. Brambillasca, F. Villa, A. Baccarelli, S. Bouthors, H. T. Stelfox, L. M. Bigatello, J. Moss and A. Pesenti, Internally coated endotracheal tubes with silver sulphadiazine in polyurethane to prevent bacterial colonisation: a clinical trial, *Intensive Care Med.*, 2008, **34**, 1030.

114. T. Sollemann, Silver, in *A Manual of Pharmacology and its Applications to Therapeutics and Toxicology*, Saunders, Philadelphia, 6th edn., 1942, pp. 1102–1109.

115. US Department of Health and Human Services, *12th Annual Report on Carcinogens*, National Toxicology Program, Research Triangle Park, NC, 2005.

116. G. Pershagen, The carcinogenicity of arsenic, *Environ. Health Perspect.*, 1981, **40**, 93.

117. Anon, Silver arsphenamine, *Cal. State. J. Med.*, 1921, **19**, 304.

118. G. D. Clayton and F. E. Clayton, *Patty's Industrial Hygiene and Toxicology*, Wiley, New York, 1981, Vol. **2**, p. 1887.

119. R. A. Faust, *Toxicity Summary for Silver*, Prepared for the Oak Ridge Reservation Environmental Restoration Programme, Oak Ridge National Laboratory, US Department of Energy, 1992, http://rais.ornl.gov/tox/profiles/silver_c_V1.html, accessed 23 November 2009.

120. R. H. Hinton and P. Grasso, Hepatotoxicology, in *General and Applied Toxicology*, ed. B. Ballantyne, T. C. Marrs and T. Sylversen, MacMillan, London, 1999, pp. 853–892.

121. A. A. Polachek, C. B. Cope, R. F. Willard and T. Enns, Metabolism of radioactive silver in a patient with carcinoid, *J. Lab. Clin. Med.*, 1960, **56**, 499.

122. D. Newton and A. Holmes, A case of accidental inhalation of zinc-65 and silver-110m, *Radiat. Res.*, 1966, **29**, 403.

123. K. F. Bader, Organ deposition of silver following silver nitrate therapy for burns, *Plast. Reconstr. Surg.*, 1966, **37**, 550.

124. M. Trop, M. Novak, S. Rodl, B. Hellbom, W. Kroell and W. Goessler, Silver-coated dressing Acticoat caused raised liver enzymes and argyria-like symptoms in a burn patient, *J. Trauma*, 2006, **60**, 648.

125. A. O. Gettler, C. P. Rhoads and S. Weiss, A contribution to the pathology of generalised argyria with a discussion on the fate of silver in the human body, *Am. J. Pathol.*, 1927, **3**, 631.

126. International Commission on Radiological Protection, *Recommendations of the International Commission on Radiological Protection*. Report of committee II on permissible dose for internal radiation, Pergammon Press, Oxford, 1959, ICRP Publication 2.

127. J. P. Marshall and R. P. Schneider, Systemic argyria secondary to topical silver nitrate, *Arch. Dermatol.*, 1966, **113**, 1077.

128. J. W. Pifer, B. R. Friedlander, R. T. Kintz and D. K. Stockdale, Absence of toxic effects in silver reclamation workers, *Scand. J. Work Environ. Health*, 1989, **15**, 210.
129. T. Tsipouiras, C. R. Rix and P. H. Brady, Solubility of silver sulphadiazine in physiological media and relevance to treatment of thermal burns with silver sulphadiazine cream, *Clin. Chem.*, 1995, **41**, 87.
130. J. Alexander and J. Aaseth, Hepatobiliary transport and organ distribution of silver in the rat as influenced by selenite, *Toxicology*, 1981, **21**, 179.
131. A. Kartal, Y. Tatkan, M. Belviranli, M. Sahin, S. Duman, A. Karahan, M. Gurbilek and S. Temur, Serum and tissue silver levels after burns treated with silver compounds, *J. Chir. (Paris)*, 1989, **126**, 676.
132. P. Tichy, J. Rosina, K. J. R. Blaha and J. Cikrt, Biliary excretion of super ^{110m}Ag and its kinetics in isolated perfused liver in rats, *J. Hyg. Epidemiol. Microbiol. Immunol.*, 1986, **30**, 145.
133. F. Walker, The deposition of silver in glomerular basement membrane, *Virchows Arch., B Cell Pathol.*, 1972, **11**, 90.
134. N. Suguwara and C. Suguwara, Effect of silver on caeruloplasmin synthesis in relation to low molecular weight protein, *Toxicol. Lett.*, 1984, **20**, 99.
135. E. M. Mogilnicka, M. Milaszewicz and J. K. Piotrowski, Binding of silver in the liver of the rat, *Bromat. Chem. Toksykol.*, 1978, **11**, 59.
136. L. Friberg, G. F. Nordberg, E. Kessler and V. B. Vouk, *Handbook of the Toxicology of Metals*, Elsevier, Amsterdam, 1986, Vol. **2**, p. 527.
137. N. Williams and I. Gardener, Absence of symptoms in silver refiners with raised blood silver levels, *Occup. Med.*, 1995, **45**, 205.
138. J. P. Berry, R. Dennebouy, M. Chaintreau, F. Dentin, G. Slodzian and P. Galle, Scanning electron microscopy mapping of basement membrane elements and arterioles in the kidney after selenium–silver interaction, *Cell Mol. Biol. (Noisy le Grand)*, 1995, **1**, 265.
139. C. J. Owens, D. R. Yarbrough and N. C. Brackett, Nephrotic syndrome following topically applied sulfadiazine silver therapy, *Arch. Intern. Med.*, 1974, **134**, 332.
140. S. Maitre, K. Jaber, J. L. Perot and F. Cambazard, Increased serum and urinary levels of silverduring treatment with topical silver sulphadiazine, *Ann. Dermatol. Venereol.*, 2002, **129**, 217.
141. S. R. Vijan, M. A. Keating and A. F. Althausen, Ureteral stenosis after silver nitrate instillation in the treatment of essential hematuria, *J. Urol.*, 1988, **139**, 1015.
142. A. Mandhani, A. Kapoor, R. K. Gupta and H. S. Rao, Can silver nitrate instillation for the treatment of chyluria be fatal?, *Br. J. Urol.*, 1998, **82**, 926.
143. C. M. Su, Y. C. Lee, W. J. Wu, H. L. Ke, Y. H. Chou and C. H. Chou, Acute necrotising ureteritis with obstructive uropathy following instillation of silver nitrate in chyluria, a case report, *Kaohsiung J. Med. Sci.*, 2004, **20**, 512.
144. A. A. Kulkarni, M. S. Pathak and R. A. Sirsat, Fatal renal and hepatic failure following silver nitrate instillation for treatment of chyluria, *Nephrol. Dial. Transplant*, 2005, **20**, 1276.

145. M. S. Gulati, R. Sharma, A. Kapoor and M. Berry, Pelvi-calyceal cast formation following silver nitrate treatment for chyluria, *Aust. Radiol.*, 1999, **43**, 102.
146. Y. Kojima, K. Uchida, H. Takiuchi, A. Wakatsuki, T. Sakurai, Y. Fujita, D. Shirai and Y. Kobayashi, Argyrosis of the urinary tract following silver nitrate instillation: report of a case, *Hinyokika Kiyo*, 1993, **39**, 41.
147. S. C. Dash, Y. Bhargav, S. Saxena, K. Agarwal, S. C. Tiwari and A. Dinda, Acute renal failure and renal papillary necrosis following instillation of silver nitrate for treatment of chyluria, *Nephrol. Dial. Transplant.*, 1996, **11**, 1841.
148. C. D. Klaassen, Biliary excretion of silver in the rat, rabbit and dog, *Toxicol. Appl. Pharmacol.*, 1979, **50**, 49.
149. Z. Gregus and C. D Klaassen, Disposition of metals in rats: a comparative study of fecal, urinary and biliary excretion and tissue distribution or eighteen metals, *Toxicol. Appl. Pharmacol.*, 1986, **85**, 24.
150. D. B. Moffat and M. Creasey, Distribution of ingested silver in the kidney of the rat and of the rabbit, *Acta Anat.*, 1972, **83**, 346.
151. US Environmental Protection Agency, *Ambient Water Quality Criteria Document: Silver*, USEPA, Washington DC, 1980, EPA 440/5-80-071.
152. A. B. G. Lansdown, Animal models for the study of skin irritants, *Curr. Probl. Dermatol.*, 1978, **7**, 26.
153. R. J. Scheuplein, Permeability of the skin: a review of major concepts, *Curr. Probl. Dermatol.*, 1978, **7**, 172.
154. W. S. Bullough, Mitotic and functional homeostasis: a speculative review, *Cancer Res*, 1965, **25**, 1683.
155. R. K. Freinkel and D. T. Woodley, *The Biology of the Skin*, Parthenon Publishing, New York, 2001.
156. G. S. Schultz, R. G. Sibbald, V. Falanga, E. A. Ayello, C. Dowsett, K. G. Harding, M. Romanelli, M. C. Stacy, L. Teot and W. Vanscheidt, Wound bed preparation: a systematic approach to wound management, *Wound Rep. Regen.*, 2003, **11**, 1.
157. V. Falanga, Classifications for wound bed preparation and stimulation of chronic wounds, *Wound Rep. Regen.*, 2000, **8**, 347.
158. S. Enoch and K. G. Harding, Wound bed preparation: the science behind the removal of barriers to healing, *Wounds*, 2003, **15**, 213.
159. A. B. G. Lansdown, Silver : toxicity in mammals and how its products aid wound repair, *J. Wound Care*, 2002, **11**, 173.
160. A. B. G. Lansdown, A. Williams, S. Chandler and S. Benfield, Silver absorption and antibacterial efficacy of silver dressings, *J. Wound Care*, 2005, **14**, 155.
161. E. Skog and J. E. Wahlberg, A comparative investigation of the percutaneous absorption of metal compounds in the guinea pig skin by means of radioactive isotopes, ^{51}Cr, ^{58}Co, ^{65}Zn, ^{110m}Ag, ^{115m}Hg, *J. Invest. Dermatol.*, 1964, **43**, 187.
162. O. Nørgaard, Investigations with ^{111}Ag into the resorption of silver through human skin, *Acta Dermatol. Venereol.*, 1954, **34**, 415.

163. G. Perelli and G. Piolatto, Tentative reference values for gold, silver and platinum: literature data and analysis, *Sci. Total Environ.*, 1992, **120**, 93.
164. Bray Healthcare, Toughened silver nitrate caustics in medicine: warts, verrucae, cautery, granuloma, 1997, www.bray.co.uk.
165. H. Schlepler, J. Kesseler and B. Hartmann, Abuse of silver nitrate solution for planing periorbital folds, *Burns*, 2002, **28**, 90.
166. A. B. G. Lansdown, B. Sampson and A. Rowe, Sequential changes in trace metal, metallothionein and calmodulin concentrations in healing skin wounds, *J. Anat.*, 1999, **195**, 375.
167. M. Karasawa, N. Nishimura, H. Nishimura, C. Tohyama, C. Abe and T. Kroki, Localisation of metallothioneins in hair follicles, of normal skin and the basal layer of hyperplastic epidermis, possible role association with cell proliferation, *J. Invest. Dermatol.*, 1991, **96**, 97.
168. I. Bremner, Interactions between metallothioneins and trace elements, *Prog. Food Nutr. Sci.*, 1987, **11**, 1.
169. I. Bremner and J. H. Beattie, Metallothioneins and the trace minerals, *Annu. Rev. Nutr.*, 1990, **10**, 63.
170. J. J. Van den Oord and M. DeLey, Distribution of metallothioneins in normal and pathological human skin, *Arch. Dermatol.*, 1994, **286**, 62.
171. J. L. Reymond, P. Stoebner and P. Amblard, Cutaneous argyria: an electron microscope study of four cases with microanalysis X study of one case, *Ann. Dermatol. Venereol.*, 1981, **107**, 251.
172. C. R. Baxter, Topical use of 1% silver sulfadiazine, in *Contemporary Burn Management*, ed. H. C. Polk and H. H. Stone, Little Brown, Boston, 1971, pp. 217–225.
173. P. Melotte, B. Hendrickx, P. Melin, A. Mullie and C. M. Lapiere, Efficiency of 1% silver sulphadiazine dream in treating bacterial infection of leg ulcers, *Curr. Ther. Res.*, 1985, **3**, 197.
174. H. E. Stern, Silver sulphadiazine and healing of partial-thickness burns: a prospective clinical trial, *Br. J. Plast. Surg.*, 1989, **42**, 581.
175. D. Wyatt, D. N. McGowan and M. P. Najarian, Comparison of a hydrocolloid dressing and silver sulphadiazine cream in the outpatient management of second degree burns, *J. Trauma*, 1990, **30**, 857.
176. J. B. Bishop, L. G. Phillips, T. A. Mustoe, A. J. Van der Zee, L. Weirsema, D. E. Roach, J. P. Heggers, D. P. Hill, E. L. Taylor and M. C. Robson, A prospective randomised evaluator-blinded trial of two potential wound healing agents for treatment of venous stasis ulcers, *J. Vasc. Surg.*, 1992, **16**, 251.
177. C. S. Chu, N. P. Matylevitch, A. T. McManus, C. W. Goodwin and B. A. Pruitt, Accelerated healing with a mesh autograft/allodermal composite skin graft treated with silver nylon dressings with and without direct current in rats, *J. Trauma*, 2000, **49**, 115.
178. A. B. G. Lansdown and A. Williams, How safe is silver in wound care, *Wound Care*, 2004, **13**, 131.

179. M. A. Watcher and R. G. Wheeland, The role of topical agents in the healing of full thickness skin wounds, *J. Dermatol. Surg. Oncol.*, 1989, **15**, 1188.

180. I. O. Leitch, A. Kucukcelebi and M. C. Robson, Inhibition of wound contraction by topical antimicrobials, *Aust. N. Z. J. Surg.*, 1993, **63**, 289.

181. D. Kjolseth, J. M. Frank, J. H. Barker, G. L. Anderson, A. I. Rosenthal, R. D. Acland, D. Schulke, F. R. Campbell, G. R. Tobin and L. J. Weiner, Comparison of the effects of commonly used wound agents on epithelialization and neo-vascularization, *J. Am. Coll. Surg.*, 1994, **179**, 305.

182. M. J. Hoekstra and M. P. Rogmans, Experimental burns in pigs for the evaluation of topical wound therapy, in *Proceedings of the 1st European Conference on Advances in Wound Management*, ed. K. G. Harding, D. L. Leaper and T. D. Turner, MacMillan, London, 1992, pp. 11–13.

183. G. L. Brown, L. B. Nanny, J. Griffin, A. B. Cramer, J. M. Yancey, L. J. Curtsinger, L. Holtzin, G. S. Schultz, M. J. Jurkiewicz and J. B. Lynch, Enhancement of wound healing by topical treatment with epidermal growth factor, *New Engl. J. Med.*, 1989, **321**, 76.

184. L. S. Tanner and D. J. Gross, Generalized argyria, *Cutis*, 1990, **45**, 237.

185. E. R. Farmer and A. F. Hood, Systemic argyria associated with ingestion of colloidal silver, in *Pathology of the Skin*, Appleton & Lange, Norwalk, CT, 2000, pp. 507–508.

186. L. E. Gaul, Incidence of sensitivity to chromium, nickel, gold, silver and copper compared to reactions to their aqueous salts including cobalt sulphate, *Ann. Allergy*, 1954, **12**, 429.

187. L. E. Gaul and G. B. Underwood, Effects of aging a solution of silver nitrate on its cutaneous reaction, *J. Invest. Dermatol.*, 1948, **11**, 7.

188. M. A. Hollinger, Toxicological aspects of topical silver pharmaceuticals, *CRC Crit. Rev. Toxicol.*, 1996, **26**, 255.

189. P. Hultman, S. Enestrom, S. J. Turley and K. M. Pollard, Selective induction of anti-fibrillarin auto-antibodies by silver nitrate in mice, *Clin. Exp. Immunol.*, 1994, **96**, 285.

190. I. L. Steffensen, O. J. Mesha, E. Andruchow, E. Namork, K. Hylland and R. A. Andersen, Cytotoxicity and accumulation of Hg, Ag, Cd, Pb, and Zn in human peripheral T and B lymphocytes and monocytes *in vitro*, *Gen. Pharmacol.*, 1994, **25**, 1621.

191. R. C. Howard, Allergy to argyrol in a patient with chronic purulent otitis media, *Laryngoscope*, 1930, **40**, 215.

192. L. H. Criep, Allergy to Argyrol, *J. Am. Med. Assoc.*, 1943, **121**, 421.

193. P. V. Marcussen, Eczematous allergy to metals, *Acta Allergol.*, 1962, **17**, 311.

194. H. De Greef and A. Dooms-Goosens, Patch testing with silver sulpha-diazine cream, *Contact Derm.*, 1985, **12**, 33.

195. A. Fraser-Moodie, Sensitivity to silver in a patient treated with silver sulphadiazine (Flamazine), *Burns*, 1992, **18**, 74.

196. I. Rasmusson, Patch test reactions to Flamazine, *Contact Derm.*, 1984, **11**, 133.

197. A. M. Clarke, Febrile reaction to silver sulphadiazine cream, *Med. J. Aust.*, 1981, **2**, 456.

198. S. D. Clarke, L. Gallur and G. N. Threlfall, Febrile reaction to silver sulphadiazine cream, *Med. J. Aust.*, 1981, **2**, 208.

199. L. L. Dupuis, N. H. Shear and R. M. Zucker, Hyperpigmentation due to topical application of silver sulphadiazine cream, *J. Am Acad. Dermatol.*, 1985, **12**, 1112.

200. S. Agarwal and D. J. Gawkrodger, Occupational allergic contact dermatitis to silver and colophonium in a jeweler, *Am. J. Contact Dermatol.*, 2002, **13**, 74.

201. J. Vilaplana, Jewellers in Kanerva, in *Handbook of Dermatology*, ed. P. Elsner and J. E. Wahlberg, Springer, Berlin, 2000.

202. T. Heyl, Contact dermatitis from silver coat, *Contact Derm.*, 1979, **5**, 197.

203. P. Sarsfield, J. E. White and J. M. Theaker, Silver worker's finger: an unusual occupational hazard mimicking a melanocytic lesion, *Histopathology*, 1992, **20**, 73.

204. I. R. White and R. J. Rycroft, Contact dermatitis from silver fulminate-fulminate itch, *Contact Derm.*, 1982, **8**, 159.

205. R. Marks, Contact dermatitis due to silver, *Br. J. Dermatol.*, 1966, **78**, 606.

206. S. Kaaber, Allergy to dental materials with special reference to the use of amalgam and polymethylmethacrylate, *Int. Dent. J.*, 1990, **40**, 359.

207. E. C. Munksgaard, Toxicology versus allergy in restorative dentistry, *Adv. Dent. Res.*, 1992, **6**, 17.

208. J. Laine, K. Kalimo and R. -P. Happonen, Contact allergy to dental restorative materials in patients with oral lichenoid lesions, *Contact Derm.*, 2006, **36**, 141.

209. M. Shirkhanzadeh and M. Azadegan, Formation of carbonate apatite on calcium phosphate coating containing silver ions, *J. Mater. Sci. Mater. Med.*, 1998, **9**, 385.

210. G. W. Gould, J. Colyer, J. M. East and A. G. Lee, Silver ions trigger Ca^{2+} release by interaction with the $(Ca^{2+}-Mg^{2+})$-Ag^+ in reconstituted systems, *J. Biol. Med.*, 1987, **262**, 7676.

211. G. Salama and J. Abramson, Silver ions trigger Ca^{2+} release by acting at the apparent physiological release site in sarcoplasmic reticulum, *J. Biol. Chem.*, 1984, **259**, 13363.

212. S. D. Prabhu and G. Salama, The heavy metal ions Ag^+ and Hg^{2+} trigger calcium release from cardiac sarcoplasmic reticulum, *Acta Biochem. Biophys.*, 1990, **277**, 475.

213. R. Tupling and H. Green, Silver ions induce ca^{2+} release from the SR *in vitro* by acting on the Ca^{2+} release channel and the Ca^{2+} pump, *J. Appl. Physiol.*, 2002, **92**, 1603.

214. V. Alt, T. Bechert, P. Steinruche, P. Seidel, E. Dingeldein, E. Domann and R. Schnettler, *in vitro* testing of antimicrobial activity of bone cement, *Antimicrob. Agents Chemother.*, 2004, **48**, 4084.

215. V. Alt, T. Bechert, P. Steinruche, M. Wegener, P. Seidel, E. Dingeldein, E. Domann and R. Schnettler, An *in vitro* assessment of the antibacterial properties and cytotoxicity of nanoparticulate silver bone cement, *Biomaterials*, 2004, **25**, 4383.

216. R. Dueland, J. A. Spadaro and B. A. Rahn, Silver antibacterial bone cement, *Clin. Orthop. Relat. Res.*, 1982, **169**, 264.

217. U. Lindh, D. Brune, G. Nordberg and P. O. Webster, Levels of antimony, arsenic, cadmium, copper, lead, mercury, selenium, silver, tin and zinc in bone tissue of industrially exposed workers, *Sci. Total Environ.*, 1980, **16**, 109.

218. M. C. Cortizo, M. Fernández, L. de Mele and A. Cortizo, Metallic dental material biocompatibility in osteoblast-like cells, *Biol. Trace Elem. Res.*, 2004, **100**, 151.

219. J. Hardes, A. Streitburger, H. Ahrens, T. Nusselt, C. Gebert, W. Winkelmann, A. Battmann and G. Gosheger, The influence of elementary silver versus titanium on osteoblasts behaviour *in vitro* using human osteosarcoma cell lines, *Sarcoma*, 2007, **10**, 1.

220. J. A. Spadaro, D. A. Webster and R. O. Becker, Silver polymethylmethacrylate antibacterial bone cement, *Clin. Orthop. Relat. Res.*, 1979, **143**, 266.

221. M. Bosetti, A. Masse, E. Tobin and M. Cannas, Silver-coated materials for external fixation devices; *in vitro* biocompatibility and genotoxicity, *Biomaterials*, 2002, **23**, 887.

222. E. Sudemann, H. Vik, M. Rait, K. Todnam, K. J. Anderson, K. Juhlsham, O. Fresland and J. Rungby, Systemic and local silver accumulation after total hip replacement using silver-impregnated bone cement, *Med. Prog. Technol.*, 1994, **20**, 179.

223. A. B. G. Lansdown, Critical observations on the neurotoxicology of silver, *CRC Crit. Rev. Toxicol.*, 2007, **37**, 237.

224. J. R. Jones, L. M. Ehrenfried, P. Saravanapavan and L. L. Hench, Controlling ion release from bioactive glass foam scaffolds, *J. Mater. Sci. Mater. Med.*, 2006, **17**, 989.

225. P. D. Thomson, N. P. More, T. L. Rice and J. K. Prasad, Leukopenia in acute thermal injury: evidence against silver sulphadiazine as a causative agent, *J. Burn Care Rehabil.*, 1998, **10**, 418.

226. F. W. Fuller and P. E. Engler, Leukopenia in non-septic burn patients receiving topical 1% silver sulphadiazine cream therapy: a survey, *J. Burn Care Rehabil.*, 1988, **9**, 606.

227. J. Viala, L. Simon, C. Le Pommelet, L. Philippon, D. Devictor and G. Huault, Agranulocytosis after application of silver sulphadiazine in a two month old infant, *Arch. Pediatr.*, 1997, **4**, 1103.

228. J. Doull, C. D. Klassen and M. D. Amdur, in *Casarett and Doull's Toxicology*, MacMillan, New York, 1986, p. 625.

229. G. D. Clayton and F. E. Clayton, Silver, in *Patty's Industrial Hygiene and Toxicology*, Wiley, 1981, pp. 1881–1894.
230. R. H. Dreisbach, in *Handbook of Poisoning*, ed. R. H. Dreisbach and W. O. Robertson, Appleton & Lange, Norwalk, CT, 1987, p. 373.
231. G. Reinhardt, G. Von Mallink, H. Kittel and O. Opitz, Acute fatal poisoning with silver nitrate following an abortion attempt, *Arch. Kriminol.*, 1971, **148**, 69.
232. D. G. Lowe, D. A. Levison, P. R. Crocker and J. H. Shepherd, Silver deposition in the cervix after application of silver nitrate as a cauterising agent, *J. Clin. Pathol.*, 1988, **41**, 871.
233. C. A. D. Ringrose, Office tubal sterilization, *Obstet. Gynecol.*, 1973, **42**, 151.
234. US Department of Health and Human Resources, *12th Report on Carcinogens*, National Toxicology Program, Research Triangle Park, NC, 2007.
235. B. S. Oppenheimer, E. T. Oppenheimer, I. Danishefsky, A. P. Stout and M. Willhite, Carcinogenic effect of metals in rodents, *Cancer Res.*, 1956, **16**, 439.
236. A. Furst and M. C. Schlauder, Inactivity of two noble metals as carcinogens, *J. Environ. Pathol. Toxicol.*, 1978, **1**, 51.
237. A. Furst, Bioassay of metals for carcinogenesis: whole animals, *Environ. Health Perspect.*, 1981, **40**, 83.
238. H. Nothdurft, Experimental production of sarcomas in rats following implantation of round discs of gold, platinum, silver or ivory, *Naturwissenschaften*, 1955, **42**, 75.
239. D. Schmähl and D. Steinhoff, Versuche zur Krebserzeugung mit kolloidalen Silber-und Goldlösungen an Ratten, *J. Cancer Res. Clin. Oncol.*, 1960, **53**, 586.
240. N. H. Dubin, T. H. Pamley, R. T. Cox and T. M. King, Effect of silver nitrate on pregnancy termination in cynomolgus monkeys, *J. Fertil. Steril.*, 1981, **36**, 106.
241. J. Rungby and G. Danscher, Neuronal accumulation of silver in brains of progeny from argyric rats, *Acta Neuropathol.*, 1983, **61**, 258.
242. J. Rungby, An experimental study on silver in the nervous system and on aspects of its general cellular toxicity, *Dan. Med. Bull.*, 1990, **37**, 442.
243. J. Rungby, L. Slomianka, G. Danscher, A. H. Andersen and M. J. West, A quantitative evaluation of the neurotoxic effect of silver on the volumes of the components of the developing rat hippocampus, *Toxicology*, 1987, **43**, 261.

CHAPTER 9
Silver and Organs of Special Sense

9.1 Introduction

The brain and central nervous system, peripheral nerves, eye, ear and sensory nerves of the tongue, and nasal mucosae, and nerve endings in the skin are justifiably included under organs of special sense. In the study of silver toxicology, the adverse effects in these tissues are least well researched and published reviews are frequently misleading or controversial. It is notable that the German obstetrician, Carl Credé (1819–1892), claimed that 2% silver nitrate instilled into the conjunctival sac was efficacious in treating neonatal eye diseases (ophthalmia neonatorum),[1] but that the procedure may be hazardous and injurious to the tissues.[2]

Credé's method

1. A method for preventing ophthalmia neonatorum by administering one drop of a 2 percent solution of silver nitrate into each eye of a newborn infant.
2. A method for expelling the afterbirth by resting the hand on the fundus of the uterus after expulsion of the fetus, gently rubbing the fundus with the hand in case of hemorrhage or failing contractions, then, when the afterbirth is loosened, expelling it by firmly squeezing the fundus with the hand.
3. A method for expressing urine by pressing the hand on the bladder, especially a paralyzed bladder.

The American Heritage® Medical Dictionary Copyright © 2007, 2004 by Houghton Mifflin Company. Published by Houghton Mifflin Company. All rights reserved.

Issues in Toxicology No. 6
Silver in Healthcare: Its Antimicrobial Efficacy and Safety in Use
By Alan B. G. Lansdown
© Alan B. G. Lansdown 2010
Published by the Royal Society of Chemistry, www.rsc.org

In more recent times, topical colloidal silver therapies for such conditions as shingles, Meniere's disease, epilepsy, infections of the mouth, nose, ears, eyes, sinuses and skin infections have been implicated in neurological damage.[3–5] It may be difficult to determine the scientific credibility of the various claims for colloidal silver remedies in view of the lack of published clinical observations, but there is abundant evidence that metallic silver, silver alloys and numerous silver compounds used in clinical medicine do come into contact with body surfaces to some extent through occupational or therapeutic circumstances. Much of the Ag^+ released precipitates as silver sulfide, silver selenide, silver chloride or silver proteinates, but some is metabolised *via* the systemic circulation to bone and soft tissues. During the World Wars and at earlier times, severe injuries to the skull were "repaired" by the use of silver plates, with claims of silver toxicity to the nerve cells in the brain and spinal column. Silver has been used in craniofacial surgery since at least the 18th century, and such cases as "the gunner with the silver mask" and elaborate facial reconstruction with silver plates are recorded.[6]

The eye, external ear and sensory receptors on the tongue, oral and nasal membranes, skin and urinogenital surfaces will be exposed to silver directly through intentional therapeutic application or accidental exposures, whereas the brain and central nervous system may be influenced by silver deposition *via* the systemic circulation (Figure 9.1). The vulnerability of the brain, spinal cord and nerve fibres to silver or other heavy metals has been researched and the importance of the blood–brain barrier established.[7,8] Silver-coated catheters inserted for intraventricular drainage in cases of congenital or infection-related hydrocephaly might be expected to release Ag^+ along the insertion tract of the catheters, and if silver is toxic to neurological tissues, then a band of pathological change is expected.[9,10]

9.2 Brain and Peripheral Nervous System

9.2.1 Neurotoxicity of Metals

Metals differ greatly in their transport mechanisms in the human body and their accumulation in the central nervous system. Acknowledged neurotoxic metals differ greatly in their ability to accumulate in soft tissues and their mechanisms of action and pathogenicity, but share the characteristics listed in Table 9.1.[11,12] At least nine metals are known to penetrate into neurological tissues[7,8] including the toxic metals, lead, cadmium and mercury with no known trace metal value. Sodium, potassium, calcium, iron, copper, zinc, manganese, cobalt and molybdenum perform essential physiological functions in the human body or serve as enzyme co-factors, transcriptional factors or modulators of gene expression but may exert pathophysiological changes if present in supra-optimal quantities.[13] The position of silver as a neurotoxic metal is equivocal at the moment.[7,8] Early evidence of neurotoxicity was provided by a simple experimental study in which tadpoles exposed to dilute silver nitrate developed a "white matter oedema", with water-filled vacuoles between

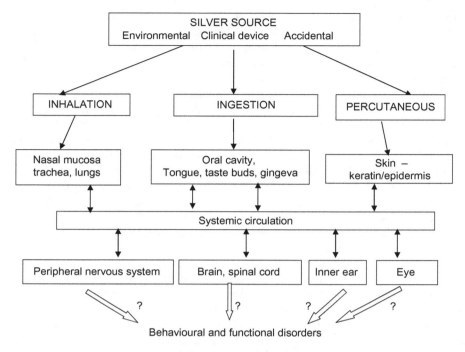

Figure 9.1   Routes of exposure of neurological tissue to silver.

Table 9.1   General characteristics of neurotoxic metals.

1. Penetrate the blood–brain barrier and the blood cerebrospinal fluid barrier to enter tissues of the central nervous system.
2. Exhibit a predilection for specific cell or tissue types (neurones, glial cells).
3. Impair essential metabolic pathways in target cells leading to functional disturbances and/or progressive degenerative and changes resulting in cell death.
4. The severity of responses seen show a direct correlation with the amount of toxin present and the duration of its action.

extracellular membrane surfaces of myelin lamelli.[14–16] The mechanism is unclear but comparable observations have not been observed in mammals.

9.2.2   Blood–Brain Barrier

The capacity of heavy metals or other xenobiotics to injure tissues of the brain and central nervous system correlates well with their capacity to cross the blood–brain barrier (BBB) and enter neurological tissues.[7–8,17] The blood–brain barrier performs a central function of maintaining chemical homeostasis within the central nervous system, but is particularly vulnerable to agents that damage cell membranes. Classical studies by Broman, Lindberg-Broman and Rapoport in the 1940s revealed that the function of the blood–brain barrier lay within the capillary endothelial cells lining the capillary network of the brain and spinal cord, and not basement membranes as previously thought.[18–20] By modulating the uptake of nutrients and electrolytes from the circulation and regulating the egression of metabolites, the blood–brain barrier controls brain chemistry and limits minor changes that may be expressed in terms of learning difficulties, memory loss and behavioural dysfunction.

This research illustrates how chemically induced defects in the blood–brain barrier may be a cause of oedema, aberrant brain development and neurodegeneration.[11–12,15] Zheng *et al.* demonstrated that the blood–brain barrier has special significance in regulating the uptake and neurotoxic action of metals and exhibits a limited capacity to metabolise certain lipophilic materials that influence carrier-mediated processes.[7,8] The neurotoxicity of metals and other xenobiotic materials is largely determined by the protective efficiency of the blood–brain barrier in different regions of the central nervous system.[17] Thus, whereas lead tends to accumulate in the blood–brain barrier before permeating the brain, mercury in its methylated form is highly lipophilic and penetrates the blood–brain barrier and nervous tissues more readily.[11] An understanding of the critical role of the blood–brain barrier is essential is appreciating the putative neurotoxic action of silver and the increased vulnerability of certain areas of the brain to injury.

The blood–brain barrier is a complex system comprising the interface between the blood and the brain, and that separating the blood and cerebrospinal fluid (CSF).[12,17,21] Endothelial cells lining the extensive vascular network of the brain and sub-arachnoid space provide a major component of the blood–brain barrier, whereas the blood–CSF barrier resides largely in the choroid plexus and the ependymal cells lining the cavity of the CSF. Peripheral nerves have analogous barrier systems comprising the vascular network and connective tissues of the endoneurium and the perineurium surrounding nerves and nerve bundles, respectively. In humans, the blood–brain barrier is established at birth, but in the choroid plexus and circumventricular organs (median eminence, sub-fornical organ, area postrema and neurohypophysis) it becomes less well developed. The endothelial cells lining vascular channels are of a fenestrated type even though they maintain tight gap junctions.[11] The permeability of the blood–brain barrier varies according to age and the region of the brain implying that certain areas of the brain are more vulnerable to metal-induced injury than others.[20,21] In the rat, the structure of the blood–brain barrier and its relationship to surrounding astrocytes has been investigated using a silver protein complex as a marker.[22] This has demonstrated that the sub-fornical organ is largely devoid of blood–brain barrier and that microvessels are separated from surrounding astrocytes only by a basement membrane.

Early evidence of the protective role of the blood–brain barrier in controlling the penetration of xenobiotic materials was provided by experiments in which the intra-vital dye, trypan blue, was injected intravenously into rabbits.[23,24] The dye bound to plasma protein was not absorbed into the tissues of the brain, but sequestered and bound lysosomally in endothelial cells. It stained other soft tissues but was excluded from the brain, unless injected intracerebrally. These observations promoted the concept that the blood–brain barrier was unique to the central nervous system and acted in the form of an "exclusionary interface" separating the brain from circulating blood. According to Rapoport, cerebral capillaries facilitate diffusion and regulate exchange of metabolites between blood and brain.[17] The ependymal surfaces of the cerebral ventricles and the pia-glial surfaces of the brain do not impede transfer of substances between the cerebrospinal fluid and the brain, and do not constitute a sub-barrier. Recent experimental models in rabbits have demonstrated a new method for implantation of intraventricular catheters using a flexible electromagnetic navigation and dynamic reference frame.[9] The method is claimed to allow insertion of catheters with minimal displacement or mechanical damage, but cannot predict changes attributable to the elution of silver or other materials from the catheter into the brain.

Zheng *et al.* studied the morphology of the blood–CSF barrier with particular reference to the role of the choroid plexus in modulating metal-induced neuro-toxicities.[7,8] The choroid plexus is a highly vascular villous structure extending from the ventricular surfaces of the brain into CSF-like coral fronds.[25,26] Although it represents less than 5% of total brain weight,[27] the choroid plexus has a proportionately higher surface area permitting greater exposure to the circulating CSF. Compelling evidence illustrates the critical role of the choroid plexus in sequestering toxic heavy metals like lead and mercury, and may regulate the neurotoxic action of silver.[7] The blood flow in the choroid plexus is high and exposes it to a greater influx of toxic materials and efflux of metabolites than elsewhere in the brain.[17]

Ependymal cells lining the CSF surface are densely packed with tight junctions providing a modest barrier to the transfer of metal ions. In contrast, the fenestrated endothelial cells lining choroidal capillaries are more porous or "leaky", thereby permitting greater exchange of solutes and metal ions between the blood and connective tissue matrix.[5,17] The ependymal cells regulate the production and composition of the CSF including the interchange of metal ions, but their mechanisms of action (possibly involving sodium and potassium ATP-ase pump mechanisms) are imperfectly understood. Transport through the blood–brain barrier is limited to non-polar substances and several nutrients for which special carrier mediated pathways exist.[11] Experimental studies have demonstrated the ability of the connective tissue of the choroid plexus to concentrate metal ions including organic mercury, cadmium, arsenic and lead, and to regulate their penetration into the neural tissues to evoke pathological damage.[28–30] Metal ions may be conveniently classified according to their specific action on the choroid plexus[8] (Table 9.2).

Silver-induced neurotoxicity is rare,[8] even though some experimental studies in the rat claim that silver ions do penetrate the blood–brain barrier and the blood–placental barrier to locate heterogeneously throughout the central

Table 9.2 Classification of metal ions according to their toxic action on the choroid plexus.

1. General toxicants	Metals which accumulate within the tissues of the plexus and cause substantial structural and functional damage.	mercury, arsenic and cadmium
2. Selective toxicant	Metals which do not alter the permeability of the plexus or evoke significant patho-physiological changes, but influence critical regulatory functions as a prelude to neurological damage.	lead, manganese, copper and tellurium
3. Sequestered toxicants	Metals that are sequestered by the choroid plexus as an integral part of its neuroprotective role.	silver, gold, zinc and iron

nervous system.[28–29,31,32] Predictive experimental studies conducted in animal models are expected to provide more accurate and reproducible information on the neurotropic action of silver than is possible with human post-mortem material, where neurological tissues obtained from patients dying with argyria or supposed silver intoxication autolyse readily after death and visualisation of sites of silver deposition is obscured. Electron microscopy has been widely used in examining the deposition of silver in the region of the brain and other tissues, but X-ray microanalysis and autometallography now provide a more accurate means of observing fine silver deposits in the choroid plexus, neurons, glial cells and extra-neural tissues of the blood–brain barrier.[32–36] Analysis of silver in "the brain" using [111]Ag-tracer studies, atomic absorption spectrometry and neutron activation analysis is insufficient by itself to discriminate between silver deposited within tissues of the brain and that contained within tissues of the blood–brain barrier.[37]

9.2.3 Experimental Studies

Early studies conducted in rats exposed chronically to silver nitrate in drinking water failed to substantiate that silver passes the blood–brain barrier in any form to accumulate in neural tissues of any part of the central nervous system. When silver nitrate was employed as an "intra-vital dye" to demonstrate the integrity of the blood–brain barrier in the rat, silver was precipitated preferentially in basal laminae and perivascular spaces of the choroid plexus, hypophysis, pineal body, area postrema and sub-fornical organ.[38–42] It could not be identified outside circumventricular areas or around cerebral capillaries, even in severely argyric animals. Although silver is readily metabolised from tissues such as liver and kidney in humans,[43] it exhibited a longer half-life in endothelial cells of the BBB site than in other soft tissues in the rat. Later more elaborate studies by Scott and Norman confirmed the inability of silver to cross the blood–brain barrier and demonstrated fine electron-dense silver granules (10–15 nm diameter) in the basal laminae of arterioles of the parietal cortex and subcortical white matter.[44] Accumulation in these sites "maximised" by 241 days and did not change in concentration or distribution up to 455 days after exposure. In an attempt to increase the vulnerability of the brain to silver, Scott and Norman induced surgical intracerebral stab wound injury.[44] This had the effect of increasing silver accumulation in the laminae of small blood vessels, tissue fragments of the blood–brain barrier and associated macrophages, but deposits were not identified within the brain parenchyma. They confirmed earlier studies demonstrating that silver protein complexes do not penetrate the gap junctions of cerebral endothelia, even though some silver might dissociate at cell membranes and penetrate cells by an undefined mechanism other than pinocytosis.[36] Alternatively, silver ion bound strongly to collagen and glyco-proteins of the blood–brain barrier.[45]

Two Russian studies (unseen) cited by the Joint FAO/WHO Expert Committee on Food Additives (JEFCA) in 1977[i] may provide evidence of a direct toxic effect on silver in the brain.[46,47] The first claimed to show decreased brain RNA and DNA and dystrophic changes in rats given 0.2% silver nitrate in drinking water for 12 months or 2.0% for six months, whilst in the second study histopathological changes were reported in neuronal, glial and vascular tissues of the encephalon and medulla of rabbits dosed with 0.025 or 0.25 mg kg^{-1} silver (?intravenous injection). Further details were not available.

Evidence that silver penetrates the blood–brain barrier and blood placental barrier relies heavily on a comprehensive series of anatomical, histochemical and electron microscopy studies conducted in rats at the University of Aarhus in Denmark. The researchers considered that the symptomatic effects of paralysis, loss of co-ordination, cerebella ataxia, convulsions and electroencephalogram (EEG) changes seen in patients with severe argyria and chronic silver exposure were difficult to explain if silver had not penetrated the blood–brain barrier.[46] They administered silver nitrate or silver lactate to rats and mice orally or by intraperitoneal injection and employed autometallographic methods to demonstrate silver penetration of the blood–brain barrier and claimed deposition in all parts of the central nervous system.[34,48–52] Animals given silver nitrate or silver lactate chronically in drinking water (0.01%) or injected intraperitoneally with silver lactate (3–55 mg for up to 13 months) or a colloidal silver preparation (Protagol (0.1–0.5 mL for 2–5 days) exhibited intracellular and extracellular silver sulfide deposition throughout the brain, dorsal root ganglia, enteric ganglia, peripheral nervous system, anterior pituitary gland and neural retina of the eye.[53] These distribution patterns were heterogeneous, but particularly heavy in large motor neurones and protoplasmic astrocytes. The silver

[i] WHO Food Additive Series, No. 12, *Summary of Toxicological Data of Certain Food Additives*, WHO Technical Report Series, 1977, Geneva.

granules were bound specifically in secondary lysosomes.[35] In keeping with earlier observations,[44] BBB tissues were heavily stained with silver sulfide deposits but much silver was located as extracellular deposits on basement membranes of cerebral blood vessels and on elastic fibres.[29,35,36] The intensity of silver deposition in each case was proportional to the amount of silver administered and the duration of exposure, although subtle differences were evident between administration of silver nitrate and silver lactate. Importantly, silver deposits were transitory in these locations and declined when silver treatment was withdrawn. Macrophages engorged with silver deposits have been consistently reported in the region of the blood–brain barrier. Interestingly, when silver was injected into the lateral ventricles of the brain, it was absorbed into ependymal cells of the blood–brain barrier rather than locating in neurones or glial cells. Rungby and Danscher conceded that the paralysis reported in earlier studies in rats dosed with silver nitrate might be attributable to the toxic effects of silver accumulating in capillaries associated with the central nervous system,[34,54] but gave no details. No specific neurobehavioural tests of the type promulgated by Roper[55] were conducted.

Although Rungby and Danscher claimed that silver does cross the blood–brain barrier to accumulate in specific locations in the central nervous system, basing their evidence upon electron microscopy, photochemical and auto-metallographic techniques, they failed to correlate silver accumulation in neurological tissues with frank neurological damage in any tissue or provide evidence for behavioural changes indicating silver neurotoxicity.[34,49,50] They claimed that silver accumulated in lysosomal vacuoles in neurones and glial cells of the olfactory lobes, cerebral cortex, hippocampus, substantia innominata and hypothalamus in a dose-related fashion irrespective of route of administration (oral or intravenous), but that the thalamus, substantia nigra and nuclei pontis were not noticeably affected. Neurons of the globus pallidus, brain stem, spinal cord and basal root ganglia, cerebellum (deep nuclei) and the trigeminal nerve also showed a strong tendency to concentrate silver. Silver deposits in the rat hippocampus and in the peripheral nervous system remained stable for at least 45 days.[49] The brains of young post-natal animals appeared to be more vulnerable to the effects of silver as suggested by a significant reduction the pyramidal cell layer of the hippocampus.[50] This may be an indication of a cytostatic effect, or other toxic or pathological effect of silver on developing hippocampal cells.

Unequivocal toxic damage has not been seen in neurological tissues of rats exposed by various routes to silver nitrate, silver lactate or Protargol, but mice exposed chronically to very low levels (0.015%) of silver nitrate or silver lactate in drinking water became argyric and hypoactive in open field behavioural studies.[56] If this represents evidence of neurobehavioural change attributable to silver accumulation, further pathophysiological studies are required to investigate the mechanism of action. More recent work by Pelkonen *et al.* reported silver accumulation in the cerebellum and the soleus muscles of young adult mice given $0.03\,\text{ml L}^{-1}$ silver nitrate to drink for 1–2 weeks, but failed to show changes in behavioural activity or disturbed health patterns (Table 9.3).[57] Some neurobehavioural studies would have been useful to explore the possible health

Table 9.3   Mean tissue silver concentration ($ng\,g^{-1}$ wet weight) in mice given $0.03\,mg\,L^{-1}$ silver nitrate to drink for one or two weeks.[57]

Tissue	1 week	2 weeks
Cerebellum	9.39 + 1.90	7.45 + 4.4.9
Cerebrum	2.46 + 0.57	1.72 + 0.48
Musculus soleus	23.63 + 9.79	28.57 + 15.35
Musculus gastrocnemus	1.28 + 0.41	1.55 + 0.22
Blood	0.97 + 0.30	0.95 + 0.27

implications of these findings. Their pooled data indicate that tissue concentrations of silver do not vary significantly within two weeks and that concentrations of silver in blood, cerebellum, soleus and gastrocnemius muscles stabilise.

An unpublished communication purporting to show permanent facial paralysis in human patients following post-operative mastoid surgery and application of silver nitrate to dehiscent facial nerves suggests that silver or nitrate ions may exert direct toxic damage on exposed nerves.[58] This view is supported by experimental studies in Sprague–Dawley rats where silver nitrate cautery for one second was shown to evoke some axonal injury but allowed modest neuronal recovery. In contrast, exposures of 5 or 10 seconds resulted in a 50% axonal loss and impaired mobility in the 14 days after operation. It is expected that the corrosive action of the nitrate anion is largely responsible for the observed changes.

Experimental models have been developed to study patterns of silver release into the brain and central nervous system from medicated catheters designed for in-dwelling use in patients with hydrocephaly.[59] An experimental study designed to evaluate the release of silver ion from a silver-iontophoretic catheter in rabbits showed that a total of $0.4–2.5\,mg$ was liberated within 12 weeks. Blood silver levels increased to $1.0\,\mu g\,L^{-1}$ over seven days, but declined to $0.28\,\mu g\,L^{-1}$ after 8–12 weeks. It is unclear from this study to what extent the silver released accumulated in the central nervous system or evoked behavioural or other pathological changes. *In vitro* experiments have demonstrated that polyurethane catheters (Erlanger silver catheters) impregnated with finely dispersed metallic silver particles (2% w/w) were biocompatible with human fibroblasts and lymphocytes and not cytotoxic. Joyce-Wöhrmann and Münstedt (1999) demonstrated that release of Ag^+ from silver impregnated catheters *in vitro* under simulated clinical conditions was equivalent to $0.1\,\mu g\,L^{-1}$.[60] When samples of catheter material coated with $0.1–30\,\mu g\,cm^{-2}$ nanoparticles of silver were implanted subcutaneously in mice, no signs of toxicity were seen within 10 days.[61] Approximately 15% of the coated silver was released (ionised) within the 10-day period with 8% excreted in faeces and 3% deposited at the implantation site. Silver ion release from a new polymeric medical device composed of silicone impregnated with silver nanoparticles was studied under simulated in-use conditions by immersion in human plasma for up to five days.[62] Whereas minimal amounts of Ag^+ were released into a control aqueous medium, greatest amounts of silver were released into the human plasma extracts within three days, *i.e.* 4 parts per million (ppm) (Table 9.4).

Table 9.4 Comparison of the mean Ag isotope released by a silver-impregnated medical catheter material immersed in human plasma or water (control) for up to five days. Samples analysed inductively coupled plasma mass spectrometry, in water and plasma extracts.[62]

Sample day	^{107}Ag isotope concentration (ppm)	^{109}Ag isotope concentration (ppm)
Water		
3	0.270 ± 0.004^a	0.267 ± 0.010
4	0.147 ± 0.006	0.144 ± 0.006
5	0.131 ± 0.003	0.129 ± 0.004
Plasma		
3	4.132 ± 0.010	4.135 ± 0.033
4	0.859 ± 0.016	0.854 ± 0.021
5	0.568 ± 0.003	0.562 ± 0.006

^{a}Results are means±SD.

Furno concluded that this pattern of release reflecting the affinity of Ag^+ for protein has clear clinical implications. In catheter implantation surgery, the surfaces of implanted devices rapidly become coated with glycoproteins from tissue and plasma.[62] In part, this reduces the Ag^+ available for bactericidal action but it provides a safety mechanism for immobilising a potentially toxic xenobiotic ion. Many approaches have been made in recent years to the production of silver-coated or silver impregnated catheters; in each case a balance is sought between Ag^+ release for antimicrobial purposes and the biocompatibility of the materials used.[63,64]

9.2.4 Clinical Experience

Clinical definition of neurotoxic risks associated with occupational or environmental exposure to silver is complicated by wide variations in patterns of exposure, quantitative analysis of silver in blood and tissues, and the scientific detail presented. Clinical studies on silver nitrate and silver sulfadiazine in treating patients with severe burns injury provide fundamental information on silver absorption and tissue distribution, but accurate information on the accumulation or distribution in the central nervous system is still urgently required.[43,65,66] Wan *et al.* critically examined methods available for quantifying silver in body tissues and fluids, and provided useful "baseline" or control silver levels in key tissues of patients with no known exposure to silver occupationally or therapeutically.[43] Flameless thermal atomic absorption spectrometry was far more accurate than older and more commonly reported techniques including spectrophotometry and flame atomic absorption.[67–69] The level of silver found in the cerebral grey matter of patients not knowingly exposed to silver and analysed by high resolution spectroscopy has been given as $0.029\,\mu g\,g^{-1}$.[70] More recent analyses of patients dying in North America has shown the silver content of tissues to be:

- brain: 0.5–$0.8\,\mu g\,g^{-1}$
- skin: $1.3\,\mu g\,g^{-1}$ (range 0.8–2 to $5\,\mu g\,g^{-1}$)

- liver: $0.7–1.0\,\mu g\,g^{-1}$
- adrenal: <0.1 to $2\,\mu g\,g^{-1}$.[71]

It is expected that silver concentrations recorded for the brain probably represent that contained within the blood–brain barrier and in serum protein complexes rather than neural tissue *per se*.

The uptake and accumulation of silver in tissues is illustrated by a cancer patient injected intravenously with 0.1 mg [111]Ag-tracer.[72] Much of the silver absorbed into the circulation within the first three days (77%) was bound to plasma macroglobulins, 15% to albumin and 8% to fibrinogen, but this declined rapidly following injection and only 10% remained after two hours and thereafter declining to 2% for the next 20 days as it was redistributed to other tissues. Silver remained high in the skin and liver until this patient died 195 days later. The silver deposition in the brain of this patient is not known.

Available information on the clinical use of silver-impregnated polyurethane catheters for external ventricular drainage suggests that risks attributable to the elution of Ag^+ are very low. In most published case studies, greatest attention has been given to the efficacy of the catheters in eliminating catheter-related infections rather than determining the influence of the ion in neurological tissues along the implantation tract. Rickham claimed that, in three years experience, an "admixture of 15% precipitated silver" did not influence the physicochemical properties of silastic catheters, but improved their radio-opacity for radiography. He did question whether silver-impregnated catheters were as inert as untreated silastic catheters.[73] Histological studies conducted in 19 patients with acute occlusive hydrocephaly implanted with external ventricular drain catheters impregnated with nanosilver particles to control infective ventriculitis demonstrated that tissue reactions around the silver-impregnated catheters were not significantly different from those associated with non-silvered silastic catheters.[74] Catheter-related infective ventriculitis is a recurrent cause of patient distress, CSF pleocytosis and malfunction of catheters. This small pilot study failed to implicate silver ion release as a possible cause of toxic changes and provided evidence that silver-impregnated catheters present a "new option" in controlling catheter-related ventriculitis in neurocritical care patients.

9.2.5 Argyria, Argyrosis Associated with Silver Deposition in Neurological Tissues

Argyria and argyrosis are the commonest observations reported in patients exposed to silver occupationally or therapeutically.[75–79] Light and electron microscopic examination of argyric patients has demonstrated electron-dense granules of 30–100 nm in the skin and other tissues; these granules being composed largely of silver sulfide with traces of selenium, mercury, titanium and iron. The electron microscope X-ray analyser is capable of detecting silver sulfide deposits in tissue at concentrations as low as $1 \times 10^{-14} \mu g\,m^{-2}$.[79] In each case, the granules have been observed mostly within secondary lysosomes of the basal lamina of the epidermis, small dermal blood vessels, Schwann cells, basement membranes of eccrine glands, and dermal elastic and collagen fibres; and not associated with pathological changes.[78] Other clinical studies examining the chemical constitution of so called "silver deposits" in brain, liver and other tissues have confirmed these observations.[80–83]

The "blue man" of Barnum and Bailey's Circus in 1927 is possibly the earliest recorded evidence of silver in the brain. Of an estimated total body silver of 90–100 g, 0.011% was located in "the brain" associated with connective tissues and macrophages.[84] The reliability of these estimates might be questioned on account of the accuracy of the investigative techniques and analytical procedures available at the time.

Occupational exposure to silver in refining, metalwork, photography and preparation of silver compounds for industry was commonly associated with argyria and argyrosis.[77–79] Blood silver concentrations were more than twice as high as in unexposed individuals ($11.0 \mu g L^{-1}$), showing high urinary ($5 \mu g g^{-1}$) and faecal silver excretion ($15 \mu g g^{-1}$).[67] The majority of reported case studies focus on the deposition of silver precipitates in the skin and eye, with rare reference to neurological abnormalities. However, occupational health studies do suggest that the cornea can serve as a sensitive indicator of silver exposure. Moss *et al.* examined 30 employees in an industrial plant involved in the manufacture of silver nitrate and silver oxide, and identified corneal and conjunctival pigmentation in 20, the severity of the discolouration being directly related to duration of employment.[85] Ten workers with impaired night vision attributable to the silver deposits failed to show electrophysiological or psychophysiological evidence of functional deficits. Although direct evidence of silver-induced neurobehavioural changes has not been seen, Rosenman *et al.* observed that most of the 20 New York factory workers showing occupational argyrosis complained of headaches, tiredness and nervousness.[86] They emphasised the importance of monitoring silver in the work environment and regularly examining staff with a slit lamp to determine ocular function. In a more recent case of occupational argyrosis, multifocal degenerative epithelial changes in the cornea were associated with a diffuse deposition of silver in the corneal stroma and Descamet's membrane and tissue debris.[87]

9.3 Silver Nitrate and Colloidal Silver in Oral Hygiene

Silver nitrate and colloidal silver preparations have been used in the treatment of mucus membrane infections and infective rhinitis for many years.[2,88] Although not now legally available in the USA and some other countries, colloidal silver is widely available in various forms for treating miscellaneous ailments. Both therapies are commonly associated with overt signs of argyria and on occasions have been implicated in neurological disorders.[89–92] In his *Manual of Pharmacology*, Sollemann[88] listed recommendations for the use of silver nitrate and colloidal silver for nose and throat infections as 2–10% silver nitrate, 0.5–10% strong silver proteins (Protargol) and 10–30% sprays of mild silver proteins (Argyrol). In practice, it is almost impossible to calculate the amount of silver consumed in long-term therapies, and blood silver levels are a poor guide to silver absorption in the chronic consumption or inhalation of over-the-counter silver products. Silver accumulates in the blood initially, but rapidly declines as some is excreted in urine and faeces and the balance is distributed to soft tissues throughout the body.

A recent case reported as the "silver man" concerned a 42-year-old patient with severe argyria resulting from chronic use of a silver protein containing vasoconstrictor preparation (Coldargan, SigmaPharm, Vienna) for treating allergic rhinitis.[93] He consumed 10–20 mL weekly of Coldargen drops containing 0.85 mg silver protein and punch biopsies showed perivascular deposits characteristic of argyria in muscle, skin and nerves but no other undesirable effects. In contrast, argyria reported in a fatal case of a 72-year-old lady with carcinoma of the stomach and uterus was associated with a deposition of silver sulfide in the basal lamina of her choroidal epithelium.[94] This patient had consumed an unknown amount of Argyrol in nose drops over 2–5 years. Although her tissues were badly autolysed, the authors claimed that silver sulfide granules (70–220 nm) were not membrane bound (lysosomal) and mostly associated with collagen fibrils and stroma of blood vessel walls. Silver granules were not contained within leptomeninges, ependymal cells or subependymal regions or in the cells of the choroid plexus, and minimal amounts were present within the area postrema. Elsewhere, florid argyria reported in a 78-year-old lady following chronic administration of over-the counter nasal drops was associated with widespread silver sulfide deposits in skin, liver, kidney, arteries, pituitary and choroid plexus.[95,96] The authors employed scanning electron microscopy with energy-dispersive spectroscopy (X-ray microanalysis) (EDAX) to characterise the chemical composition of the deposits. A later analysis of this case suggested that silver deposits were predominantly in those parts of the brain having higher regional blood flow and possibly greater permeability to environmental chemicals.[17,21,75]

A case of myoclonic status epilepticus is reported following repeated oral administration of colloidal silver in the form of a home-made "silver drink".[70] Myoclonic status epilepticus has not previously been associated with silver toxicity, but this case suggests irreversible neurologic toxicity with poor prognosis. The 71-year-old male had used this homeopathic remedy containing colloidal silver for four months, along with an anti-androgen for treating prostatic cancer and various nutritional supplements. He developed paralysis with high levels of silver in blood and CSF, and markedly elevated urinary silver excretion. Although his blood and CSF silver levels declined following plasmaphoresis, the patient showed no improvement in his neurological condition and he lapsed into a coma. His EEG revealed 14–18 Hz electropositive central-frontal polyspikes during myoclonic jerks and he died 5.5 months after the onset of his seizures. Post-mortem examination revealed evidence of diffuse Alzheimer type 2 astrocytosis and microglial activation, but no evidence of neuronal loss or focal pathology. High resolution spectroscopy demonstrated a twofold increase in silver deposition in cerebral grey matter (0.068 μg g^{-1} wet weight). Silver deposits were not specifically associated with Alzheimer-related changes and their distribution in the choroid plexus and the blood–brain barrier is not known.

More tangible evidence of neurotoxicity resulting from silver nitrate administration was reported in a 59-year-old lady using self-administered drops for ulcers of her tongue.[80] She developed cutaneous argyria and a manic depressive psychosis, but died six years later from a ruptured aortic aneurism. At autopsy silver deposits were identified in skin, mucus membranes and in many aspects of her central nervous system—notably leptomeninges, choroid plexus, basal ganglia, hypothalamus, substantia nigra and cerebellum. The lysosomally bound silver deposits were located within intraparenchymal regions and not neurones or glial cells. Progressive glial cells changes and cellular gliosis were evident in many areas of the brain. In a similar way, generalised argyrosis was reported in a 52-year-old man treated with 35 mg of an unidentified silver preparation for 18 years (estimated total intake of 35 g silver).[97] This patient died of cardiac failure, but dense silver sulfide deposits were observed in blood vessels, kidney, liver and choroid plexus at postmortem.

Westhofen and Schafer considered that silver exhibits a strong predilection for membrane and neuronal structures in severe cases of argyria with neurological involvement, but that silver sulfide deposition advanced the "progression of clinical disease".[98] They used light and electron microscopy to demonstrate silver sulfide granules in the perineurium of peripheral nerves of a severely argyric patient following chronic self-administration of an unidentified silver product. Symptoms of progressive taste and smell disorders, vertigo and hypesthesia were confirmed by chemosensitivity tests and electrophysiological investigations. Blood and brain silver levels in this patient were not given and it is unclear whether the symptoms (other than argyria) receded following withdrawal of silver therapy. (Silver-induced alterations in zinc metabolism and metallothionein induction may underlie changes in smell and taste perception.[99])

9.4 Neurological Implications from Silver in Medical Devices

9.4.1 Wound Care

In Sollemann's *Manual of Pharmacology*,[2] inorganic salts of silver (notably the nitrate) were recorded as being astringent, caustic and antiseptic but that their local action is easily controlled by their precipitation as protein complexes at the site of application. The literature is replete with case studies of silver nitrate in treating neonatal eye disease, abrasion of warts, ulcers and excessive granulations, and in the cauterisation of chronic catarrhal infections, but limited evidence is available to show it to be a potential cause of neurological damage.[75–76,100] In a fatal case of a 60-year-old man exposed to silver nitrate dressings 8 h daily for 30 days, argyria developed and skin silver levels of 2800 mg kg^{-1} and plasma silver 0.12 mg L^{-1} recorded, but no silver was seen in his brain.[101] In an 18-year-old man receiving silver nitrate for only six days, plasma silver was 0.12 mg L^{-1} and skin silver 1250 mg L^{-1}.[101] Neither patient was reported as showing neurological or behavioural changes.

Risk of argyria and silver deposition in the central nervous system have been recorded when silver nitrate has been used cauterise the cervix or as an abortifacient agent.[102] A forensic case is recorded of a German woman given a highly corrosive 7% silver nitrate solution to induce abortion.[54] She died from extreme trauma but her brain and other tissues were heavily congested.

Silver sulfadiazine as a 1% formulation in amphiphilic cream is appreciably less of an irritant than silver nitrate, but in a similar way, topical therapy leads to accumulation of Ag$^+$ in skin wounds.[103–105] Up to 10% of topically applied silver sulfadiazine absorbed from deep partial thickness burn wounds in patients with extensive burns can lead to blood concentrations 300 μg L^{-1}. Although early reports emphasise the low risks of silver sulfadiazine toxicity in routine wound care,[106] its more extensive use and incorporation in medical devices for long-term implantation these days indicate greater caution than at one time considered, including the risk of neurological damage.

Argyria is a rare complication of silver sulfadiazine therapy for severe burn injuries, but cases are recorded where a marked elevation of blood silver has been associated with deterioration in mental state.[107] In one case, blood silver levels of 291 μg L^{-1} were associated with greatly raised brain silver (617.3 ng g^{-1} cerebrum; 823.7 ng g^{-1} cerebellum wet weight). Haemodialysis, haemofiltration and plasma exchange were effective in reducing blood silver, but the patient died. Although this case might implicate silver *per se* as a neurotoxin, the information presented fails to demonstrate silver within neurological tissues or its association with neurodegenerative changes. The study does not preclude infection or immuno-suppression as a possible cause of fatality. [Flammacerium® (Solvay Pharmaceuticals) (containing 1% silver sulfadiazine and 2.2% cerium nitrate) was introduced to alleviate problems of immunosuppression attributable to products forming in burn wounds as a result of thermal energy;[108] see Chapter 5].

Sustained silver-release wound dressings are used widely these days but evidence of significantly raised blood silver or neurological change has not been seen, even with those products containing total silver of between > 10 mg per 100 cm^2 and 100 mg per 100 cm^2.[109–112] Where behavioural changes have been seen in patients with severe indolent wounds, the disability has invariably been linked to the severity of the lesion rather than to any neurotoxic action of the silver released.

9.4.2 Miscellaneous Medical Devices

Medical devices including catheters, bone cements, orthopaedic fixation pins, and cardiac prostheses and valves are notoriously prone to bacterial adhesion, colonisation and biofilm formation. Recent advances in silver nanotechnology, materials science and ion beam silver (IBAD) coating techniques have been increasingly employed in an attempt to engineer out these risks of infection and improve patient comfort and survival.[113,114] Adverse effects (including argyria) attributable to silver leaching from medical devices over a long period are exceedingly rare, but a lot more information and clinical and experimental study is urgently required. It should be anticipated that Ag^+ released in a sustained fashion from in-dwelling devices into the circulation or sites favouring metal ion absorption has a high chance of being deposited in or around neurological tissues. However, available reports fail to provide satisfactory evidence of silver absorption from intraurethral catheters,[114–117] even those with elaborate hydrophilic coatings impregnated with silver metal (including nanocrystalline forms), silver oxide and silver sulfadiazine to maximise antimicrobial efficacy.[118,119] New technology allows accurate calibration of silver ionisation patterns and release of free Ag^+ for absorption and/or carrier protein binding, but concentrations of silver in the circulation are rarely recorded except in cases of fatality or overt ill-health.

Cymet questioned whether silver "alloy" catheters might increase the inherent risks of systemic argyria and risks of silver toxicity,[120] but no satisfactory responses have been received. An unseen Russian study did report urethral argyria following use of silver nitrate,[121] but details of blood silver levels were not available. Cymet further emphasised that argyraemia could be accurately monitored and that this information is clinically important, if merely to ascertain whether or not levels were consistent with regulatory requirements in the USA and elsewhere.[120]

Saint *et al.* conducted five randomised trials using silvered catheters for short-term use but failed to observe evidence of local argyria,[122] but acknowledged that the risk did exist in long-term urethral drainage with the consequence of silver deposition in internal organs. Clinical studies with haemodialysis and intravascular catheters have similarly failed to produce evidence of silver toxicity or brain involvement, even though silver ion released directly into the circulation would be expected to lead to increased plasma-bound silver and greater tissue deposits. Tobin and Bambauer reviewed clinical studies designed to assess the efficacy and biocompatibility of silver-coated dialysis catheters.[123] They noted that blood silver levels increased from a mean of 1.3–$6.9\,\mu g\,L^{-1}$ for acute catheters and from 3.4–$19.6\,\mu g\,L^{-1}$ for long-term catheters, in each case with plasma levels returning to normal on removal of the catheters. No data were provided for tissue silver deposition or evidence of toxic side effects. Maki *et al.* evaluated triple lumen catheters (ARROWgard) designed for intravenous insertion.[124] These contained $0.70\,mg$ of silver sulfadiazine and evoked a mild local erythema at insertion points and plasma silver levels of 45–$73\,ng\,mL^{-1}$ in the 12 patients tested; these incredibly low concentrations are of minimal toxicological significance. The toxicological implications of silver alloy or silver oxide coating of catheters for intracerebral/intraventricular implantation are presently unclear.

Mechanical heart valves containing silver have been associated with greater hazard than in-dwelling catheters. Thus, St Jude Medical applied a silver coating to the sewing cuff of its range of Silzone® range of heart valves with the objective of reducing risks of infective endocarditis. Over 30 000 of these valves were distributed after 1997, but the valve was withdrawn following thromboembolic complications.[125] In one such case, a St Jude Medical Silzone® valve was implanted into a 72-year-old lady suffering from mitral valve disease. Her fatality was attributed to chronic inflammatory disease but the implication of silver in this case is unclear. Experimental studies in sheep implanted with the Silzone® valve showed plasma silver of 40 parts per billion (ppb) within ten days and mean brain silver of $4.32 + 0.28 \, \mu g \, g^{-1}$ dry tissue weight after up to 20 days.[126] Liver concentrations were a lot higher at $16.75 + 5.18 \, \mu g \, g^{-1}$, but changes were not reported in other tissues.

Acupuncture needles should be included amongst the medical devices containing silver. Their use has been associated with macular or widespread argyric changes with occasional neurological involvement.[127] This "Hari" therapy conducted in Japan for many centuries for relief of fatigue and headache involves long-term intracutaneous insertion of silver–gold needles. A 21-year-old Japanese lady given this Hari therapy over two years to relieve asthma developed a profound macular argyria and chrysiasis on her neck, face and chest.[82] (Chrysiasis results from precipitation of dark coloured insoluble gold compounds in tissue.) Minute silver–gold particles (20–60 nm) were deposited along the outer edge of basement membranes of blood vessel and sweat glands, and in lines around but not in nerve fibres. Small amounts of silver–gold were evident also in basement membrane collagen associated with myelinated and non-myelinated nerves, but nerve damage was not reported. Silver acupuncture needles are used treating for sterility and general fatigue conditions, but have argyric implications.[83] In a particularly severe case of a lady using up to 2500 needles over 13 years, macular argyria with irregularly shaped silver sulfide granules of 40–500 nm diameter was observed distributed mainly in extracellular dermal sites and around nerve fibres and elastic tissues, but overt neurological changes were not recorded. Suggestions that argyria developing through implantation of acupuncture needles might impair tissue function have not been substantiated.[75] Discolourations of the face and body have been associated with acupuncture needles containing up to 69% silver, but neurological damage has not been reported.[83,127] Blood silver levels are not known in these cases but are expected to be well below the supposed toxic range of $50–500 \, mg \, kg^{-1}$ body weight and in which patients were claimed to show abnormal encephalographic changes and brain scan findings.[128]

Antismoking remedies containing silver are included in medical devices, although there is no evidence that they present a neurotoxic hazard. A healthy 47-year-old lady who showed profound argyria following excessive oral dosage of silver acetate as an antismoking remedy for six months accumulated as much as 6.4 g of silver in her body but only 1.8% of the original dose was retained within her circulation.[129] Silver absorption and retention analysed by radio-active trace administration showed that, after an initial decrease, 18% silver tracer remained in her body for up to 30 weeks although it is not known whether any of this was deposited in her brain or peripheral nervous system. She remained in overt good health throughout.

Strong evidence implicating silver as a cause of neurological toxicity and behavioural changes derives from use of a silver as an antimicrobial agent in arthroplasty cement. Bone cement containing an unknown quantity of silver was used to anchor a Christiansen prosthesis in a 78-year-old lady.[81,130] Five years after insertion of the prosthesis, the patient became unstable and exhibited muscle weakness in her left leg. Electromyography revealed no activity in those muscles innervated by her left tibial and femoral nerves, and a total paralysis of her quadriceps muscle. This was related to exceptionally high levels of silver in her hip joint fluid ($956 \, nmol \, L^{-1}$), blood ($58 \, nmol \, L^{-1}$) and biopsies of acetabulum. Biopsies of soft tissue revealed granules characteristic of argyria in the region of elastic fibres and in numerous macrophages, but not in peripheral nerves. The right leg was entirely normal. The prosthesis was removed and blood silver levels declined to $15 \, nmol \, L^{-1}$ within 12 months, with conducting activity restored fully to her tibial muscle and partially to her femoral nerve. Motor activity was also improved in her quadriceps muscle. The patient was closely monitored over ten years, by which time the paralysis had receded and the patient was able to walk unaided. Although transitory electrophysiological changes were observed several months after removal of the prosthesis, neurotoxic action of silver was not substantiated.

9.5 The Eye

The eye is highly vulnerable to the effects of silver through occupational exposure or accidents in silver plating, silver nitrate manufacture, photographic procedures and other industries using metallic silver, silver alloys and silver compounds (Table 9.5).[85] However, apart from discolourations of the conjunctiva and cornea, few pathogenic lesions have been associated with generalised argyria, although rare cases of lachrymal gland and lens involvement have been reported.[85] Silver nitrate and colloidal silver preparations have been used therapeutically for many years in the treatment of perioptic infections and facial infections with contamination of the eye and optic membranes. Argyrosis resulting from the precipitation of insoluble silver compounds in the cornea is the principal effect of silver in the eye, but as in argyria, argyrosis has not been shown to impair the retinal structures or be associated with irreversible toxic damage in any region of the eye.[85]

9.5.1 Argyrosis (*Argyrosis conjunctivae, Argyrosis oculi*)

Argyrosis is recognised as a dusky blue-grey discolouration of the conjunctival membranes, cornea and lachrymal tissues through accumulation of insoluble precipitates or silver granules.[76] It is distinct from but commonly associated with argyria, which mainly affects the skin but can involve other tissues (Figure 9.2).

Table 9.5 Causes and conditions of silver exposure leading to manifestations of argyrosis.

Situation	Silver exposure	Clinical
Occupational	Silver smelting and refining Preparation of silver oxide, silver nitrate and alloys Silver craftsmen, silver artwork Soldering	Silver nitrate is irritant and can cause burns of the conjunctiva Corneal and conjunctival argyrosis Impaired night vision Corneal opacity Explosive injuries → necrosis and visual impairment
Therapeutic	Colloidal silver remedies, eye drops, anti-infectives Silver "stick" for relieving eyelid lesions 1-2% silver nitrate in ophthalmia neonatorum Silver clips and sutures in eye surgery	Corneal and conjunctival argyrosis Lunal caustic is corrosive (possible cataract, blindness, and scarring)
Miscellaneous	Jewellery Silver acetate in antismoking remedies Eyelash tints	Corneal and conjunctival argyrosis

It may result from chronic endogenous or exogenous exposure to silver compounds (notably colloidal silver preparation) and occupational exposures.[85,131–134] Other aetiologies leading to manifestations of argyrosis include use of silver-containing cosmetics (eyelash tints), silver clips and sutures used in eye surgery, and eye drops to control ocular infections, use of silver acetate in antismoking remedies, and in the manufacture and wearing of jewellery.[135–137]

Present views are that silver absorbed into the body as Ag^+ by inhalation, ingestion, percutaneous or other route, complexes with albumins and macroglobulins in the systemic circulation, and is transported to all parts of the body including the fine capillary network of the eye region. Through the action of solar irradiation, silver is precipitated in soft tissues as silver sulfide or silver selenide in much the same way as seen in the skin, hair and nails. Greatly improved methods of investigation have shown that these physiologically inert silver deposits and conglomerates can occur in such tissues as eyelids, lid margins, cornea, caruncle, conjunctiva and Descamet's membrane, with occasional deposits in basement membranes and superficial conjunctival substantia propria.[85,131,138] Electron microscopy with energy dispersive X-ray analysis has shown that silver conglomerates seen in the lachrymal sac of patients treated with colloidal silver occur bound in secondary lysosomal vacuoles in the extracellular matrix, predominantly on elastic fibres and within fibroblastic cells having prominent rough endoplasmic reticula.[138]

Conjunctival involvement is most commonly seen at the caruncle and semilunar fold.[85] Occasionally, the eye lens is involved. Electron microscopy

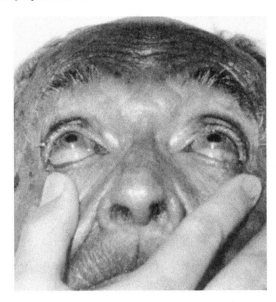

Figure 9.2   Argyrosis: discolouration of the cornea and conjunctiva and severe argyria in a patient following long-term consumption of colloidal silver. (By courtesy of Dr. B. A, Bouts, Ohio, U.S.A.,

has revealed that deposits may develop in the extracellular matrix in the vicinity of elastic fibres or in intracellular regions encapsulated in secondary lysosomes of fibroblasts and epithelial cells.[138] Corneal changes investigated using ultrasound biomicroscopy for determining corneal topography have further shown that diffuse opacities in stroma relate to diffuse epithelial deposits in the deep corneal stroma, with more granular deposits in the anterior region of the endothelium.[85] On rare occasions, discolourations of the orbit have been mistaken at initial ophthalmological examinations for conjunctival melanomatous lesions but with biopsy analysis confirming the changes to be attributable to Ag^+ leaching from suture material or silver particles rubbed into the eye from work practice.[136,137] A rare case of localised argyrosis seen in a 70-year-old lady 58 years following surgery for a strabismus condition revealed pigmented episcleral areas in the region of remnants of silver-containing suture fragments.[137] She was free from ocular symptoms and initially conjunctival malignant melanoma was suspected, but biopsy confirmed argyrosis. Detailed studies of the incidence of argyrosis and conjunctival pigmentation in silver workers show a strong relationship between the duration of silver exposure and the severity of the condition[67,85] (Table 9.6).

Argyroses are not commonly seen these days in view of greatly improved attention to health and safety in workplaces and wider control of therapeutic practices using silver and silver-containing medical devices in healthcare. Nevertheless, the literature is replete with illustrations of argyrosis, conditions under which it occurs, and its influence on general health and eyesight. Few

Table 9.6 Relationship between the severity of conjunctival discolouration and duration of work in the silver industry.[85]

Years of Work in the Industry	Number of Silver Workers	Severity Rating of Discoloration			Total Number of Workers showing Evidence of Argyrosis
		+	+ +	+ + +	
≤ 1	6	2	0	0	2
1–5	8	3	1	0	4
5–10	8	2	3	1	6
≥ 10	8	1	0	7	8
TOTAL	30	8	4	8	20

+ Slight, + + modest , + + + severe (as assessed using a slit lamp)

cases have been investigated in detail.[139] DiVincenzo *et al.* monitored 37 workers employed in silver smelting and preparation of silver products for the photographic industry and noted that argyrosis and more generalised argyria were consistent with blood, urine and faecal silver of $11 \, \mu g \, L^{-1}$, $< 0.005 \, \mu g \, g^{-1}$ and $15 \, \mu g \, g^{-1}$, respectively, compared with $< 5 \, \mu g \, L^{-1}$, $< 0.005 \, \mu g \, g^{-1}$ and $1.5 \, \mu g \, g^{-1}$ in non-silver exposed workers.[67] At the time, threshold limit values for silver exposure of $0.1 \, \mathrm{mg \, m^{-3}}$ were expected to lead to faecal silver excretion of about 1 g daily.

The overt changes attributable to occupational silver exposure reflect the type of work and the duration of exposure, but commonly argyroses are regarded as more sensitive indicators of silver exposure than more generalised argyria which appears gradually as a darkening of the skin in exposed areas.[85] Williams evaluated the case of a 51-year-old silver refiner engaged in production of silver ingots who developed corneal and conjunctival argyrosis in both eyes but with normal vision after seven years in the industry, where occupational health standards of exposure are $0.01 \, \mathrm{mg \, mg^{-3}}$ (eight hours time-weighted average) for silver compounds and $0.1 \, \mathrm{mg \, m^{-3}}$ for metallic silver.[133] Over a five-year period during which the individual continued to work, his blood silver levels declined from $74.0 \, \mu g \, L^{-1}$ to a mean of $11.2 \, \mu g \, L^{-1}$ but symptoms of argyrosis did not change appreciably. Other refinery workers were unaffected at environmental silver levels ranging from $0.11–0.17 \, \mathrm{mg \, m^{-3}}$, equivalent to near the occupational exposure standard for metallic silver of $0.1 \, \mathrm{mg \, m^{-3}}$ (229 minutes sampling period). The patient exhibited normal vision in both eyes despite the argyrosis. Recent searches of databases (1949–1999) showed total of 214 case reports, occupational health studies, and miscellaneous accounts of argyria and argyrosis.[133] None provide useful information regarding minimal levels of argyraemia consistent with recognisable argyrosis. Earlier studies of two patients subject to occupational silver exposure showed argyraemias of 49 and $74 \, \mu g \, L^{-1}$, respectively, but were asymptomatic and exhibited no signs of argyric neuropathy; however, the authors described one man as showing "non-characteristic clinical signs of argyrosis".[140]

More severe occupational risks are encountered in cases of explosion injuries.[87] Thus a patient with multifocal argyroses showed discolouration of eyelids, periocular skin, episclera, conjunctiva and cornea consistent with deposition of silver precipitates in epithelial basement membranes, Bowman's layer, Descamet's membranes and corneal stroma.[87] Although lysosomally bound, silver was identified in association with collagen fibres and epidermal cell debris; the observations indicated that intracellular silver may progress to more severe cell damage and necrotic changes leading to impairment in vision.

Optic changes resulting from the therapeutic or surgical use of silver mainly relate to the use of colloidal silver preparations such as Argyrol as eye drops or anti-infective agents for periorbital lesions.[76] Mild silver protein eye drops (Argyrol) (1%) used to treat herpetic keratitis in a 63-year-old lady led to corneal argyrosis, with biopsy showing high levels of silver, sulfur and selenium similar to that seen in skin.[78,131] No evidence has been seen of functional visual defects following the use of colloidal silver, but there is a strong tendency for silver deposits and discolouration of many parts of the eye.[131,132]

The extent to which silver-coated clips or sutures are used in eye surgery these days is not known, but interesting and diagnostically challenging cases of argyrosis show that Ag^+ leaching from the materials over many years can give rise to unsightly lesions, sometimes masquerading as melanoma. A 70-year-old lady with a history of strabismus developed a pigmented episcleral lesion in one eye 58 years after undergoing surgery.[137] Fragments of silver and suture material were found and the original diagnosis malignant melanoma discounted by biopsy examination. A second case involved an 82-year-old lady who presented with localised conjunctival argyrosis simulating melanoma 76 years after strabismus surgery; biopsies revealed silver precipitates in her conjunctiva and lateral rectus muscle, but she had no visual impairment.[138]

A more complex case of conjunctival argyrosis simulating melanoma was reported in a photographic technician who had a habit of wiping his eyes whilst working in the darkroom.[136] His eye was scarred following a road accident 50 years previously, but he developed argyrotic discolourations of the conjunctiva, cornea and eyelids. Brownish-black particles were demonstrated in biopsies to be extracellular and lysosomally bound in the region of elastic fibres immediately beneath the epithelium. It seemed likely that, in this case, the lesions were attributable more to exogenous exposure to silver through working practice than silver leaching from surgical material. Argyrosis may involve only the cornea on occasions; but in the opinion of some, intensity of distribution of the minute deposits of silver (or silver precipitates) intracellularly and extracellularly in the connective tissues of the conjunctiva and Descamet's membrane of the cornea are a reflection of the duration and magnitude of exposure.[141]

More serious optic damage is reported following the use of silver nitrate for infective conjunctivitis, haemostasis in periorbital surgery, eyelid and conjunctival bleeding, and limbic keratoconjunctivitis. Silver nitrate (20%) astringents were disallowed in the USA on account of the inherent risks of chemical conjunctivitis, corneal opacity and even blindness.[142] Although mild burn-like wounds "recovered" with minimal corneal scarring, it is recommended that therapies using 0.5–1% silver nitrate be used with extreme caution in treating conjunctivitis.[143] In a similar way, lunar caustic (silver nitrate) sticks are used to stem haemorrhages with extreme caution to avoid permanent eye damage.[71,142] Accidental splashing of >5% silver nitrate into the eye can be expected to cause severe inflammatory change, oedema, cataractous changes and permanent damage leading to blindness.[144,145]

The practice of using dilute silver nitrate to alleviate neonatal eye disease supposedly introduced by Credé in the late 1800s may have been superseded by safer and more efficacious therapies these days,[88] but a case of corneal argyrosis presenting as greyish-brown lime-like plaques was presented to the Aachen eye clinic recently.[146] Scanning electron microscopy of biopsy samples showed silver deposits 100–300 nm diameter deep within the corneal stroma, illustrating that silver applied to the surface of the cornea has the capacity to penetrate with possible long-term implications in eyesight.

Silver may not be used much in eye cosmetics these days,[147] but self-application of an eyelash and brow tint (Revlon Professional Roux Lash and Brow Tint, Colomer USA Corp., New York) was associated with ocular argyrosis in three patients.[135] Silver deposits on the upper eyelid, lid margin, caruncle, conjunctiva and Descamet's membrane extending to the level of the basement membrane and superficial substantia propria of the conjunctiva occurred after prolonged use of this product. This is surely a case of an undesirable cosmetic effect with an expensive cosmetic!

Calvery *et al.* were possibly first to conclude that ocular deposition of silver precipitates was the first and most objective sign of generalised argyria,[148] even though the technology for detecting silver in tissues at the time was limited. With more sophisticated equipment, later workers confirmed that not only was discolouration of the conjunctiva the most sensitive clinical index of argyria but that it was clearly visible without need of a biomicroscopic examination.[85] Moss *et al.* examined the condition of the eyes and visibility of 30 silver workers using slit lamp and electrophysiological methods and confirmed that the conjunctiva is a sensitive index of occupational silver exposure, but were unable to establish minimal threshold exposure levels consistent with incipient argyrosis. No correlation was apparent between detectable silver in the blood and recognisable symptoms of argyrosis using electroretinography, visual acuity tests and Schiötz tonometry. Ten silver workers examined by Moss experienced impaired night vision (nyctalopia) with extensive deposition of silver precipitates in conjunctiva and cornea after 5–25 years in the industry. Although his visual acuity tests failed to show evidence of impaired visual function, it is conceivable that with further advances in optical instrumentation, some subtle and progressive alterations in the cornea, conjunctiva or lens may be appreciated.

9.5.2 Experimental Argyrosis

The human eye is unique in many ways in its structure and function, but experimental studies in laboratory animals exposed to silver compounds in drinking water, diet and by sub-conjunctival and intraperitoneal injection have thrown some light on its possible toxicological implications.[49,145] As in human patients, the eye is highly vulnerable to periorbital instillation of silver nitrate in all mammals and instillation of >6% are liable to cause extensive scarring or blindness.[145] Grant reviewed the case of a grey cat poisoned by administration of 3 g of silver nitrate by stomach tube and which developed transient blindness within two days with swelling of optic nerves and dilatation of retinal blood vessels.[145] In a less traumatic experiment, Rungby administered silver nitrate or silver lactate to rats in order to investigate the deposition patterns of argyrophilic granules in the eye.[49] Using autometallographic techniques, he showed that silver granules were lysosomally bound and located in most cell types other than in the neural retina. As in studies in the human eye, silver granules located in vascular basal laminae and in association with connective tissue fibres. The neurophysiological implications are still unclear.

9.6 The Ear

The external ear is exposed to metallic silver, silver alloys and silver compounds through occupational exposures and it is expected that some ionised silver will complex with and be deposited in outer layers of keratin in the external auditory meatus and possibly the tympanic membrane. Possibly in rare cases of use of silver-plated hearing aids (Figure 9.3), some silver will be released to interact with epidermal surfaces. In cases of generalised argyria, discolouration of the ear is expected as in other exposed tissues. No cases of impaired hearing or complications arising from silver deposition have been reported.

Silver oxide impregnated silastic tympanostomy tubes have been found beneficial in reducing post-operative otorrhoea with minimal clinical complications. Gourin and Hubbell implanted tubes in the ears of patients exhibiting mucoid or purulent effusions or blood at the myringotomy sites and did not identify silver-related complications in a total of 1254 implants.[149] In a study of

Figure 9.3 An 18th century silver-plated ear trumpet.

26 children aged 16–127 months, silver-impregnated silastic tympanostomy tubes were compared with gold-plated silver tubes.[150] No differences were noted in the rate of otorrhoea, but the longer survival rate of the silver-impregnated tubes seemed to be independent of infectious otorrhoea and possibly dependent on the effects of biocompatibility.

The biocompatibility of silver oxide impregnated tympanostomy tubes has been established in experimental studies in gerbils and guinea pigs.[151] The devices were well tolerated when implanted for 12 months in the middle ear of gerbils. When applied to the round window, there was no evidence of ototoxicity attributable to silver release and electrocochleography and cytocochleography (hair cell counts) were normal.

9.7   Taste and Smell

Disturbances in taste and smell attributable to silver ingestion or inhalation may result from the deposition of silver precipitates in or around sensory nerves endings in the nasal cavity or in the buccal/lingual mucosae.[152] In their review of the literature 1980–1996, Ackerman and Kasbekar appreciated that elderly and chronically sick patients were more prone to develop signs of anosmia, and ageusia following exposure to drugs and suggested that these sensory disturbances may result from silver sulfate deposition, imbalances in trace metal ions (notably zinc) and disturbed pharmacological mediators, but that much research was still necessary.[152]

Greater information has accumulated on the use of silver acetate as an antismoking remedy. Mouthwashes, lozenges and chewing gum containing silver acetate (possibly with an enzyme such as co-carboxylase to enhance its effect) were introduced in the early 1970s and numerous studies have shown their efficiency in controlling long-term smoking tendencies. Whilst there seems to be no clear evidence that silver (as Ag^+) accumulating in mucosal surfaces is detrimental to sensory nerve function, silver in the presence of cigarette smoke creates an unpleasant taste in the mouth and has a deterrent effect on smoking.[153–156] The silver acetate content of the various formulations range from 2.5–6 mg, but since the remedies are frequently available on an over-the-counter basis, the actual amount of silver consumed is not accurately known. Based on large-scale surveys, most authors are of the opinion that silver acetate in gums, lozenges and sprays is modestly effective as antismoking remedies and that side effects were generally mild and transient. Mild buccal irritancy due to silver acetate had a negative influence on the success of clinical studies.[154] The results of a more detailed appraisal of side effects from a chewing gum deterrent containing 6 mg silver acetate with small amounts of co-carboxylase and ammonium chloride (three times daily for 21 days) are shown in Table 9.7.[129,155] The authors reported that patients in the silver acetate and placebo treated groups complained of a bad taste with foods, dry mouth, green tint on their tongues and adhesion of the antismoking

Table 9.7 Adverse effects of consuming silver acetate in chewing gum anti-smoking remedy.[155]

Side effect	Patients affected in Silver Acetate Group	Patients affected in Placebo Group
Oral cavity irritancy	15	5
Gastrointestinal complaints (nausea, heartburn, cramps *etc.*)	9	0
Dermatological symptoms	6	2
Central nervous system symptoms (not defined)	4	1

preparations to their dentures. One patient using the silver acetate chewing gum developed an erythematous tender tongue.

The US Food and Drug Administration Advisory finding in 1982 was that silver acetate is safe as a non-prescription treatment for smoking cessation and should be classified as an over-the-counter product (OTC).[157] The side effects are viewed by some as minor, but in the clinical case reviewed by East *et al.,* long-term obsessive use of silver to prevent smoking can lead to argyria and irreversible skin and mouth discolourations.[129] Argyria as a consequence of long-term usage of silver acetate smoking remedies like Respaton is an expected complication.[158,159]

References

1. C. S. F. Credé, *Die Verhütung der Augenentzündung der Neugeborenen, der häufigsten und wichtigsten Ursache der Blindheit*, Hirschwald, Berlin, 1895.
2. T. Sollemann, *A Manual of Pharmacology and its Applications to Therapeutics and Toxicology*, Saunders, Philadelphia, 6th edn., 1942, p. 1108.
3. J. Turner, *Colloidal silver*, in *The Gale Encyclopedia of Alternative Medicine*, ed. K. Krapp and J. L. Long, Gale Group, Detroit, 2001.
4. S. Barrett, *Colloidal Silver: Risk without Benefit*, 2005, Quackwatch, www.quackwatch.org/01QuackeryRelatedTopics/PhonyAds/silverad.html, accessed 24 November 2009.
5. US Food and Drug Administration, Department of Health and Human Services, Over-the-counter drug products containing colloidal silver ingredients or silver salts [proposed rule], *Fed. Regist.*, 1996, **61**, No. 200 (October 15, 1996), 53685.
6. M. H. Kaufman, J. McTavish and R. Mitchell, The gunner with the silver mask: observations on the management of severe maxillofacial lesions over the last 160 years, *J. R. Coll. Surg. Edinburgh*, 1997, **42**, 367.

7. W. Zheng, M. Aschner and J.-F. Ghersi-Egea, Brain barrier systems: a new frontier in metal neurotoxicological research, *Toxicol. Appl. Pharmacol.*, 2003, **192**, 1.
8. W. Zheng, Toxicology of the choroid plexus: special reference to metal-induced toxicities, *Microsc. Res. Tech.*, 2001, **52**, 89.
9. T. Rodt, G. Köppen, M. Lorenz, O. Majdani, M. Leinung, S. Bartling, J. Kaminski and J. K. Krauss, Placement of intraventricular catheters using flexible electromagnetic navigation and a dynamic reference frame: a new technique, *Stereotact. Funct. Neurosurg.*, 2007, **85**, 243.
10. R. Bayston, A. Mills, S. M. Howdle and W. Ashraf, Comment on the increasing use of silver-based products as antimicrobial agents: a useful development or a cause for concern?, *J. Antimicrob. Chemother.*, 2007, **59**, 587.
11. F. Fonnum, Neurotoxicology, in *General and Applied Toxicology*, ed. B. Ballantyne, T. Marrs and T. Sylversen, MacMillan Reference, London, 1999, Vol. **2**, pp. 631–647.
12. J. M. LeFauconnier and C. Bouchard, Neurotoxicology, in *General and Applied Toxicology*, ed. B. Ballantyne, T. Marrs and P. Turner, MacMillan Press, 1993, Vol. **1**, pp. 469–487.
13. A. B. G. Lansdown, Physiological and toxicological changes in the skin resulting from the action and interaction of metal ions, *CRC Crit. Rev. Toxicol.*, 1995, **25**, 397.
14. A. Hirano, S. Levine and H. M. Zimmerman, Experimental cyanide encephalopathy: electron microscopic observation of early lesions in white matter, *J. Neuropathol. Exp. Neurol.*, 1967, **26**, 200.
15. A. Hirano, Edema damage, *Neurosci. Res. Prog. Bull.*, 1967, **9**, 493.
16. H. De F. Webster, A. G. Ulsamer and M. F. O'Connell, Hexachlorophene induced myelin lesions in developing nervous system of Xenopus tadpoles: morphological and biochemical observations, *J. Neuropathol. Exp. Neurol.*, 1974, **33**, 144.
17. S. I. Rapoport, *The Blood–Brain Barrier in Physiology and Medicine*, Raven Press, New York. 1976.
18. T. Broman, *The Permeability of the Cerebrospinal Vessels in Normal and Pathological Conditions*, Monksgaard, Copenhagen, 1949.
19. T. Broman and A. M. Lindberg-Broman, An experimental study of disorders in the permeability of the cerebral vessels ("the blood–brain barrier") produced by chemical and physico-chemical agents, *Acta Physiol. Scand.*, 1945, **10**, 102.
20. Y. Olsson, Studies on the vascular permeability of peripheral nerves, IV. Distribution of circulating fluorescent serum albumin in the rat sciatic nerve after injection of 5-hydroxytryptamine, histamine and compound 48/80, *Acta Physiol. Scand.*, 1966, **69**(Suppl. 284), 1.
21. H. Davson, *The Physiology of the Cerebrospinal Fluid*, Churchill, London, 1967.
22. C. Bouchaud, M. Le Bert and P. Dupouey, Are close contacts between astrocytes and endothelial cells a prerequisite condition of a blood–brain

barrier? The rat sub-fornical organ as an example, *Biol. Cell*, 1989, **67**, 159.

23. E. E. Goldman, Die aüssere und innere Sekretion des gesunden und kranken Organismus im Lichte der vitalen Färbung, *Beitr. Z. Klin. Chir.*, 1909, **64**, 192.

24. E. E. Goldman, Vitalfärbung am Zentralnerven System, in *Beitrage zur Physiologie des Plexus Choroideus und der Hirnhaute*, Berlin, 1913.

25. J. W. Millen and D. M. H. Woolham, *Anatomy of the Cerebrospinal Fluid*, Oxford University Press, 1962.

26. E. M. Wright, Mechanisms of ion transport across the choroid plexus, *J. Physiol.*, 1972, **226**, 545.

27. H. F. Cserr, Physiology of the choroid plexus, *Physiol. Rev.*, 1971, **51**, 273.

28. E. Freidheim, C. Corvi, J. Graziano, T. Donnelli and D. Breslin, Choroid plexus as a protective sink for heavy metals, *Lancet*, 1983, **1**(8331), 981.

29. E. Takeuchi, K. Eto and H. Tokunaga, Mercury level and histochemical distribution in the human brain with Minamata disease following a long term clinical course of 26 years, *Neurotoxicology*, 1989, **10**, 651.

30. W. I. Manton, J. B. Kirkpatrick and J. D. Cook, Does the choroid plexus really protect the brain from lead?, *Lancet*, 1984, **2**(8398), 351.

31. G. Danscher and J. Rungby, Differentiation of histochemically visualised mercury and silver, *J. Histochem. Cyochem.*, 1986, **18**, 109.

32. M. Stoltenberg, S. Juhl, E. H. Poulsen and E. Ernst, Auto-metallographic detection of silver in hypothalamic neurones of rats exposed to silver nitrate, *J. Appl. Toxicol.*, 1994, **14**, 275.

33. G. Danscher, Light and electron microscopic localisation of silver in biological tissue, *Histochemistry*, 1981, **71**, 177.

34. J. Rungby and G. Danscher, Localisation of exogenous silver in brain and spinal cord of silver-exposed rats, *Acta Neuropathol. (Berlin)*, 1983, **60**, 92.

35. G. Danscher, M. Stoltenberg and S. Juhl, How to detect gold silver and mercury in human brain and other tissues by auto-metallographic silver amplification, *Neuropathol. Appl. Neurobiol.*, 1994, **20**, 454.

36. M. Stoltenberg and G. Danscher, Histochemical differentiation of auto-metallographically traceable metals (Au, Ag, Hg, Bi, Zn): protocols for chemical removal of separate autometallographic metal clusters on Epon sections, *Histochem. J.*, 2000, **32**, 645.

37. W. L. Scott, Silver uptake in the brains of chronically gamma-irradiated rats: a study of neutron activation analysis, *Radiat. Res.*, 1967, **31**, 522.

38. E. W. Dempsey and G. B. Wistlocki, An electron microscopic study of the blood–brain barrier in the rat employing silver nitrate as a vital stain, *J. Biophys. Biochem. Cytol.*, 1955, **1**, 245.

39. V. L. Van Breemen and C. D. Clemente, Silver deposition in the central nervous system and the hematoencephalic barrier studies with the electron microscope, *J. Biophys. Biochem. Cytol.*, 1955, **1**, 161.

40. K. G. Scott and J. G. Hamilton, The metabolism of silver in the rat with radio-silver used as an indicator, *Univ. Cal. Publ. Pharmacol.*, 1955, **1**, 161.
41. J. E. Furchner, C. R. Richmond and G. A. Drake, Comparative metabolism of radio-nuclides in mammals. IV. Retention of silver-111n in the mouse, rat, monkey and dog, *Health Phys.*, 1968, **15**, 505.
42. G. B. Wistlocki and E. H. Leduc, Vital staining of the hematoencephalic barrier by silver nitrate and trypan blue, and cytological comparisons of the neurohypophysis, pineal body, area postrema, inter-columnar tubercle and supra-optic crest, *J. Comp. Neurol.*, 1952, **96**, 371.
43. A. T. Wan, R. A. Conyers, C. J. Coombs and J. P. Masterton, Determination of silver in blood, urine and tissues of volunteers and burn patient, *Clin. Chem.*, 1991, **37**, 1683.
44. T. Scott and P. M. Norman, Silver deposition in arteriolar basal laminae in the cerebral cortex of argyric rats, *Acta Neuropathol.*, 1980, **52**, 243.
45. F. Walker, The deposition of silver in the glomerular basement membrane, *Virchows Arch. B Cell Pathol.*, 1972, **11**, 90.
46. P. D. Kharchenko, G. D. Berdyshev, P. Z. Stepanenko and A. A. Velikoivaneko, Changes in nucleic acid level in rat brain and liver following prolonged administration of silver ions in drinking water, *Fiziol. Zh.*, 1973, **19**, 362.
47. G. D. Barkov and L. I. El'Piner, The need for limiting the silver content of drinking water, *Gig. Sanit.*, 1968, **33**, 16.
48. J. Rungby and G. Danscher, Neuronal accumulation of silver in brains of progeny from argyric rats, *Acta Neuropathol. (Berlin)*, 1983, **61**, 258.
49. J. Rungby, Exogenous silver in dorsal root ganglia, peripheral nerve, enteric ganglia and adrenal medulla, *Acta Neuropathol. (Berlin)*, 1986, **69**, 45.
50. J. Rungby, L. Slomianka, G. Danscher, A. H. Andersen and M. J. West, A quantitative evaluation of the neurotoxic effect of silver on the volumes of the components of the developing rat hippocampus, *Toxicology*, 1987, **43**, 261.
51. J. Rungby, An experimental study of silver in the nervous system and aspects of its general cellular toxicity, *Dan. Med. Bull.*, 1990, **37**, 442.
52. O. Thorlacius-Ussing and J. Rungby, Ultrastructural localisation of exogenous silver in the anterior pituitary gland of the rat, *Exp. Mol. Pathol.*, 1984, **41**, 58.
53. J. Rungby, Experimental argyrosis: ultrastructural localisation of silver in rat eye, *Exp. Mol. Pathol.*, 1986, **45**, 22.
54. G. Reinhardt, M. Geldmacher-v-Mallinkrodt, H. Kittel and O. Opitz, Akute tödliche Vergiftug mit Silbernitrat als Folge eines Abreibungs versuches, *Arch. Kriminol.*, 1971, **148**, 69.
55. W. L. Roper, *Toxicological Profile for Silver*, Agency for Toxic Substances and Disease, Registry US Public Health Service, Atlanta, GA, 1990.

56. J. Rungby and G. Danscher, Hypoactivity in silver exposed mice, *Acta Pharmacol. Toxicol. (Copenh.)*, 1984, **55**, 398.

57. K. H. O. Pelkonen, H. Heinonen-Tanski and O. O. P. Hänninen, Accumulation of silver from drinking water into cerebellum and musculus soleus in mice, *Toxicology*, 2003, **186**, 151.

58. B. G. Wachter, J. P. Leonetti, J. M. Lee, R. D. Wurster and M. R. Young, Silver nitrate injury in the rat sciatic nerve: a model of facial nerve injury, *Otolaryngol. Head Neck Surg.*, 2002, **127**, 48.

59. R. Y. Hachem, K. C. Wright, A. Zermeno, G. P Bodey and I. I. Raad, Evaluation of the silver iontophoretic catheter in an animal model, *Biomaterials*, 2003, **24**, 3619.

60. R. Joyce-Wöhrmann and R. Münstedt, Determination of the silver ion release from polyurethanes enriched with silver, *Infection*, 1999, **27**(Suppl), S46.

61. D. A. Roe, B. Karandikar, N. Bonn-Savage, B. Gibbins and J.-B. Roullet, Antimicrobial surface functionalization of plastic catheters by silver nanoparticles, *J. Antimicrob. Chemotherap.*, 2008, **61**, 869.

62. F. Furno, K. S. Morley, B. Wong, B. L. Sharp, P. L. Arnold, S. M. Howdle, R. Bayston, P. D. Brown, P. D. Winship and H. J. Reid, Silver nanoparticles and polymeric medical devices: a new approach to prevention of infection, *J. Antimicrob. Chemother.*, 2004, **54**, 1019.

63. B. Jansen, M. Rinck, P. Wolbring, A. Strohmeier and T. Jahns, *In vitro* evaluation of the antimicrobial efficacy and biocompatibility of a coated central venous catheter, *J. Biomater. Appl.*, 1994, **9**, 55.

64. J. M. Schierholz, A. F. Rump and G. Pulverer, Clinical and preclinical efficiency of antimicrobial catheters, *Anasthesiol. Intensivmed. Notfallmed. Schmerzther.*, 1997, **32**, 298.

65. C. J. Coombs, A. T. Wan, J. P. Masterton, J. Pedersen and Y. T. Chia, Do burn patients have a silver lining?, *Burns*, 1992, **18**, 179.

66. X. Wang, N. Z. Wang, O. Z. Zhang, R. L. Zapata-Sirvent and J. W. Davies, Tissue determination of silver following topical use of silver sulphadiazine in extensive burns, *Burns Incl. Therm. Inj.*, 1985, **11**, 197.

67. G. D. DiVincenzo, C. J. Giordano and L. S. Schreiver, Biologic monitoring of workers exposed to silver, *Int. Arch. Occup. Environ. Health*, 1985, **56**, 207.

68. M. G. Boosalis, J. T. McCall, D. H. Ahrenholz, L. D. Solem and C. J. McClain, Serum and urinary silver levels in thermally injured patients, *Surgery*, 1987, **101**, 40.

69. S. Sano, R. Fujimori, M. Takashima and Y. Itokawa, Absorption, excretion and tissue distribution of silver sulphadiazine, *Burns*, 1981, **8**, 278.

70. S. M. Mirsattari, R. R. Hammond, M. D. Sharpe, F. Y. Leung and G. B. Young, Myoclonic status epilepticus following repeated oral ingestion of colloidal silver, *Neurology*, 2004, **62**, 1408.

71. G. D. Clayton and F. E. Clayton, *Patty's Industrial Hygiene and Toxicology*, Wiley, New York, 1981, Vol. 2, p. 1881.

72. A. A Polachek, C. B. Cope, R. F. Williard and T. Enns, Metabolism of radioactive silver in a patient with carcinoid, *J. Lab. Clin. Med.*, 1960, **56**, 499.

73. P. P. Rickham, A new silver-impregnated silastic type C catheter for use with the Holter valve in the treatment of hydrocephalus, *Dev. Med. Child. Neurol.*, 1970, **12**(Suppl), 14.

74. P. Lackner, R. Beer, G. Broessner, R. Helbok, K. Galiano, C. Pleifer, B. Pfauser, C. Brenneis, C. Huck, K. Engelhardt, A. A. Obwegeser and E. Schmutahard, Efficiency of silver nanoparticles-impregnated external ventricular drain catheters in patients with acute occlusive hydrocephalus, *Neurocrit. Care*, 2008, **8**, 360.

75. S. D. M. Humphreys and P. A. Routledge, The toxicology of silver nitrate, *Adverse Drug React. Toxicol. Rev.*, 1998, **17**, 115.

76. M. C. Fung and D. L. Bowen, Silver products for medical indications: risk-benefit assessment, *Clin. Toxicol.*, 1996, **34**, 119.

77. R. J. Pariser, Generalised argyria: clinic-pathological features and histochemical studies, *Arch. Dermatol.*, 1978, **114**, 373.

78. S. S. Bleehan, D. J. Gould, C. I. Harrington, T. E. Durrant, D. N. Slater and J. C. E. Underwood, Occupational argyria: light and electron microscopic studies and X-ray microanalysis, *Br. J. Dermatol.*, 1981, **104**, 19.

79. W. R. Buckley and C. J. Terhaar, The skin as an excretory organ in argyria, *Trans St John's Dermatol. Soc.*, 1973, **59**, 39.

80. H. W. Dietl, A. P. Anzil and P. Mehraein, Brain involvement in generalised argyria, *Clin. Neuropathol.*, 1984, **3**, 32.

81. E. Sudman, H. Vik and M. Rait, Systemic and local silver accumulation after total hip replacement using silver-impregnated bone cement, *Med. Prog. Technol.*, 1994, **20**, 179.

82. H. Suzuki, S. Baba, S. Uchigashi and M. Murase, Localised argyria with chrysiasis caused by implanted acupuncture needles. Distribution and chemical forms of silver and gold in cutaneous tissue by electron microscopy and X-ray microanalysis, *J. Am. Acad. Dermatol.*, 1993, **29**, 833.

83. S. Sato, H. Sueki and A. Nishijima, Two unusual cases of argyria: the application of an improved tissue processing method for X-ray microanalysis of selenium and sulphur, *Br. J. Dermatol.*, 1999, **140**, 15.

84. A. O. Gettler, C. P. Rhoads and A. Weiss, A contribution to the pathology of generalised argyria with a discussion on the fate of silver in the human body, *Am. J. Pathol.*, 1927, **3**, 631.

85. A. P. Moss, A. Sugar, N. A. Hargett, A. Atkin, M. Wolkstein and K. D. Rosenman, The ocular manifestations and functional effects of occupational argyrosis, *Arch. Ophthalmol.*, 1979, **97**, 906.

86. K. D Rosenman, A. Moss and S. Kon, Argyria: clinical implications of exposure to silver nitrate and silver oxide, *J. Occup. Med.*, 1979, **21**, 430.

87. U. Schlotzer-Schrehardt, L. M. Holbach, C. Hofmann-Rummelt and G. O. Nauman, Multifocal corneal argyrosis after explosion injury, *Cornea*, 2001, **20**, 553.
88. M. C. Bowen, M. Weintraub and D. L. Bowen, Colloidal silver proteins marketed as health supplements, *J. Am. Med. Assoc.*, 1995, **274**, 1196.
89. A. B. G. Lansdown, Controversies over colloidal silver, *J. Wound Care*, 2003, **12**, 120.
90. National Center for Complementary and Alternative Medicine, Colloidal Silver Products, National Institutes of Health, Bethesda, MD, 2004.
91. J. P. Marshall and R. P. Schneider, Systemic argyria secondary to topical silver nitrate, *Arch. Dermatol.*, 1977, **113**, 1077.
92. A. C. Timmins and A. R. Morgan, Argyria or cyanosis?, *Anaesthesia*, 1988, **43**, 755.
93. N. S. Tomi, B. Kränke and W. Aberfer, A silver man, *Lancet*, 2004, **363**, 532.
94. H. H. Goebel and J. Muller, Ultrastructural observations on silver deposition in the choroid plexus of a patient with argyria, *Acta Neuropathol. (Berlin)*, 1973, **26**, 247.
95. S. Landas, J. Fischer, L. D. Wilkin, L. D. Mitchell, A. K. Johnson, M. Theriac and A. C. Moore, Demonstration of regional blood brain barrier permeability in the human brain, *Neuroscience Lett.*, 1985, **57**, 251.
96. S. Landas, S. M. Bonsib, R. Ellerbroek and J. Fischer, Argyria, Microanalytic-morphologic correlation using paraffin embedded tissue, *Ultrastruct. Pathol.*, 1986, **10**, 129.
97. H. Steininger, E. Langer and P. Stommer, Generalised argyrosis, *Deutsch Med. Wochenschr.*, 1990, **115**, 657.
98. M. Westhofen and H. Schafer, Generalised argyria in man: neurological, ultrastructural and X-ray microanalytical findings, *Arch. Otorhinolaryngol.*, 1986, **243**, 260.
99. A. B. G. Lansdown, N. Stubbs, E. Scanlon and M. S. Ågren, Zinc in wound healing: theoretical, experimental and clinical aspects, *Wound Rep. Regen.*, 2006, **15**, 2.
100. Bray Healthcare, Toughened silver nitrate caustics in medicine: warts, verrucae, cautery, granuloma, 1997, www.bray.co.uk
101. K. F. Bader, Organ deposition of silver following silver nitrate therapy, *Plast. Reconstruct. Surg.*, 1966, **37**, 550.
102. D. G Lowe, D. A Levison, P. Crocker and J. H. Shepherd, Silver deposition in the cervix following application of silver nitrate as a cauterizing agent, *J. Clin. Pathol.*, 1988, **41**, 871.
103. C. Dollery, Silver sulphadiazine, in *Therapeutic Drugs*, Churchill Livingstone, Edinburgh, 1991, Vol. **2**.
104. D. S. Cook, Crystal and molecular structure of silver sulphadiazine (N^I-pyrimidin-2-yl sulphanilamide), *J. Chem. Soc. Perkin Trans. 2*, 1975, **1021**.

105. C. R. Baxter, Topical use of 1% silver sulphadiazine, in *Contemporary Burns Management*, ed. H. C. Polk and H. H. Stone, Little Brown, Boston, 1971, pp. 217–225.
106. A. B. G. Lansdown and A. Williams, How safe is silver in wound care?, *J. Wound Care*, 2004, **13**, 131.
107. S. Iwasaki, A. Yoshimura, T. Ideura, S. Koshikawa and M. Sudo, Elimination study of silver in a hemodialyzed burn patient treated with silver sulphadiazine cream, *Am. J. Kidney Dis.*, 1997, **30**, 287.
108. A. B. G Lansdown, S. R. Myers, J. Clark and P. O'Sullivan, A reappraisal of the role of cerium in burn wound management, *J. Wound Care*, 2003, **12**, 113.
109. A. B. G. Lansdown, Silver 2: toxicity in mammals and how its products aid wound repair, *J. Wound Care*, 2002, **11**, 173.
110. A. B. G. Lansdown, A review of the use of silver in wound care: facts and fallacies, *Br. J. Nurs.*, 2004, **13**(Tissue Viability Suppl), S6.
111. A. B. G. Lansdown, *Silver in Wound Care and Management*, Wound Care Society, 2003, Educational Supplement.
112. A. B. G. Lansdown, Silver in healthcare: antimicrobial effects and safety in use, *Curr. Probl. Dermatol.*, 2006, **33**, 17.
113. D. G. Maki and P. A. Tambyah, Engineering out the risk of infection with urinary catheters, *Emerg. Infect. Dis.*, 2001, **7**, 342.
114. T. S. J. Elliott, Role of antimicrobial central venous catheters for prevention of associated infections, *J. Antimicrob. Chemother.*, 1999, **43**, 441.
115. S. Saint, R. H. Savel and M. A. Mathay, Enhancing safety of critically ill patients by reducing urinary and central venous catheter-related infections, *Am. J. Resp. Crit. Care Med.*, 2002, **165**, 1475.
116. P. Thibon, X. Le Coutour, R. Leroyer and J. Fabry, Randomised multicentre trial of the effect of a catheter coated with hydrogel and silver salts on the incidence of hospital acquired infections, *J. Hosp. Infect.*, 1999, **45**, 117.
117. U. Sammuel and J. P. Guggenbichler, Prevention of catheter-related infections: the potential of a new nano-silver impregnated catheter, *Int. J. Antimicrob. Agents*, 2004, **23**(Suppl. 1), S75.
118. J. P. Guggenbichler, M. Böswald, S. Lugauer and T. Krall, A new technology of micro-dispersed silver in polyurethane induces antimicrobial activity in central venous catheters, *Infection*, 1999, **27**(Suppl. 1), S16.
119. J. Brosnahan, A. Jull and C. Tracy, Types of urethral catheters for management of short-term problems in hospitalised adults, *Cochrane Database System Review*, 2004, **1**, CD004013.
120. T. Cymet, Do silver alloy catheters increase the risk of systemic argyria?, *Arch. Intern. Med.*, 2001, **161**, 1014.
121. N. S. Liakhovitaskii, Argyria of the urethra, *Urol. Nefrol. (Mosk.)*, 1968, **33**, 59.
122. S. Saint, D. L. Veenstra, S. D. Sullivan, C. Chenoweth and M. Fendrick, The potential clinical and economic benefits of silver alloy urinary

catheters in preventing urinary tract infection, *Arch. Intern. Med.*, 2000, **160**, 2670.

123. E. J. Tobin and R. Bambauer, Silver coating of dialysis catheters to reduce bacterial colonisation and infection, *Ther. Apher. Dial.*, 2003, **7**, 504.

124. D. G. Maki, S. M. Stolz, S. Wheeler and L. A. Mermel, Prevention of central venous catheter-related bloodstream infection by use of an anti-septic-impregnated catheter. A randomised controlled trial, *Ann. Intern. Med.*, 1997, **127**, 257.

125. Medicines and Healthcare Products Regulatory Agency, *AN 1999(06) Thromboembolic Complications involving Silzone Mechanical Heart Valves*, MHRA, London, 1999, Ref. 04/01/98122131, www.mhra.gov.uk/Publications/Safetywarnings/MedicalDeviceAlerts/Advicenotices/CON008871, accessed 19 November 2009.

126. D. Langanki, M. F. Ogle, J. D. Cameron, R. A. Litzman, R. F. Schroeder and M. W. Mirsch, Evaluation of a novel bioprosthetic heart valve incorporating anti-calcification and antimicrobial technology in a sheep model, *J. Heart Valve Dis.*, 1998, **7**, 633.

127. Y. Tanita, T. Kato, K. Hanada and H. Tagami, Blue macules of localised argyria caused by implanted acupuncture needles. Electron-microscopy and roentgenographic microanalysis of deposited metal, *Arch. Dermatol.*, 1985, **121**, 1550.

128. M. J. Rosenblatt and T. C. Cymet, Argyria: report of a case associated with abnormal encephalographic and brain scan findings, *J. Am. Osteopath. Assoc.*, 1987, **87**, 509.

129. B. W. East, K. Boddy, E. D. Williams, D. MacIntyre and A. L. C. McLay, Silver retention, total body silver and tissue silver concentrations in argyria associated with exposure to an anti-smoking remedy containing silver acetate, *Clin. Exp. Dermatol.*, 1980, **5**, 305.

130. H. Vik, K. J. Andersen, K. Juhlsham and K. Todnem, Neuropathy caused by silver absorption from arthroplasty cement, *Lancet*, 1985, **1**(8433), 872.

131. Z. A. Karclioglu and D. R. Caldwell, Corneal argyrosis: histologic ultrastructural and microanalytic study, *Can. J. Ophthalmol.*, 1985, **20**, 257.

132. W. H. Spencer, L. K. Garron, F. Contreras, T. L. Hayes and C. Lai, Endogenous and exogenous ocular and systemic silver deposition, *Trans. Ophthalmol. Soc. UK*, 1980, **100**, 171.

133. N. Williams, Longitudinal medial surveillance showing lack of progression of argyrosis in a silver refiner, *Occup. Med.*, 1999, **49**, 397.

134. G. Pala, A. Fronterré, F. Scala, R. Ceccuzzi, E. Gentile and S. M. Candura, Ocular argyrosis in a silver craftsman, *J. Occup. Health*, 2008, **50**, 521.

135. M. J. Gallardo, B. Randleman, K. M. Price, D. A. Johnson, S. Acosta, H. E. Grossniklaus and R. D. Stutling, Ocular argyrosis after long-term self-application of eyelash tint, *Am. J. Ophthalmol.*, 2006, **141**, 198.

136. L. Zografos, S. Uffer and L. Chamot, Unilateral conjunctival-corneal argyrosis simulating conjunctival melanoma, *Arch. Ophthalmol.*, 2003, **121**, 1483.

137. J. Frei, B. Scröder, J. Messerli, A. Probst and P. Meyer, Localised argyrosis 58 years after strabismus operation—an ophthalmological rarity, *Klin. Monbl. Augenheilkd.*, 2001, **218**, 61.

138. D. F. E. Holck, D. F. Klintworth, J. J. Dutton, G. N. Foulks and and F. J. Manning, Localized conjunctival argyrosis: a late sequala of strabismus surgery, *Ophthalmic Surg. Lasers*, 2000, **31**, 495.

139. K. U. Leoffler and W. R. Lee, Argyrosis of the lachrymal sac, *Graefes Arch. Clin. Exp. Ophthalmol.*, 1987, **225**, 146.

140. N. Williams and I. Gardner, Absence of symptoms in silver refiners with raised blood silver levels, *Occup. Health (Lond.)*, 1995, **45**, 205.

141. C. Hanna, T. F. Fraunfelder and J. Sanchez, Ultrastructural study of argyrosis of the cornea and conjunctiva, *Arch. Ophthalmol.*, 1974, **92**, 18.

142. E. B. Shaw, Questions need for prophylaxis with silver nitrate, *Pediatrics*, 1977, **59**, 792.

143. P. A. Laughrea, J. J. Arentsen and P. R. Laibson, Iatrogenic ocular silver nitrate burn, *Cornea*, 1985–1986, **4**, 47.

144. R. M. Stein, W. M. Bourne and T. J. Liesegang, Silver nitrate injury to the cornea, *Can. J. Ophthalmol.*, 1987, **22**, 279.

145. W. M. Grant, *The Toxicology of the Eye*, Charles C. Thomas, Springfield, IL, 1986.

146. G. Schirner, N. F. Schrage, S. Salla, C. Tepling, M. Reim, W. G. Burchard and B. Schwab, Silver nitrate burn after Credé's preventive treatment. A roentgen analytic and scanning electron microscopy study, *Klin. Monbl. Augenheilkd.*, 1991, **199**, 283.

147. H. Butler, *Poucher's Perfumes, Cosmetics and Soaps*, Kluwer Academic Publishers, Dordrecht, 2000.

148. H. O. Calvery, H. D. Lightbody and B. Rones, Effects of some silver salts on the eye (silver nitrate, silver ammonium nitrate, silver ammonium sulphate, silver ammonium lactate, and a mixture of silver ammonium nitrate and silver ammonium sulphate), *Arch. Ophthalmol.*, 1941, **25**, 839.

149. C. G. Gourin and R. N. Hubbell, Otorrhea after insertion of silver oxide-impregnated silastic tympanostomy tubes, *Arch. Otolaryngol. Head Neck Surg.*, 1999, **125**, 4446.

150. D. Ulrich and S. Kreutzer, Comparison of gold-plated silver and silver oxide-impregnated silastic tympanostomy tubes: a randomised prospective clinical trial, *Laryngorhinootologie*, 2006, **85**, 501.

151. R. A. Chole, R. E. Brummett and S. P. Tinling, Safety of silver oxide-impregnated silastic tympanostomy tubes, *Am. J. Otol.*, 1995, **16**, 722.

152. B. H. Ackerman and N. Kasbekar, Disturbances of taste and smell induced by drugs, *Pharmacotherapy*, 1997, **17**, 482.

153. N. Hymowitz and H. Eckholdt, Effects of a 2.5-mg silver acetate lozenge on initial and long term smoking cessation, *Prevent. Med.*, 1996, **25**, 537.
154. E. J. Jensen, E. Schmidt, B. Pedersen and R. Dahl, Effect of nicotine, silver acetate and ordinary chewing gum in combination with group counselling on smoking cessation, *Thorax*, 1990, **45**, 831.
155. R. Malcolm, H. S. Curry, M. A. Mitchell and J. E. Keil, Silver acetate gum as a deterrent to smoking, *Chest*, 1986, **90**, 107.
156. D. MacIntyre, A. MacLay, B. W. East, E. Williams and K. Boddy, Silver poisoning associated with an antismoking lozenge, *Br. Med. J.*, 1978, **2**, 1749.
157. US Food and Drug Administration, Smoking deterrent drug products for over-the-counter human use: establishment of a monograph, *Fed. Regist.*, 1982, **47**, 490.
158. D. Shelton and R. Goulding, Silver poisoning associated with an anti-smoking lozenge, *Br. Med. J.*, 1979, **1**, 267.
159. E. J. Jensen, J. Rungby, J. C. Hensen, E. Schmidt, B. Pedersen and R. Dahl, Serum concentrations and accumulation of silver in skin during three months treatment with an antismoking chewing gum containing silver acetate, *Human Exp. Toxicol.*, 1988, **7**, 535.

CHAPTER 10

A Final Thought: How Much Silver is Too Much?

Silver is ubiquitous in the human environment. Most people are expected to show some silver in their blood or body tissues, but there is no tangible evidence that the metal fulfils any trace metal value or exhibits any physiological function. However, whereas metallic silver is inert in the presence of human tissues, body fluids and secretions, silver ion (probably Ag^+) is biologically active and readily binds cell surface receptors, metal carrier proteins like metallothioneins, and albumins and macroglobulins in serum, and is metabolised to most parts of the body by well-established mechanisms. Significantly higher blood and tissue levels of silver were common following occupational exposure to silver or silver compounds in dust, fumes and atmospheric pollution, but introduction of environmental safety precautions has greatly reduced these risks. Prolonged exposure can lead to overt symptoms of argyria and argyrosis but these conditions, although cosmetically undesirable and seriously disfiguring, are not life-threatening. With the current widespread and increasing use of ionising silver compounds as antibiotic agents in wound therapies, medical devices, textiles, water purification systems and consumer products, higher levels of body silver can be expected but experience has shown that these argyraemias are rarely significantly higher than normal, and are of no clinical or toxicological consequence. Invariably, questions do arise concerning how much silver can the human body tolerate without showing "adverse" reactions, and what are the minimal concentrations of silver in the circulation or in tissues consistent with signs of toxicity? On the basis of existing knowledge reviewed in this publication, neither question can be answered adequately.

Unlike many xenobiotic metals, silver is not a cumulative poison and few examples exist to show that silver accumulation in any tissue is a cause of cytogenicity, mutagenicity or functional disturbances. My present knowledge of the literature indicates that silver *per se* is not life-threatening in man or other

Issues in Toxicology No. 6
Silver in Healthcare: Its Antimicrobial Efficacy and Safety in Use
By Alan B. G. Lansdown
© Alan B. G. Lansdown 2010
Published by the Royal Society of Chemistry, www.rsc.org

mammalian species, and on rare occasions where patients with profound argyria have died, fatality has been attributable to unrelated causes (*i.e.* underlying medical problems such as cardiovascular disease, renal failure, diabetes, *etc.*). Reports of toxicity have been discussed in relation to high doses of silver compounds like silver nitrate and silver arsphenamine, where adverse changes are attributable to the anions and not Ag^+ or other silver ion released in the presence of moisture or tissue fluids. Regulatory authorities have attempted to formulate maximum exposure levels of silver in environmental exposure (water, air), occupational safety limits, and silver ion emission from products using manifestations of argyria or argyrosis as criteria for "silver toxicity". However, it should be recognised that:

1. There is no good evidence at the moment to show the minimal amount of silver present in the circulation consistent with unequivocal signs of long-lasting discolouration of the skin or eyes.
2. In view of wide variations in silver uptake from ionisable silver compounds consumed or inhaled, there is a poor relationship between silver exposure level and serum silver concentration.
3. The human body exhibits well-defined pathways of silver metabolism and excretion. Silver excreted in the urine or faeces can be used as a guide to silver uptake by various routes of exposure. [Silver in faeces represents that excreted *via* the liver/biliary pathway and that not absorbed from the diet or drinking water (?up to 90%).]
4. A number of early research papers are regularly cited as supporting evidence for estimations of MTD (minimal toxic dose) or permissible exposure limits. Many of these are incompatible with present day standards of quality control in experimentation (including expectations of control and randomisation), or analysis of silver in biological tissues. Considerable advances have been made in the chemistry and bio-technology of silver in the past 20 years and much learned on the cellular and biochemical management of Ag^+ by human tissues.

Current guidelines indicating permissible exposure limits for metallic silver and soluble silver compounds are somewhat ambiguous. Thus the US Occupational Safety and Health Administration and the National Institute of Occupational Safety and Health currently specify permissible exposure limits of 0.01 mg m^{-3} for metallic and soluble silver compounds in an 8-hour workday, 40-hour work week,[1] without seemingly recognising the importance of the ionisable concentration of the materials. Metallic silver ionises slowly in the presence of moisture at room temperature and the small amount absorbed into the body even from implanted catheters or orthopaedic devices is unlikely to significantly change blood silver levels. On the other hand, silver nitrate and colloidal silver preparations ionise rapidly and can lead to greatly increased argyraemias, and are a cause of profound argyria following chronic exposure. Occupational exposure to silver dust, silver oxide and silver nitrate over many years is a well-established cause of argyria, but the extent to which it occurs these days is not known. Recommended exposure limits of $0.01\,\mathrm{mg\,m^{-3}}$ may be a prudent safety margin but this might usefully be revised in the light of more recent clinical and occupational health information. The US Environmental Protection Agency (EPA) gave a reference dose for oral "silver" exposure of $5\,\mu\mathrm{g\,kg^{-1}}$ body weight for an average adult, based on clinical studies with the highly toxic material silver arsphenamine conducted 63 years earlier and other isolated patient studies for which some critical data are lacking.[2,3] Confidence in these assessments must be low, as criteria for silver "effects" are entirely subjective.

Other organisations including the American Conference of Governmental Industrial Hygienists, the Health and Safety Executive (UK) and the European Commission have in part revised their thinking on occupational exposure to metallic silver and silver compounds and recognised that metallic silver poses a lesser risk than "soluble" compounds (metallic silver and all inorganic compounds of silver ionise to some extent). Thus, they recommend exposure limits of $0.1\,\mathrm{mg\,m^{-3}}$ (eight hour time-weighted average) for metallic silver in dust and fumes, and $0.01\,\mathrm{mg\,m^{-3}}$ for "soluble compounds" which are held to pose a greater risk of causing argyria.[1] All regulatory data seen fail to recognise the impact of nanotechnology and its implications in the much higher ionisation of metallic silver in atmospheric dust and the impact on human health including the induction of argyria and argyrosis. Particle size is as critical as the actual concentration of silver in assessing risk.

A nanotechnology law report published in 2008 for The International Council of Nanotechnology noted that, in 1999, the worldwide production of silver amounted to 15.5 million kilograms and that 2.5 million kilograms were disseminated into the environment.[4] Humans will be exposed to silver in food chains and through drinking water, but there is no evidence that exposure through either route is injurious in any way. Substances present in food will absorb or otherwise precipitate silver as insoluble or inert compounds. The US EPA has established National Primary Drinking Water Regulations that set mandatory maximum concentrations of silver in drinking water of $0.1\,\mathrm{mg\,L^{-1}}$.[5] The World Health Organisation in 2003 concluded that consumption of $0.1\,\mathrm{mg\,L^{-1}}$ of silver in drinking water (equivalent to a total dose over 70 years of half the estimated no effect level in man) could be tolerated without risk.[6] To my knowledge, no cases of argyria or argyrosis have been reported following ingestion of silver in drinking water or food, and where these conditions have been created experimentally, silver concentrations administered to test animals were unrealistically high, even in relation to the most highly polluted areas such as the San Francisco Bay area.[7] Further clinical toxicology data are urgently required.

The human body exhibits a number of protective mechanisms against the "toxic" effects of silver as a xenobiotic element, which must be taken into account in assessing safety thresholds. In summary these include:

1. Barriers limiting silver absorption through gastrointestinal and respiratory mucosae, and the skin.

2. Silver ion binding and precipitating as inert complexes with proteins in blood, exudates, excretions and intracellular metal-binding proteins (metallothioneins). (Silver ion also binds with and precipitates with a wide range of anions and cell surface protein residues rendering it "unavailable for absorption".)

3. Silver ion metabolised to tissues such as the kidney, liver, blood–brain-barrier and macrophages by pinocytic action has been shown by electron microscopy to be deposited in lysosomal vacuoles as inert precipitates of silver sulfide. Silver does not enter neurological tissues and is not neurotoxic as supposed by several case studies.

4. Argyria and argyrosis as routinely quoted in regulatory assessments comprise innocuous depositions of insoluble silver sulfide and silver selenide in lysosomally bound vacuoles or as intercellular deposits affording brown black discolourations. There is no evidence that silver is released into the circulation in any form from these deposits which can remain unmetabolised for many years to produce profound disfigurement.

Silver "overload" situations will develop following massive intake of illegal colloidal silver remedies or chronic exposure to silver in an occupational setting. Such situations would be expected to lead to a saturation of these protective mechanisms. Alternatively, high levels of tissue free silver ion competing with essential trace minerals as enzyme co-factors, matrix metalloproteinases or structural sites may lead to toxic signs. However, although this situation has been demonstrated in the case of cytotoxic metals such as lead and cadmium, it has not been seen with silver.

The antibiotic efficacy of silver ion is determined by:

- the environment into which it is released;
- the rate of ionisation of the silver source;
- the range of microflora present in the medium (*in vitro* or *in vivo*);
- interaction between micro-organisms, propensity to biofilm formation;
- the sensitivity of micro-organisms to silver ion, and their molecular and genetic intracellular and cell membrane protein expression;
- existence of silver resistant strains;
- the availability of "sufficient" free Ag^+ to interact with target bacteria, yeasts, fungi, *etc.* (Figure 10.1).

No device containing silver and releasing Ag^+ on a sustained basis as an antibiotic is entirely satisfactory in achieving a germ-free state following topical, intraparenteral or intra-urethral administration. However, numerous claims have been made that silver controls pathogenic organisms and improves the quality of life in patients where infections such as *Pseudomonas aeruginosa, Escherichia coli, Enterobacteriaceae, Staphylococcus aureus* and methicillin-resistant *S. aureus* (MRSA) were potentially detrimental in wound repair, function of devices and general health. In our experience, patients treated with

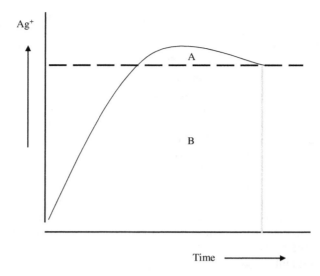

Figure 10.1   Free silver ion as the antimicrobial fraction of that released from medical devices. A = propotion of silver ion released available for antibiotic action. B = silver ion released bound irreversibly in protein complexes and as precipitates with inorganic anions ($^-$Cl, $^-$PO$_4$, *etc.*).

high quality, sustained silver-release wound dressings have shown residual infections after many months therapy; however, silver is effective in controlling bacterial balances in the wound bed, limiting the antisocial aspects of infection, and aiding clinical objectives in wound bed preparation. Bacteria persisting in wounds, catheter-related infections, prosthetic devices, *etc.* may be resistant to silver or not exposed to silver released from devices. There may be insufficient free Ag$^+$ available for antibiotic purposes, or biofilm formation with bacterial/ fungal cell transformation and acquired resistance to many antibiotics including silver.

No evidence has been produced to show that prolonged exposure to infections to silver favours the emergence of silver-resistant strains, although this subject requires more study *in vitro* and *in vivo*. The true incidence of silver-resistant bacteria and fungal infections in human infections is not known. Test kits and primers for identifying the genetically modulated *Sil*-gene complex and the characteristic protein complex are available for determining silver resistant organisms but are rarely utilised, possibly for reasons of cost, lack of suitable qualified technical staff, or the "time" factor. Nevertheless, the full benefit of silver technologies and maximisation of the advances made in the development of silver as a broad spectrum antibiotic in healthcare require clinical and nursing staff to record when silver fails to achieve desired expectations and the clinical circumstances.

References

1. P. L. Drake and K. J. Hazelwood, Exposure-related health effects of silver and silver compounds: a review, *Ann, Occup. Hyg.*, 2005, **49**, 575.
2. Environmental Protection Agency, Colloidal silver safety and toxicity: the EPA Guidelines 2008, http://www.silvermedicine.org/safety.html
3. Integrated Risk Information System (IRIS), Silver (CASRN 7440-22-4), 2008, http://www.epa.gov/iris/subs/0099.htm
4. N. R. Panyala, E. M. Pena-Mendz and J. Havel, Silver or silver nano-[articles: a hazardous treta to the environment, *J. Appl. Biomedicine*, 2008, **6**, 117.
5. Environmental Protection Agency, Secondary drinking water regulations: guidance on nuisance chemicals, EPA, 1992, Ref. 810/K-92-001.
6. World Health Organisation, Silver in Drinking Water, WHO/SDE/WSH/ 3.04.2003, Geneva.
7. A. R. Flegal, C. L. Brown, S. Squire, J. R. Ross, G. M. Scelfo and S. Hibdon, Spatial and temporal variations in silver contamination and toxicity in San Francisco Bay, *Environ. Res.*, 2007, **105**, 34.

Subject Index

Note: Where multiple entries occur, page numbers in **bold** indicate a more comprehensive coverage. Numbers in *italic* refer to figures.

CPSIA information can be obtained
at www.ICGtesting.com
Printed in the USA
LVHW08*2230011018
592086LV00007B/53/P